AF166350

BORON IN SOILS AND PLANTS

Developments in Plant and Soil Sciences

VOLUME 76

The titles published in this series are listed at the end of this volume.

Boron in Soils and Plants

Proceedings of the International Symposium on Boron in Soils and Plants
held at Chiang Mai, Thailand, 7–11 September, 1997

Edited by

R. W. BELL

School of Environmental Science, Division of Sciences, Murdoch University, WA 6150, Australia

and

B. RERKASEM

Multiple Cropping Centre, Chiang Mai University, Chiang Mai, Thailand 50200

SPRINGER-SCIENCE+BUSINESS MEDIA, B.V.

Library of Congress Cataloging-in-Publication Data

ISBN 978-0-7923-4705-7   ISBN 978-94-011-5564-9 (eBook)
DOI 10.1007/978-94-011-5564-9

Printed on acid-free paper

All rights reserved

©1997 Springer Science+Eusiness Media Dordrecht
Originally published by Kluwer Academic Publishers in 1997

No part of the material protected by this copyright notice may be reproduced
or utilized in any form or by any means, electronic or mechanical,
including photocopying, recording or by any information storage and
retrieval system, without written permission from the copyright owner.

List of referees

R. L. Aitken
R. N. Noppakoonwong

R. W. Bell
Z. Rengel

D. G. Blevins
B. Rerkasem

I. Cakmak
R. N. Sah

D. G. Edwards
B. J. Shelp

H. Goldbach
V. M. Shorrocks

S. Goldberg
M. Thellier

U. C. Gupta
D. Turner

E. J. Hanson
I. R. Willet

L. B. Huang
J. W. C. Wong

S. Jamjod
J. R. Woodruff

T. Matoh
X. Yang

J. McGrath
B. A. Zarcinas

List of sponsors

Principal sponsor

Borax International

Major sponsors

The Australia International Development Assistance Bureau (AusAID)
Australian Centre for International Agricultural Research (ACIAR)
XIII th International Plant Nutrition Colloquium
Thai Soils and Fertiliser Society

Symposium organisers

Chiang Mai University
Murdoch University
Thai Soils and Fertiliser Society

Executive Committee for Boron in Soils and Plants

Richard Bell
Patrick Brown
Bernie Dell
Robin Graham
Longbin Huang
Irb Kheoruenromne
Jack Loneragan
Toru Matoh
Benjavan Rerkasem
Victor M. Shorrocks

International Advisory Group

Pax Blamey, University of Queensland, Australia
Mitsuo Chino, Tokyo University, Japan
the late H. Marschner, Hohenheim University, Germany
Ankasith Pongsakul, Dean, Faculty of Agriculture, Chiang Mai University, Thailand
A. D. Robson, University of Western Australia, Australia
K. Shivashankar, University of Agricultural Science, Bangalore, India
Sompong Thirawong, President, Soils and Fertiliser Society of Thailand Ross Welch, Cornell University, PR China
I. R. Willett, Australian Centre for International Agricultural Research, Australia
Yuai Yang, Zhejiang Agricultural University, PR China
Fusuo Zhang, China Agricultural University, PR China

Preface

The economic significance of boron (B) in agriculture, horticulture, and forestry has been beyond dispute for several decades. Even in the last two decades, the areas where B deficiency limits plant production has grown with increased reports from China, south Asia and southeast Asia. The present volume is reflective of the growing awareness of the significance of low soil B with reports from Australia, Bangladesh, Brazil, north, central and southern China, India, Nepal, and the North West Frontier Province of Pakistan contained herein. Boron deficiency also continues to be a problem for crop yield and quality in areas where B deficiency has been known for some time, for example in Germany and the USA. The problem of low soil B is not limited to effects on field crop yield, with papers reporting on depressed wood yield and quality in timber trees (Lambert et al.), and depressed fruit quality (Dong et al.; Smith et al.; Zude et al.) also appearing in the present volume. Globally, Shorrocks (1997)[1] estimates that ?? tonnes of B fertiliser is applied annually in agriculture. The economic benefits from the use of B fertiliser have not been quantified but are clearly enormous.

Paradoxically, the clear economic imperatives for using B fertiliser on low B soils are not matched by a similar clarity of understanding of the role and functions of B in plants. Several decades of research on B have been largely descriptive and contained significant speculation about the putative roles of boron. A consensus has been emerging that the most fundamental property of B in plants is the tendency of boric acid and the borate ion to form complexes with cis-diol functional groups. The reversible nature of these complexes and their pH dependence are severe impediments to the in vivo characterisation of B in plant cells. The reversibility of the chelation of B is illustrated by Findeklee et al. who showed an alteration of cell wall elasticity and hydraulic conductivity within minutes when roots are transferred to solution lacking B.

The state-of-the art with B research is reviewed in the accompanying monograph of invited reviews, 'Boron in Soil and Plants'[2]. Evidence is now clearly behind the cell wall being the locus of the primary role of B in plants. Several new technical developments have positioned B research to move to a new level of understanding. The most significant of these developments is the increased use of inductively coupled plasma- mass spectrometry (Kerrien et al.; Shu), immunocyto-chemistry (Takasaki et al.), and fluorescence probes (Glusenkamp et al.) all of which provide opportunities for locating B in cells and the compounds with which it is associated in vivo. Papers in the present volume illustrate the use of all of these techniques in relation to location of B in cell walls, and the remobilisation of B.

Two contentious issues have repeatedly dogged the discussion of B in plants: the extent of B remobilisation in plants, and the extent of active uptake of B. The situation with B retranslocation is now substantially clearer. The phloem mobility of B has now been shown to be unique amongst other elements in that its phloem mobility is species dependent. In species which transport B in the phloem, the efficacy of B foliar application is greatly enhanced compared to other species. However, even in species which only weakly re-translocate B, considerable interest is attached to differences in re-translocation of different forms of applied B (Kerrien et al.).

Whilst there is clear evidence that B uptake is largely a passive process, nevertheless at low external B in solution sunflower absorbed B at a rate faster than explained by simple diffusion (Pfeffer et al.; Dannel et al.). It is suggested that B uptake takes place by a facilitated diffusion but further research is needed to identify the mechanisms by which such a facilitated diffusion of B can occur.

Boron toxicity is a significant problem in several cereal producing areas of the world, and has prompted breeding and selection programmes to identify tolerant germplasm (Chantachume et al.; Jamjod et al.; Yau et al.). Concerns also surround the possibility of B toxicity from the use of wastewater containing B, and from the use of waste products like flue gas desulfurization products (Tsadilas; Dowdy et al.). However, two papers in the present volume challenge a long held dogma of B research that a narrow range exists between deficiency and toxicity of B (Blamey et al.; Chapman et al.). Blamey and colleagues present evidence from carefully controlled solution culture studies that several species tolerate a wide range of external B concentrations be tween deficiency and toxicity. Further

[1] Shorrocks V. M., 1997. In Boron in Soils and Plants. Eds B. Dell, P. H. Brown and R. W. Bell. Kluwer Academic Publishers, Dordrecht, The Netherlands.

[2] Dell B., Brown P. H. and Bell R. W. (Eds) 1997 Boron in Soils and Plants. Kluwer Academic Publishers, Dordrecht, The Netherlands.

viii

consideration of this issue is warranted because at face value the results suggest that the concerns about B toxicity from overuse of B fertiliser may be overstated.

Many people deserve special thanks for enabling us to bring to a conclusion the task of editing the present volume. We pay special tribute to two people who have directly and indirectly contributed to our involvement as joint-editors of this volume. Professor Alan Robson offered the initial encouragement to us to hold the International Symposium on Boron in Soils and Plants. Professor Jack Loneragan provided the circumstances that bought us together to work initially on B in northern Thailand in the mid- 1980s. More importantly, both of them are our teachers and have continued to inspire us through their leadership, clear thinking, energy and enthusiasm for research on micronutrients.

Twenty seven referees generously reviewed papers for us and adhered closely to the tight schedules set. Most of them are listed below, and whilst some have chosen to remain anonymous we thank them all for their rigour, punctuality and constructive criticism of manuscripts.

It would not have been possible to undertake the task of mounting the International Symposium on Boron in Soils and Plants (Boron97) without the assistance of our sponsors. Initially, the generosity of the XIII th International Plant Nutrition Colloquium under the chair of Professor Mitsuo Chino in providing sponsorship gave us the seed funding to begin planning the Symposium and the encouragement to proceed. The Borax Group have been very generous in funding the symposium, the attendance of several invited speakers, and of three delegates. Without their substantial contribution, it is doubtful that we would have continued with the task of organising the Symposium. We thank Terry Lynch and Martin Phillips for their interest in Boron97 and for their flexibility in assisting the Symposium. We also acknowledge the assistance of Dr Vic Shorrocks in negotiating the sponsorship arrangements with the Borax Group.

The Australian Agency for International Development (AusAID) provided funding for two invited speakers from Australia to participate in Boron97, and for the attendance of another eight authors of contributed papers from Third World countries. AusAID's support is gratefully acknowledged because one of the goals of the Boron97 was to promote the presentation and publication of good quality research on B occurring in Third World countries. For many of the authors, it was their first experience of publishing in a peer reviewed, international publication. The Australian Centre for International Agricultural Research (ACIAR) generously sponsored the publication of the present volume and its companion, the invited review papers of Boron in Soils and Plants. Participants from several ACIAR-funded projects in Thailand and China have contributed papers in the proceedings and written invited reviews also. In some respects the holding of the Symposium in Thailand can be directly attributed to the previous support of research on B by ACIAR. With ACIAR sponsorship of the proceedings and invited reviews, copies of the two volumes will be distributed to interested scientists in Third World countries as well as participants. We are confident that the support from AusAID and ACIAR will not only ensure that scientists from all over the world become familiar with some high quality research in Third World countries but also that the learning experience will assist in the career development of Third World scientists, not to mention the increase in agricultural production on low B soils.

The Thai Soils and Fertilisers Society (TSFS) has enthusiastically worked to support Boron97 by obtaining sponsorship within Thailand and by faciliating participation by Thai scientists. We thank especially Associate Professor Irb Kheoruenromne who was the liaison between Boron 97 and TSFS, former President of TSFS, Professor Sorasith Vacharotayan, and the currrent President, Mr Sompong Thirawong.

Chiang Mai University provided considerable support to Boron97, especially by allowing staff within the Faculty of Agriculture to contribute their time to the task of local organisation of the Symposium. We are grateful for the support of the Dean of the Faculty of Agriculture, and the Head of the Multiple Cropping Center.

The Thai Department of Agriculture, Soil Science and Horticulture Divisions assisted with the one day field trip to Hang Chat by preparing field demonstrations of B response and allocating staff to assist with the displays.

Ad Plaizier and his team at Kluwer Academic Publishers have been a pleasure to work with and are to be complemented in having the Proceedings ready for distribution to delegates at the Symposium.

At Chiang Mai University, we were ably assisted by Rojare Noppakoonwong who organised the field trip, by Sananee Jamjod who made the arrangements for the presentation of posters, and by Somjit Jina who provided valuable administrative assistance. At Murdoch University, the editing task has been assisted by Judith Adams who typed revisions and reformatted manuscripts, and by Mark Gheradi who proof read final manuscripts and prepared hard copy and disk copies of manuscripts before submission. Other members of the Executive committee of Boron97 (see above) have supported and assisted us in the task of organising the Symposium and the publication of the present volume. We offer them all our sincere thanks.

Table of Contents

PHYSIOLOGY AND FUNCTIONS OF BORON IN PLANTS

BORON UPTAKE, DISTRIBUTION AND RETRANSLOCATION

BORON IN CELL WALLS

xiv

R.W. Bell and B. Rerkasem (eds.), Boron in Soils and Plants, 1–5.
© 1997 *Kluwer Academic Publishers.*

The problem of boron deficiency in crop production in Bangladesh

Sultana Ahmed & M. B. Hossain
Bangladesh Institute of Nuclear Agriculture, PO Box 4, Mymensingh-2200, Bangladesh

Key words: agroecological zones, Bangladesh, boron deficiency, critical level

Abstract

Boron (B) deficiency problems for crop production have been identified recently in Bangladesh. Information on the subject is still limited. Deficiency symptoms typical of B have been observed in a wide range of crops including: deformed and poor root growth; death of plant apices; failure in seed setting; abortion of flowers; thick pod walls with small or no seeds; discolouration of vegetables. Deficiency of B in plants has been considered responsible for causing sterility in wheat, and mustard in the country, although it was recently reported that low soil B is not the sole factor inducing wheat sterility.

The available B level in major soil types of Bangladesh ranged between 0.1 to 1.9 mg kg^{-1} soil. Non-calcareous grey floodplain soil (Typic Fluvaquent), Terrace soil (Typic Haplaquept) and Hill soil (Lithic Udorthent), which are the principal light textured acid soils generally contain low levels of available B (0.1–0.3 mg kg^{-1}). Approximately 1 million ha of cultivable land may have B deficiency problems. Yield increases to the extent of 10–92% by B fertilization in wheat and 14–52% in different vegetables were recorded from several field trials in B deficient areas in the northern zones of Bangladesh. A sharp decline in B availability in soil is expected to occur in the near future due to both crop intensification and crop diversification as Bangladesh strives for increased agricultural production. Hence B demands more attention, along with better crop management practices, in the context of sustainable agricultural production.

Introduction

The importance of boron (B) in the nutrition of crops has been well documented (Gupta, 1979). Its deficiency impairs the quality of crops and depresses the yield. In fertilizer schedules, inclusion of B often determines the success and failure of crops (Dwivedi et al., 1990). It is reported that one of the major limiting factors decreasing wheat yield in tropical and sub-tropical Asia is due to failure in grain set caused by B deficiency (Rerkasem et al., 1991).

The problem of B deficiency in Bangladesh soils was identified through a nutrient survey conducted under a national programme of the Soil Resources Development Institute (SRDI). Crop response to application of B was also recorded in several greenhouse and field trials in different parts of the country. Soil conditions often related to B deficiency are acid sandy soils, with low organic matter content and alkaline soils.

Studies on B deficiency problems have been conducted by several agencies including the Bangladesh Agricultural Research Institute (BARI), Bangladesh Institute of Nuclear Agriculture (BINA) and Bangladesh Agricultural University (BAU). A national programme on delineation of available soil nutrient status is also in progress.

The net cultivable area of Bangladesh includes a wide range of environmental conditions on the basis of which the soils are grouped into thirty agroecological zones (AEZ). Intensive cropping is widely practiced, with 2–3 crops a year grown on most of the floodplain and a major portion of the terrace areas, which can cause severe soil nutrient depletion. Nutrients depleted by whatever means should be replenished for a sustainable crop yield. Presently, B has become a limiting nutrient in some of our agricultural lands.

The soils of Bangladesh comprise 80 percent floodplain and deltaic sediments, 8% terraces and 12% hills.

Several factors control the availability of B to crops. It is adsorbed in soil to an increasing extent as the pH is increased above 7, but there is weak adsorption by soil at lower pH. It is, therefore, leached fairly readily in acid soils, resulting in deficiency in regions of high rainfall. The mean annual rainfall in Bangladesh varies from 1400 to 5800 mm. Boron deficiency is also induced due to high rates of lime application in acid sandy soils especially in dry periods on soils with low B concentration in the subsoil (Walsh and Golden, 1952). Such situations are somewhat prevalent under different agroecological regions of Bangladesh (Anon., 1995).

Boron deficiency in soils has been reported in several parts of the country, especially in northern zones. Boron status of major soil types in Bangladesh are presented in Table 1.

The available B content of the major soil types in Bangladesh ranged between 0.1–1.9 mg kg^{-1}soil. The sandy loam soils mostly contain 0.1–0.3 mg available B kg^{-1}(Islam and Miah, 1991). Since there exists a wide range of soil heterogeneity, location specific technology can better benefit the area in terms of crop productivity.

Critical levels of boron

Determination of available B content in soil depends on the extractant used. In Bangladesh, the method outlined by Hunter (1984), using Ca (HPO$_4$)$_2$ and phenol as an extractant, is widely used to determine available B. Based on the above method, a relative rating of 0.2–0.5 mg B kg^{-1}soil is considered low and 0.5–0.9 mg kg^{-1} as optimum. Accordingly a critical limit of B was considered as 0.3 mg kg^{-1} soil, based on a scatter diagram (Cate and Nelson, 1965).

Crop response to boron application

In Bangladesh, B deficiency in crop production was first identified following nutrient omission experiments under green house and field conditions. Investigation of micronutrient deficiency symptoms was carried out using mustard (*Brassica campestris*) in a sand culture experiment where omission of B caused abortion of flowers, and poorly developed, deformed roots.

Results obtained from the greenhouse experiments on the effects of Zn and B application on the yield and uptake by wheat on four soils of different pH showed the highest yield by application of 1 kg B ha^{-1}to the soil having the lowest pH (5.2). It was also observed

that Zn and B did not have an antagonistic effect on the uptake of each other (Ahmed and Alam, 1994). The findings reported from a field experiment on the effects of micronutrients on mungbean (*Vigna radiata*) showed a 14% increase in seed yield over the control by application of 2.5 kg B ha^{-1}, grown on grey floodplain soil of Mymensingh (Typic Fluvaquent), where available B content was 0.40 mg kg^{-1}soil (Ahmed, 1987).

Field experiments were conducted with wheat (*Triticum aestivum*), to study crop response to B application in several low B soils of Bangladesh. The yield of wheat as influenced by B fertilization in different locations is presented in Table 2. A significant yield increase (10–92%) in wheat was obtained by application of complete micronutrients, including B, as compared to treatments without B. Jahiruddin et al. (1992) conducted field experiments on the response of wheat to application of B at the rate of 3 kg ha^{-1} in several locations representing grey floodplain soils (Typic Fluvaquent). It was reported that B-deficiency may be a causative factor for floret sterility, which is associated with grain set failure and low wheat yield. Increase in wheat yield (30%) was recorded by application of B. Rerkasem (1995) reported that in northern Bangladesh, sterility in wheat was observed to be associated with low B levels in soil but also related to complex environmental conditions. She suggested substantial yield response may result from B applications in such locations. It has, however, been reported that no single factor has been identified as conclusively causing wheat sterility in Bangladesh (Saifuzzaman, 1995). Further, soil application of B was observed to be a more efficient method than foliar spray (Jahiruddin et al., 1992). Mixed responses in application of B by spraying was also reported (Saifuzzaman and Meisner, 1995). Increased grain yield (19%) in wheat by application of 4 kg B ha^{-1}was observed on a low B (0.25 mg kg^{-1}) silt loam soil (Abedin et al., 1994).

Several field experiments have been conducted to study the effect of B fertilization on some upland crops in non-calcareous grey floodplain soil of Rangpur (Typic Fluvaquent) (Islam and Talukder, 1991). They observed remarkable yield increases in most of the crops by fertilization with B, especially in chickpea and in mustard (Table 3).

Field experiments were laid-out on terrace soils in Gazipur (Typic Haplaquept) to study the response of vegetable crops to B application (Table 4). Yield increase in different vegetables ranged between 9–

Table 1. Boron status with some important soil characteristics in the major soil types of Bangladesh.

General soil type	Textural class[a]	Clay (g kg⁻¹)	pH (soil:water 1:2.5)	Org. Carbon (g kg⁻¹)	CEC (cmol(+) kg⁻¹)	Available B (mg kg⁻¹)
Brown hill	SL	100	4.3	7.5	9.3	0.2
Grey piedmont	L	90	4.9	12.1	4.5	0.3
Acid basin clay	C	550	4.5	21.1	31.9	0.4
Shallow red brown terrace	SiCl	310	4.9	12.8	9.7	0.5
Deep red brown terrace	SL	70	5.6	8.2	5.2	0.3
Brown mottled terrace	L	200	5.2	7.5	11.4	0.1
Shallow grey terrace	SiL	260	5.8	11.5	7.4	0.2
Black terai	SL	140	5.0	15.1	12.9	0.3
Non-calc alluvium	SL	50	6.5	3.8	7.4	0.1
Non-calc brown FP	SL	140	5.3	6.1	6.6	0.1
Non-calc grey FP	SiL	160	5.3	8.6	4.6	0.3
Non-calc dark grey FP	SiCL	300	5.2	15.9	15.3	0.5
Calc alluvium (8.5–8.7% CaCO₃)	SiL	230	8.0	7.1	12.1	0.3
Calc brown FP (2.5–3.5% CaCO₃)	SiCL	400	7.5	12.4	17.3	0.5
Calc dark grey FP (2.5–3.5% CaCO₃)	C	700	7.1	18	20.4	1.9
Non-calc dark grey FP (saline phase 4.8–3.4%)	SiCL	360	6.7	10.3	13.2	1.0
Peat	SiL	440	6.1	24.8	31.2	0.5
Acid sulphate	C	400	4.6	13.9	10.2	1.5

[a] SL = Sandy loam, SiCL = Silty clay loam, L = Loam, Cl = Clay loam, C = Clay, SiL = Silt loam, FP = Floodplain, Calc = calcareous, Non-calc = Non- calcareous
Source: BARI, BRRI, BINA and SRDI reports.

Table 2. Effect of boron (B) on the growth and yield of wheat in B deficient areas of Bangladesh. Crops supplied with complete basal nutrients including boron (+B) or without (–B).

Location	Year	Soil texture class	Average B status (mg kg⁻¹)	Sterile spikes m⁻² +B	–B	1000 grain wt. +B	–B	Grain yield (t ha⁻¹) +B	–B
Kaunia, Rangpur *early sowing*	1988–89	Sandy loam	0.1	43	84	47	36	2.3	1.6
late sowing	”	”	0.1	47	80	48	40	2.3	1.2
Mahiganj, Rangpur *early sowing*	”	”	0.2	22	48	52	36	4.9	3.9
late sowing	”	”	0.2	50	64	50	37	3.2	2.4
late sowing	”	”	0.2	–	–	45	40	2.5	2.1
Thakurgaon	1990–91	”	0.2	–	–	50	38	3.4	2.8
Jamalpur	”	”	0.4	–	–	40	40	2.2	2.0
Nashipur, Dinajpur	”	”	0.3	–	–	–	–	4.1	3.5
Mymensingh*	1989–91		0.2	–	–	–	–	2.5	1.6
LSD ’(p=0.10)				4	5	5	3	0.23	0.27

52%. Among the vegetables, fertilization with B produced the greatest yield response (52%) in cauliflower.

A field experiment was also conducted on silt loam soil to study the effect of S, Zn and B on yield and uptake of BR-2 rice. The result indicated significant yield increase (28%) by combined application of S, Zn and B (Jahiruddin et al., 1994). Although there were significant yield increases in different crops at different locations by addition of various micronutrients, including B, as compared to without B addition, the

4

Table 3. Effect of boron (B) on the yield of different upland crops in low B soils of Rangpur. Available B status of Rangpur soils = 0.1 mg kg^{-1}.

Crop	Grain yield (tonne ha^{-1})		Yield
	Complete (all nutrients)	Complete-B	increase %
Wheat	3.1	2.4	29
Mustard	1.67	0.99	69
Chickpea	1.18	0.48	146
Potato	29.0	27.0	7
Cauliflower	20.0	13.5	48
Groundnut	1.8	1.5	20

Table 4. Effect of boron (B) on the yield of vegetables in low B Terrace soils of Gazipur. Available B status of Gazipur Soils = 0.1–0.3 mg kg^{-1}.

Crop	Grain yield (tonne ha^{-1})		Yield
	Complete (all nutrients)	Complete-B	increase %
Cauliflower	20.4	13.4	52
Cabbage	67.0	60.0	12
Tomato	75.0	69.0	9
Broccoli	18.3	16.0	14
Potato	26.0	24.5	10
Groundnut	2.0	1.8	11

Source: Islam and Talukder (1987–91)

use of 5% level of statistical significance appeared too rigid. Based on economic yield increase with certain micronutrients including B, it was suggested in a country report that a level of statistical significance may be calculated at 10% in future micronutrient research (Islam, 1986).

In Bangladesh, intensive cropping plays a significant role in increasing agricultural production. As such, micronutrients are very likely to be deficient in soil in years to come. Micronutrient deficiencies (Zn, B and Mn) are widespread and are associated with specific soil reactions. Boron deficiency in soil has been identified on the basis on initial soil analysis and crop response data. The probable B deficient areas are represented by the following soil types: Non-calcareous grey floodplain soils (Typic Fluvaquent); Piedmont soils (Aeric Fluvaquentic Haplaquept); Terrace soils (Typic Haplaquept); and Hill soils (Lithic Udorthent), mostly acid soils having medium to light texture. These areas cover about 42% of the major soil types out of which about 1 million hectare may have B deficiency problems. Important crops like rice are grown in

floodplain soil and wheat, pulses and oilseed crops are grown in the dry winter season. Further detailed studies on B are in progress. Since the range between insufficiency and excess of available B is considered to be too narrow (Cartwright et al., 1984), careful consideration must be made for judicial application of B-fertilizer, wherever it is needed.

The ultimate goal of sustainable crop productivity in Bangladesh can be achieved through constant monitoring and correction of the existing and emerging micronutrient deficiencies followed by adopting the best and most practicable soil fertility management practices. This goal can be assisted by attention to the following research in relation to B.

• Delineation of available B status in soils and upland crops of Bangladesh leading to the mapping of B deficient areas.

• Greenhouse and field experiments as a basis for soil test-crop response correlation studies covering a range of soils from different agroecological zones.

• Field and on-farm trials to be conducted to determine the response of different crops to B-fertilization and its residual effect in a cropping sequence. Such trials should examine sources, rates and methods of B application and their interaction with other nutrients.

• Identification of B efficient species and cultivars which may help overcome yield losses due to B deficiency and minimize the need for B fertilization.

Finally, analytical laboratories in the country need to be strengthened for providing micronutrient advisory services for the farmers.

Acknowledgment

The senior author is grateful to AusAID for providing sponsorship to attend the International Symposium on Boron in Soils and Plants through its International Seminar Support Scheme.

References

Abedin M J, Jahiruddin M, Hoque M S, Islam M R and Ahmed M 1994 Application of boron for improving grain yield in wheat. Progressive Agric. 5, 75–79.

Ahmed S 1987 Micronutrient studies under irrigated and non-irrigated conditions. Bangladesh Institute of Nuclear Agriculture (BINA), Annual Report 1986–87.

Ahmed S and Alam T 1994 Effect of zinc and boron on yield and nutrient content of wheat. Progressive Agric. 5, 55–59.

Anonymous 1995 The Use of Land and Soil Resources Data of Various Agroecological Zones. Soil Resources Development Institute (SRDI) Report, Dhaka.

Cartwright B, Tiller K G, Zarincas B A and Spouncer L R 1983 The chemical assessment of the boron status of soils. Aust. J. Soil Res. 21, 321–332.

Cate R B Jr and Nelson L A 1965 A rapid method for correlation of soil test analysis with plant response data: Tech. Bull. No. 1 ISFEI Series, p 2. North Carolina State University, Raleigh, North Carolina.

Dwivedi G K, Dwivedi M and Pal S S 1990 Modes of application of micronutrients in acid soil in soybean-wheat crop sequence. J. Indian Soc. Soil Sci. 38, 458–463.

Gupta U C 1979 Boron nutrition of crops. Adv. Agron. 31, 273–307.

Hunter A H 1984 Agro-Service International (ASI). Analytical methods. Soil Fertility Analytical Services in Bangladesh. Contract, (US AID/BARC), Dhaka.

Islam M S 1986 Coordinated micronutrient research under irrigated and non irrigated conditions. Terminal report (1982–85), Bangladesh Agricultural Research Institute, Joydebpur.

Islam M S and Talukder K H 1991 Research findings on the effect of boron in cereals and vegetables. Annual Report 1987–1991, Bangladesh Agricultural Research Institute, Joydebpur.

Jahiruddin M, Abedin M J and Ahmed M U 1992 Boron deficiency – A major factor for grain sterility in wheat. *In* Proc. of the Inter-

Congress Conf. Comm. IV. Eds M S Hussain, S M I Haque, M A Iqbal and T N Khan, pp 85–92. Bangladesh Agricultural Research Council, Dhaka. Soils Publ. No. 37.

Jahiruddin M, Islam M N, Hashem M A and Islam M R 1994 Influence of sulphur, zinc and boron on yield and nutrient uptake of BR2 rice. Progressive Agric. 5, 61–67.

Rerkasem B 1995 A general survey of the incidence of wheat sterility. *In* Sterility in Wheat in Sub Tropical Asia: Extent, Causes and Solutions. Eds H M Rawson and K D Subedi, pp 8–12. ACIAR Proceedings Series, no. 72.

Rerkasem B, Lodkaew S and Jamjod S 1991 Assessment of grain set failure and diagnosis for boron deficiency in wheat. *In* Proc. Wheat for Nontraditional Warm Areas. Ed D A Saunders, pp 500–504. Int. Maize and Wheat Improvement Centre, Mexico. D. F.

Saifuzzaman M 1996 Influence of seeding date, genotype and boron on sterility of wheat in Bangladesh. *In* Sterility in Wheat in Sub tropical Asia: Extent, Causes and Solutions. Eds H M Rawson and K D Subedi, pp 46–50. ACIAR Proceedings Series no. 72.

Saifuzzaman M and Meisner C A 1996 Wheat sterility in Bangladesh: an overview of the problem, research and possible solutions. *In* Sterility in Wheat in Sub tropical Asia: Extent, Causes and Solutions. Eds H M Rawson and K D Subedi, pp. 104–108. ACIAR Proceedings Series no. 72.

Walsh T and Golden J D 1987 Chemistry of essential plant micronutrients in soil. *In* Russell's Soil Condition and Plant Growth. Ed A Wild, pp. 789–801. Longman Scientific and Technical, England.

R.W. Bell and B. Rerkasem (eds.), Boron in Soils and Plants, 7–9.
© 1997 *Kluwer Academic Publishers.*

Effects of boron, potassium, sulfur, magnesium application on rapeseed and mulberry yield and quality

F. Chen, J. W. Lu, Y. F. Wan, D. B. Liu & Y. S. Xu
Hubei Academy of Agricultural Sciences, Wuhan, Hubei Province 430064, PR China

Key words: boron (B), potassium (K), sulfur (S), magnesium (Mg), rapeseed, mulberry, yield, quality

Abstract

Field trials of boron (B), potassium (K), sulfur (S) and magnesium (Mg) application on rapeseed (*Brassica napus*) and mulberry (*Morus nigra*) in eastern Hubei province, PR China were used to study the effects of these nutrients on crop yield and quality. There was a positive interaction between K and B on rapeseed yield. It is important that when using K fertilizer on B deficient soil, B must be adequate for increasing efficiency of K fertilizer. Boron, K, S, Mg application increased mulberry yield and improved its quality by increased leaf amino acid content.

Introduction

Rapeseed (*Brassica napus*) and mulberry (*Morus nigra*) are two high value cash crops in eastern Hubei, central China. The area planted to mulberry in Hubei was recently expanded, but yield is still very low (Xu and Han, 1994). Reports and research suggested that low available soil boron (B), potassium (K), sulfur (S) and magnesium (Mg) contents in soils of eastern Hubei and unbalanced fertilization were the major problems of low yield and poor quality (Liu and Chao, 1993; Liu, 1993 and Xie et al., 1993). Thus, research was started in 1988 to determine if additions of the plant nutrients B, K, S and Mg in addition to basal nitrogen (N) and phosphorus (P) fertilization would increase rapeseed and mulberry yields and quality.

Materials and methods

The field trials were located on brown red soil, brown soil and alluvial soil which are the main soils for rapeseed and mulberry planting in eastern Hubei. Soil analysis data results were: pH 4.9–6.4; organic matter 4.1–23.5 g kg^{-1}; total K 14.6–39.8 g kg^{-1}; available K 39–88 mg kg^{-1}; S 1.8–3.5 mg kg^{-1}; B 0.06–0.19 mg kg^{-1}, and; Mg 252–600 mg kg^{-1}, using the ASI method of analysis (Beijing Office of PPIC, 1992).

Fertilizer rates used were as follows: N 180–225 kg ha^{-1} (as urea and ammonium phosphate); P, 75–90 kg ha^{-1} (as ammonium phosphate); K, 67–100 kg ha^{-1} (as KCl); B 2.9 kg ha^{-1} (equivalent to 15 kg borax ha^{-1}); S, 30 kg ha^{-1} (as elemental Sulphur), and Mg 60 kg ha^{-1} (as MgCl$_2$). All fertilizers for rapeseed and mulberry were applied as a basal dressing. Mulberry fertilizer was applied on three occasions: spring, summer and autumn.

Each plot area was 20 m^2, with 3 replications in a randomized block arrangement.

Results

Effect on yield

Rape seed yield was doubled or trebled by applied B compared to NP alone on the brown red soil (Table 1). By contrast when K was applied without B, the increase of seed yield was small or negative. Hence, there was a positive interaction between B and K. This result is similar to that with some other crops (Hill and Morrill, 1975; Woodruff et al., 1987).

Yield responses from 4 sites with different soil fertility levels (Table 2) indicated that on mulberry already supplied with NP fertilizer, application of K, B, S and Mg increased average yields in the spring season by

Table 1. Effect of boron (B) and potassium (K) application on yield of rapeseed (kg ha^{-1}) fertilized with nitrogen (N) and potassium (K) on two sites in eastern Hubei province. Values are means of three replications. Fields located in Qichun county.

Soil No.	Soil B (mg kg^{-1})	Item	NP	NPK	NPB	NPKB
1	0.056	yield	490	740	1050	1580
		%	–	+50	+113	+221
2	0.128	yield	390	231	1410	1700
		%	–	–41	+262	+337

Table 2. Effect of application of potassium (K), boron (B) sulfur (S) and magnesium (Mg) on mulberry leaf yield (kg ha^{-1}) at four sites in eastern Hubei province. Values are means of three replications. Leaves were harvested in late May-early June. Fields located in Luotian county.

Treatment	Site 1	Site 2	Site 3	Site 4	Average	LSD 5%	1%
NP	7130	2240	2640	5430	4360	ef	B
NPK	9450	2510	2930	6930	5450	bcd	AB
NPB	9600	2870	2830	6000	5320	cd	AB
NPS	8010	2940	2730	6250	4980	def	B
NPMg	8930	3000	2780	6330	5260	de	AB
NPKB	11300	3160	3700	7170	6330	a	A
NPKS	10000	3350	3490	7750	6150	abc	A
NPKMg	9960	3420	3590	7920	6220	ab	A

Table 3. Effect of boron (B) and potassium (K) application on the content of oil and protein in rapeseed.

Treatment	K_0B_{15}	$K_{60}B_{15}$	$K_{120}B_{15}$	$K_{180}B_{15}$
Protein (g kg^{-1})	271	269	249	249
Oil (g kg^{-1})	356	375	376	396

Table 4. Effect of potassium (K), boron (B), sulfur (S) and magnesium (Mg) on mulberry leaf amino acid, fat and sugar content. Sampled in late May 1994, leaf number 10 from the top.

Treatment	Total amino acid (g kg^{-1})	Crude Fat (g kg^{-1})	Sugar (g kg^{-1})
NP	186	98.2	120
NPK	190	96.5	145
NPB	197	100	99
NPS	216	100	115
NPMg	204	99.5	109
NPKB	206	99.6	102
NPKS	197	96.3	130
NPKMg	191	97.4	112

11–33% with K, 7–35% with B, 3–32% with S and 5–34% with Mg. With mulberry to which NPK had been applied, applications of B, S and Mg increased yield 16%, 13% and 14%, respectively. There was a positive interaction between K and S in these experiments. However, since there was evidence that K, B, S and Mg were all deficient, it is possible that even higher yields could have been achieved at all four sites by an application of complete fertilizers comprising N, P, K, S, Mg and B.

Effect on quality

Rapeseed oil content in seed was increased by 1.9–4.0% when K was applied along with B (Table 3). Oil content had a tendency to increase as K rate increased. On the contrary, protein content declined.

With NP fertilizers applied as basal fertilizer, application of B, S and Mg increased oil content and decreased sugar content of mulberry leaf and B, S, Mg and K application increased the total amino acid content (Table 4). This results is similar to those in studies with other crops (Lu and Chen, 1994).

Discussion

Boron application significantly increased rapeseed yield on the brown red soil. If K was applied without B, visual symptoms of B deficiency increased and rapeseed yields were reduced on this soil. Therefore, application of both B and K together were essential supplements to N and P for optimum nutrition of oilseed rape. There was a significant positive interaction between K and B on rapeseed yield. Seed oil content was increased by K application with B, but protein content was reduced at higher K rates. Applications of K and B together are important since rapeseed is grown for its oil content.

Mulberry yield in eastern Hubei is low. Potassium, B, S and Mg application increased the yield of mulberry leaf by about 20% and its quality was also improved because leaf total amino acid content was increased. Sulfur application showed the most notable effect on amino acid, increasing it by 16%. Applications of K, B, S and Mg were all important for high quality mulberry leaf production.

Oil content of mulberry leaf increased and sugar content decreased with B, S and Mg application. Potassium application showed the opposite trends in oil content and sugar content to B, S and Mg applications.

Acknowledgments

The authors are grateful to the Potash and Phosphate Institute-China Program, for partial funding of the research reported in this paper.

References

Beijing Office of PPIC 1992 Soil survey of nutrient status by the method of the systematic approach. Chinese Publishing House of Agricultural Sciences and Technology, Beijing, PR China.

Hill W E and Morrill L G 1975 Boron, calcium and potassium interaction in Spanish peanuts. Soil Sci. Soc. Am. Proc. 39, 80–83.

Liu, C and Chao S 1993 Outline of sulphur in Chinese agriculture. *In* Proceedings of International Symposium of the Status and Future of China S Resource and S Fertilizer Demand. pp. 154–162 (in Chinese).

Liu Z 1993 Characterization of content and distribution of microelements in soils of China. *In* Proceedings of the International Symposium on the Role of S, Mg and Micronutrients in Balanced Plant Nutrition. Ed S Portch, pp 54–61. Publishing House of Chengdu Science and Technology University. Chengdu, P R China.

Lu J W and Chen F 1994 Effect of K, S application on crop yield and quality. Bulletin of Soil 25, 216–218 (in Chinese).

Woodruff J R, Moore F E W and Musen H L 1987 Potassium, boron, nitrogen and lime effects on corn yield and earleaf nutrient concentrations. Agron. J. 79, 520–524.

Xie J, Du C and Li F 1993 Soil magnesium status and prospects for magnesium requirement in south China. *In* Proceedings of the International Symposium on the Role of S, Mg and Micronutrients in Balanced Plant Nutrition. Ed S Portch, pp 262–272. Publishing House of Chengdu Science and Technology University, Chengdu, P R China.

Xu P Q and Han T X 1994 Effect of mulberry compound fertilizer on the yield and quality of mulberry. Hubei Agricultural Sciences 6, 34–35 (in Chinese).

R.W. Bell and B. Rerkasem (eds.), Boron in Soils and Plants, 11–15.
© 1997 *Kluwer Academic Publishers.*

A foliar boron nutrition and insecticide program for soybean

Gary J Gascho & Robert M McPherson
University of Georgia, Coastal Plain Experiment Station, Tifton, Georgia, USA

Key words: boron, fertigation, foliar, *Glycine max*, reproductive stage, soybean

Abstract

In an attempt to increase soybean (*Glycine max*) yield, studies initiated in 1986 and continued until 1996 have resulted in the recommendation that soybean growers in Georgia foliar-apply 0.28 kg boron ha^{-1} with Dimilin at 0.07 kg active ingredient ha^{-1} in reproductive stages. Such applications, at the time an insecticide is needed, have resulted in economically important yield responses of soybeans grown using other best management practices. The yield response appears due to both nutrition and protection from foliage-eating insects. Initial experiments were with applications of nitrogen (N) applied by fertigation using center-pivot sprinkler irrigation systems. Such applications increased yield and protein contents of soybeans in many cases. However, fertigation is not an option for most soybeans, therefore research on foliar effects of N applications were undertaken. Foliar nitrogen applications frequently result in leaf burns, but we found that addition of boron (B) afforded some protection from those burns and provided additional increases in yield. Additional experiments indicated that B application was resulting in yield increases similar to those attained with fertigated N at far less cost. Yield increases of up to 400 kg ha^{-1} for a single application of 0.28 kg B ha^{-1} have been attained. The major gross physiological effect of B appears to be an increase in bean size. Recent studies have been directed toward applying the boron with insecticides at the time the insecticide is needed. This method of application reduces the cost of B application to the actual cost of the B source (approximately $ 3 US ha^{-1}). Soluble disodium octaborate tetrahydrate is compatible with insecticides, including Dimilin (diflubenzuron), an insecticide which is very effective against the velvetbean caterpillar (*Anticarsia gemmatalis* Hubner).

Introduction

Boron (B) is an important element for application to many crops, but is not often recommended for soybeans. The soybean has been rated to have a low susceptibility to B deficiency (Martens and Westermann, 1991; Mortvedt and Woodruff, 1993). However, yield responses have been recorded in soils with high pH, low organic matter, low available B, coarse texture, and low cation exchange capacity (Al-Molla, 1985). Schon and Blevins (1987) found increased branching and greater numbers of pods on branches due to stem infusion of B in research plots. Since that time, several studies have indicated yield responses to foliar-applied, soil-applied, and fertigated B (Schon and Blevins, 1989; Chandel et al., 1989; Gascho, 1992; Rerkasem et al., 1993).

Table 1. Soybean yield (kg ha^{-1}) for five cultivars at three sites as affected by foliar B application (kg ha^{-1}).

B Rate	Soil - Site		
	Tifton	Greenville	Bonifay
0	3024	3696	3246
0.28	2957	3763	3599
0.56	2957	3763	3377
1.12	2822	3696	3842
LSD (P = 0.05)	NS	NS	318

Proposed mechanisms of response to B include: interactions of B with magnesium in sucrose translocation, decreased flower and pod abortion, improved membrane function, improved carbon mobilization to roots, improved nitrogen (N) fixation, and improved pollen germination and pollen tube growth (Blevins et

Table 2. Soybean cultivar yield (kg ha^{-1}) response to foliar B (kg B ha^{-1}) on Bonifay sand.

B Rate	Cultivar				
	Braxton	DP3627	NKC6738	NKC6847	P9641
0	3980	2324	3471	3764	2692
0.28	4464	2775	3950	3712	3094
0.56	4125	2680	3946	3612	2921
1.12	4323	2638	4371	3955	3920
LSD (P = 0.05)	375	NS	660	NS	1209

Table 3. Branches, pods, and bean weight related to yield on Bonifay sand, Pioneer 9641.

B rate kg ha^{-1}	Branches	Pods			Bean Weight g 100^{-1}	Yield kg ha^{-1}
		Branch	Stem	Total		
0	4.6	14.9	17.7	32.6	16.3	2692
0.28	4.3	14.1	19.0	33.1	16.5	3094
0.56	5.2	14.0	18.6	32.6	17.1	2921
1.12	4.9	15.2	24.6	39.8	17.3	3920
LSD (P = 0.05)	NS	NS	3.7	NS	0.8	1209

al., 1993; Yamagishi and Yamamoto, 1994). Rerkasem et al. (1993) described soybean seed damage induced by B deficiency. The damage, described as localized depressions on the internal surface of cotyledons of soybean seed is similar to the B deficiency symptom of peanut (Arachis hypogaea) called hollow heart which has often been found in Georgia peanuts. They indicated that susceptibility to B deficiency was variable among soybean cultivars and suggested that selection for cultivars which were not susceptible to B deficiency should be initiated in breeding programs. Alternatively, it may be suggested that some B applications may be made to cultivars which are superior or acceptable in other respects.

The Georgia research and extension program of providing supplemental nutrition to soybeans began in 1986 with vegetative N applications via fertigation using center-pivot irrigation systems. Due to our results and those of Flannery (1986) emphasis changed to N applications during pod fill stages in 1988. Boron and N+B studies began in 1990. Applications have been made to soil, but the major emphasis was using fertigation until 1990, since then major emphasis has been on foliar application as fertigation is not a possibility in most soybeans.

In this study, we investigated the relation of yield to foliar applications of B. The velvetbean caterpillar (Anticarsia gemmatalis Hubner) is an annual economic threat to Georgia's soybean crop and foliar applications of insecticide are required to prevent defoliation and yield loss (McPherson et al., 1996). We attempted to reduce costs of B application by combining the B with a needed insecticide.

Materials and methods

Foliar boron experiments

Small plot field research with spray applications of B was carried out at three sites in south Georgia in 1992 in concert with other uniform experiments in Missouri, Ohio, Wisconsin and Illinois. Two irrigated and one nonirrigated sites were selected in order to cover a range of soil conditions in Georgia soybeans. The unirrigated site was a Tifton loamy sand with a single cultivar, Delta Pine 3627. The irrigated sites were on a Greenville sandy loam and a Bonifay sand. The experiments followed bermudagrass, wheat, and fallow for the Tifton, Greenville and Bonifay sites, respectively. At the Greenville and Bonifay sites, split-plot experiments with four replications were conducted with the main-plots being cultivars and the split-plots B treatments. Split-plot size was 1.82 × 7.6 m. Plant-

Table 4. Correlations co-efficients (r) of yield components with yield for five cultivars on Bonifay sand.

Variable	Braxton	DP3627	NKC6738	NKC6847	P9641
Bean wt	0.41	0.90**	0.34	0.68**	0.65**
Branches	0.22	0.66**	0.11	−0.07	−0.06
Pods-branch	0.50*	0.75**	0.24	0.16	0.70**
Pods-main	0.25	0.76**	0.44*	0.15	0.82**
Pods-total	0.30	0.82**	0.50*	0.19	0.89**

* Significant at P = 0.05, ** Significant at P = 0.01.

Table 5. Effects of boron and Dimilin foliar applications on the population peaks of velvetbean caterpillars and soybean defoliation for three years.

Treatment	1993		1994		1995	
	count[1]	defoliation[2]	count	defoliation	count	defoliation
None	—	—	11.8	32.3	6.2	16.3
Dimilin	1.3	3.5	0.3	7.1	0.9	4.8
B + Dimilin	0.0	2.5	4.0	7.3	1.0	5.3
LSD (P = 0.05)	4.1	6.8	3.8	6.3	3.2	5.8

[1] Velvetbean caterpillars/10 sweeps at peak population, [2] Percent of total foliage removed.

ing date was 5 June at the Tifton site, 17 June at the Greenville site, and 21 May at the Bonifay site. Row spacing was 0.91 m at the Tifton and Bonifay sites and 0.76 m at the Greenville site. Boron was applied at 0.28 kg ha^{-1}, as disodium octaborate tetrahydrate ($Na_2B_8O_{13}.4H_2O$), with a compressed CO_2-back-pack sprayer, 187 l water ha^{-1}, and the surfactant Tween 20. The data of Guertal et al. (1996) indicate that the soluble source of B used in these experiments is as effective as other soluble sources currently available. Initial foliar applications at R1 (first flower) were made on 7 August, 6 August, and 17 July for the Tifton, Greenville and Bonifay sites, respectively. Rainfall during the crop totaled 424 mm at the Tifton and Bonifay sites and 361 mm at the Greenville site. Irrigation totaling 165 mm was applied at the Bonifay site. Irrigation at the Greenville site totaled 188 mm. At R6, plant populations were 225, 101 and 225 thousand ha^{-1} at the Tifton, Greenville and Bonifay sites, respectively. Branches per plant, branch pods and stem pods were counted on 5 plants per plot at R6. Yield was determined by combining the entire plot. Seed weights were obtained by weighing a sample of 100 beans from each plot.

Table 6. Soybean yield (kg ha^{-1}) as affected by foliar boron and Dimilin for four years.

Treatment	1993	1994	1995	1996
None	—	3368	3172	3588
Dimilin	2634	3786	3174	4227
B + Dimilin	2869	3759	3207	4603
LSD (P = 0.05)	NS[1]	370	NS	—[2]

[1] NS = not significant at P = 0.05.
[2] Unreplicated demonstration.

Boron-Dimilin experiments

Seven replicated randomized complete block experiments were conducted from 1993 to 1995 and a field demonstration was conducted in 1996. The experiments included treatments of foliar-applied B with and without several insecticides commonly used in Georgia soybean. For brevity sake, only a sampling of the data for the B and B-Dimilin (an insect growth regulator) treatments can be presented. Scout Xtra (tralomethrin, a pyrethroid), Dipel 4 (*Bacillus thuringiensis*), and combinations of Dimilin and Scout Xtra were also evaluated with B. The experiments were conducted on several soil-sites near Tifton, GA with methods similar to those used in the 'Foliar boron experiments'. The Dimilin rate was 0.07 kg active ingredient ha^{-1}.

14

Mixtures of the combined sprays were observed in the laboratory for physical compatability and pH. Due to some sedimentation of the mixture with time after mixing, Dimilin was added to the B solutions just prior to application. Mixture pH was elevated by addition of the disodium octaborate tetrahydrate. In some cases, a buffer was added to neutralize pH and a surfactant was added to increase residence of the sprays on the foliage. The value of such additions to the B-Dimilin combination is unproved at this time. Velvetbean caterpillar populations were monitored in all plots 7, 28, and 42 days after foliar applications. Ten sweeps were taken with a 0.38 m diameter net (Kogan and Herzog, 1980). Following the 42-day sample, plots were rated for defoliation.

Results and discussion

Foliar boron experiments

Boron application did not result in any foliage burn and had no measurable effect on senescence. Analysis of variance indicate a significant yield response to foliar B only for the Bonifay sand site. Soybean yield at that deep-sand site increased by up to 596 kg ha^{-1} due to B application (Table 1). A significant response of 353 kg ha^{-1} was found to a single application of B (0.28 kg B ha^{-1}) at flowering. These results suggest response can be expected on sands with little ability to hold the B. Yields for all cultivars appeared to respond to B on the deep sand to some extent, but the response with increasing rates (number of applications) of B was not linear and was statistically significant for only three of the five cultivars (Table 2).

Throughout these studies apparent response differences due to cultivar were noted (data not shown). In these split-plot randomized block experiments the cultivar × nutrient treatment interaction was usually not significant, even though large differences in yield response between cultivars occurred. The responses to spray B treatments (Table 2) are an example in which yield response to 1.12 kg B ha^{-1} was as little as 191 kg ha^{-1} for NKC6847 and as great as 1228 kg ha^{-1} for P9641. Success with foliar and fertigate applications may depend on the cultivar grown, as suggested by Rerkasem et al. (1993).

In an attempt to explain the effect(s) of B on the plant, Table 3 relates lateral branching, pods on lateral branches, pods on the main stem, total pods and 100 bean weight for the cultivar showing the greatest

response in yield at the Bonifay site. Yield response was the product of both increased numbers of pods and increased bean weight due to B application. Lateral branching was unaffected by B treatments. Pods on the main stem tended to increase with increased rate of B application. Bean weight increased linearly with increased B rate. Correlations of yield components with yield for all 5 cultivars on Bonifay sand (Table 4) also indicate that yield responses were the result of both pod number and seed weight, but yield was most consistently correlated with bean weight.

Boron-Dimilin experiments

Peak velvetbean caterpillar populations were similar for the B and no B treatments indicating that an application of this nutrient in the late-season does not alter the efficacy of Dimilin (Table 5). An application of Dimilin (alone or with B) significantly reduced the peak velvetbean caterpillar population and subsequently reduced defoliation. When velvetbean caterpillar populations caused around 30% defoliation, yields were also greater in the Dimilin plots where defoliation was maintained below 8% (Tables 5 and 6).

Boron applications do not always provide significant yield response. Results are year, soil, and cultivar dependent. In some cases applications have provided very significant responses and in most cases, increased yield more than justifies the cost (around $ 3 US ha^{-1}) of the application.

Conclusions

1. Foliar sprays of B during reproductive stages of soybean, as well as, fertigations and dribble applications have resulted in significant yield responses. When the B is applied with a needed insecticide the cost of B application is approximately $ 3 US ha^{-1} and the efficacy of the insecticide is maintained.

2. Responses greater than 300 kg ha^{-1} have generally been limited to irrigated sand soils, where the B moves freely and may be leached with rainfall and irrigation. Responses on loamy soils have generally been less than 300 kg ha^{-1} in studies over a period of 5 years.

3. Responses may be cultivar dependent.

4. Yield response was best correlated with seed weight and less correlated with seed number.

Acknowledgements

Financial support of these studies by TVA, FFF, FAR, U.S. Borax and the GA Agr. Com. for Soybeans is appreciated. The help of Drs. John Woodruff, Randy Hudson and several seed companies is appreciated. The authors also thank Benjie Baldree, Bert Crowe, Ellen Hall, and Del Taylor for their technical efforts on this project.

References

Al-Molla R M M 1985 Some phenological aspects of soybean development and yield as affected by boron fertilization. Ph. D. Dissertation, Univ. of Arkansas, Dissertation Abst.46-10B, 3268, Ann Arbor, MI.

Blevins D G, Reinbott T M and Boyce P J 1993 Foliar fertilization of soybeans with boron and magnesium. *In* Foliar Fertilization of Soybeans and Cotton. Ed L S Murphy. PPI/FAR Tech. Bull 1993-1. Norcross, GA.

Chandel A S, Tiwari S K and Saxena S C 1989 Effect of micronutrient application on soybean (*Glycine max*) grown in Uttar Pradesh foothills. Indian J. Agr. Sci. 59, 62.

Flannery R L 1986 Plant food uptake in a maximum yield soybean study. Better Crops with Plant Food. September, 6–7.

Gascho G J 1992 Late-season nitrogen and boron applications for soybeans. *In* Fluid Fertilizer Foundation 1992 Research Symp., Proc. pp 59–67. Fluid Fertilizer Foundation, St Louis, MO, USA.

Guertal E A, Abaye A O, Lippert B M, Miner G S and Gascho G J 1996 Sources of boron for foliar fertilization of cotton and soybean. Commun. Soil Sci. Plant Anal. 27, 2815–2828.

Kogan M and Herzog D C 1980 Sampling Methods in Soybean Entomology. Springer-Verlag, New York, 587 p.

Martens D C and Westermann D T 1991 Fertilizer applications for correcting micronutrient deficiencies. *In* Micronutrients in Agriculture, 2nd edn. Ed J Mortvedt, F R Cox, L M Sherman and R M Welch, pp 549–592. Soil Science Society America Book Series No. 4, Madison, Wisconsin.

McPherson R M, Douce G K and Riley D G 1996 Summary of losses from insect damage and costs of control in Georgia. 1995 Ga. Agric. Expt. Stn. Spec. Publ. 90. 55 p.

Mortvedt J J and Woodruff J R 1993 Technology and application of boron fertilizers for crops. *In* Boron and its Role in Crop Production. Ed. U C Gupta, pp 157–176. CRC Press, Boca Raton, Florida.

Rerkasem B, Bell R W, Lodkaew S and Loneragan J F 1993 Boron deficiency in soybean [*Glycine max* (L.) Merr.] peanut (*Arachis hypogaea* L.) and black gram [*Vigna mungo* (L.) Hepper]: Symptoms in seeds and differences among soybean cultivars in susceptibility to boron deficiency. Plant and Soil 150, 289–294.

Schon M K and Blevins D G 1987 Boron stem infusions stimulate soybean yield by increasing pods on lateral branches. Plant Physiol. 84, 969–971.

Schon M K and Blevins D G 1989 Foliar boron applications increase the final number of branches and pods on branches of field-grown soybean. Plant Physiol. 92, 602–607.

Yamagishi M and Yamamoto Y 1994 Effects of boron on nodule development and symbiotic nitrogen-fixation in soybean plants. Soil Sci. and Plant Nut. 40, 265–274.

R.W. Bell and B. Rerkasem (eds.), Boron in Soils and Plants, 17–21.
© 1997 *Kluwer Academic Publishers.*

Soil Boron Content and the Effects of Boron Application on Yields of Maize, Soybean, Rice and Sugarbeet in Heilongjiang Province, P R China

Yiying Li & Hong Liang
Soil and Fertilizer Institute, Heilongjiang Academy of Agricultural Sciences, Harbin, P. R. China

Key words: soil boron, systematic approach, boron application, maize, soybean, rice, sugarbeet

Abstract

Using the Systematic Approach methodology, 8 out of 21 soils tested from Heilongjiang province were deficient in boron (B). In field trials, using borax application as basal fertilizer with maize (*Zea mays*), soybean (*Glycine max*), rice (*Oryza sativa*) and sugarbeet (*Beta vulgaris*) on the eight soils, yields increased by an average of 8.5%, 4.0%, 6.6% and 10.2%, respectively. Boron application increased the rate of fruiting of the grain crops and reduced physiological disorders of sugarbeet. The capacity of the soils to adsorb boron was not very strong, especially when the concentration of boron added was more than 0.5 mg B kg^{-1}.

Introduction

Heilongjiang province, located in northeast China, contains more than 10 million ha cultivated land. Main crops planted are: maize (*Zea mays*), soybean (*Glycine max*), wheat (*Triticum aestivum*), rice (*Oryza sativa*), sugarbeet (*Beta vulgaris*) and potato (*Solanum tuberosum*). The main soils are: black soil, chernozemic soil, meadow black soil, lessive and meadow soil. Heilongjiang is the leading grain producing province in China.

During the 1980's the symptom of barren wheat due to boron (B) deficiency was found in large areas of the province, especially in the state farms located in the northern borders. A soil survey through the province showed that nearly 50 percent of the soils were B deficient using a critical level of 0.5 mg B kg^{-1}. On severely B deficient soils, wheat yield was very poor (Wang, 1982; Yang, 1986; Zhang, 1989). Boron application on these B deficient soils increased yields of wheat and sugarbeet by 30–50%.

This paper deals with available soil B, adsorption of applied B and the effect of B application on crops yields using the Systematic Approach introduced by Agro Services International (ASI) of Orange City, Florida, USA, through the Potash and Phosphate Institute's China Program (Beijing Office of PPIC, 1992).

The Systematic Approach uses soil sampling, routine soil analysis for all plant nutrients, soil adsorption studies on some nutrients, and greenhouse pot experiments as the basis for establishing field trials. The continuous procedures to diagnose soil nutrient status and to improve upon this, was called the Systematic Approach.

Materials and methods

Soil sample collection and analysis

In 1989, the Soil and Fertilizer Institute of Heilongjiang Academy of Agricultural Sciences established a cooperative project with the PPIC. Twenty one soils representing seven main soil types in the province were analyzed in the cooperative laboratory of the Chinese Academy of Agricultural Sciences and PPIC, in Beijing, using ASI methods (Beijing Office of PPIC, 1992). The 21 soils were: black soil (7 samples) from seven counties, with available B contents (mg B kg^{-1}) of: 0.52, 0.44, 1.09, 0.65, 0.53, 0.40, 0.34; Meadow soil (5 samples) from five counties, with available B contents of: 0.30, 0.76, 0.71, 0.1, 0.44; Chernozemic soil (4 samples) from three counties, with available B contents of: 0.55, 0.96, 0.62, 0.76; Sandy soil (2 samples)

18

Table 1. Chemical properties of the soils studied in the greenhouse and field.

Code	pH	O.M. %	N	P	K	S	B	Cu	Fe	Mn	Zn
			mg kg^{-1}								
001	5.8	2.9	24.0	34.0	78	9.0	0.52	2.8	16.0	19.8	2.8
002	5.7	4.4	14.0	17.0	74	28.0	0.25	3.9	131	18.8	3.4
003	6.0	5.3	6.0	15.0	7	16.0	0.30	2.1	27.0	17.7	1.7
004	6.2	5.1	12.0	22.0	109	16.0	0.55	2.2	11.0	13.2	1.2
005	6.2	4.6	9.5	16.1	67	6.5	0.44	1.6	64.1	18.7	1.4
006	5.1	2.8	26.6	22.8	31	26.2	0.10	4.3	486	42.3	4.2
007	6.0	6.1	14.9	14.8	105	5.6	0.65	1.7	44.8	9.2	1.5
008	5.5	5.2	30.6	20.8	82	31.5	0.71	3.0	89.2	35.1	1.9

from two counties, with available B content of: 1.65, 1.06; there was one sample each from the lessive, dark brown soil and paddy soils, with available B contents of: 0.25, 0.42 and 0.21, respectively.

The ASI method used a B critical level of 0.3 mg B mg^{-1}, but suggests that two times the critical level was suitable for crop growth and development for experimental conditions. Using these criteria, 8 soils among those studied were predicted to be B deficient. The eight soils were: black soil with thin layer in Harbin city, (code number 001); lessive soil in Lincow county (code 002); meadow black soil in Qingan county (code 003); meadow chernozem in Shuangcheng city (code 004); black soil with loess mother material at Niujia town (code 005); meadow soil at Conghe town (code 006); black soil at Zhaoguang farm (code 007); meadow soil in Fangzheng county (code 008).

The chemical characteristics of these 8 soils are presented (Table 1). Boron content ranged from 0.1-0.7 mg kg^{-1} with the lowest the meadow soil from Conghe town and the highest the meadow black soil from Fangzheng county.

Soil adsorption study

The purpose of this study was to determine B adsorption by each of the 8 low B soils. Solutions of increasing concentrations of B (0.25, 0.5, 1.0, 2.0, 4.0 mg B kg^{-1}) were added to a known volume of soil. After seven days of anaerobic/aerobic conditions, available B was extracted from the soil using the ASI extractant (0.08 M calcium phosphate solution) and measured. Native B content in soils 007 and 008 were higher than 0.6 mg B kg^{-1} soil thus, they were excluded from adsorption studies. Curves relating extractable B to B application rates are roughly divided into 2 groups. Those with low original soil B content (006, 001, and 003) and others

with relatively high original soil B content (005, 004, and 001).

The slope of the curves of 004 and 006 were flatter indicating that the B adsorption capacity of these soils was stronger than the others. The curves of 002, 003; and 004 and 005 were similar indicating similar B adsorption capacities. This information was used, with the original soil analysis data, to establish pot experiments.

Pot experiments

To obtain information on B as well as other plant nutrients that could possibly limit crop yield, pot experiments were set up. From the results of the soil analysis and adsorption studies, an optimum treatment (Opt) was selected for each soil. Plant nutrients were either added or left out of the Opt treatment depending on their need based on the original soil analysis and adsorption study. There were 13–16 treatments each with four replications.

The indicator plant, sorghum (*Sorghum bicolor*) was planted in plastic pots with a volume of 500 ml. The growing period was 30–40 days. Eight plants were harvested from each pot. Dry weight of harvested plants was measured. Only results concerning B are presented (Table 2).

Field experiments with boron application

Based on the information obtained from the soil analysis, adsorption studies and pot experiments with the 8 soils, field trials were designed where any plant nutrients considered deficient were applied at equal amounts to all treatments, with the exception of one treatment where B was omitted. According to the soil properties and local climatic conditions field trials were

Table 2. Effect of omitting boron on dry weights (g pot^{-1}) of sorghum plants in eight soils of Heilongijang province in a pot experiment

Code	Opt^a	Opt-B	5% LSD
001	3.62	3.43	NS
002	4.04	3.66	NS
003	3.97	3.81	NS
004	3.78	3.06	NS
005	5.80	5.20	0.50
006	6.20	5.67	NS
007	5.20	6.13^b	1.55
008	4.80	5.32^b	NS

^{a} Opt - complete nutrients supplied at optimal rates based on prior soil analysis
Opt - B - boron omitted from the complete nutrient mix.
NS - Not significant.
^{b} Treatment was Opt plus B in soils 007 and 008.

conducted with maize on the soils of 001, 002, 003, 004 and 005; soybeans on 001, 002, 003 and 005; and rice on 001, 002, 003, 006 and 007. These were conducted in the years 1991 and 1993. Sugarbeet trials were carried out on 001, 003, 004 and 007 in 1994. A total of eighteen field trials were completed with B application on the four different crops. The purpose was to determine if information obtained from soil analysis, adsorption and greenhouse studies was correct and how B application affected crop yields in the field. Thus, only simple comparisons were made in the experimental design.

The Opt for maize was (kg ha^{-1}): N 140, P 44, K 125, ZnSO$_4$ 20, borax 6.0; for soybean: N 85, P 55, K 104, ZnSO$_4$ 15, borax 6.0; for rice: N 125, P 50, K 125, ZnSO$_4$ 15, borax 7.5; and for sugarbeet: N 100, P 55, K 75, borax 9.0. The B portion of the experiment simply had two treatments: Opt; and Opt -B.

Plot area for maize and sugarbeet was 30 m^2, and for soybean and rice was 21 m^2. Each treatment had 4 replications except for rice with 3.

Results and discussion

Pot experiments

When B was not applied to soils 001–006, yields decreased by 4–19% (Table 2). Of these soils, 004, 005 and 002 were seriously deficient compared with

the other 3. Soils 007 and 008 had high B concentrations and their adsorption studies did not show an effect of strong adsorption of B, thus B was not added to the optimum treatments of these soils. However, to check whether B was needed or not, a treatment with B was added in the pot studies, to test against the Opt.

Results indicated that with addition of B on soils 007 and 008, dried plant weights increased by 17.9% and 10.8%, respectively, meaning that the Opt was not predicted accurately by either the soil analysis or adsorption studies. Thus, an adequate level of B was added to the Opt in the field trials.

Field experiments with boron application

With borax application on maize on 5 of the soils, yield increases ranged between 3.6–11.7%. The average increase was 566 kg ha^{-1}. Boron showed the greatest effect on maize yield on soils 002, 004 and 005 (Table 3). Borax had a definite positive effect on soybean yield on 3 of the 4 soils: average yield increase was 118 kg ha^{-1} with an average percent yield increase of 4.0% (Table 3). With borax application on rice on 5 of the soils, average yield increase was 560 kg ha^{-1}. Boron showed a consistent positive effect on rice yield with four of these soils (exception 003) (Table 3).

Boron application reduced the incidence of bare ears of maize (Table 3), empty pods of soybean (Table 3) and empty grains of rice (Table 3).

Boron had a significant effect on sugarbeet yield on all 4 soils studied. With borax application at the rate of 9.0 kg ha^{-1}, yield of sugarbeet increased from 2.3–15.8%. Sugar content increased by 0.3–0.4%. Average sugar yield increase was 3986 kg ha^{-1}, an average increase of 10.2% (Table 4). Furthermore, B application had a significant effect on reducing the incidence of physiological disorders of sugarbeet (crinkled leaves and rotten heart) reducing it to 1.8–3.6%, while without B treatment levels were 5.7–14.2%.

Conclusion

Using the Systematic Approach methodology of soil analysis, adsorption studies and potted plant trials, 8 out 21 soils were identified as potentially B deficient and studied in the field for B responses with 4 crops.

Boron application increased yields through improving grain fruiting in maize (reduced empty ears), soybean (reduced empty pods) and rice (reduced empty grain) and by reducing the incidence of physiologi-

Table 3. Effects of boron application on maize, soybean and rice yield on soils of Heilongjiang province.

Soil	Treatment	Maize		Soybean		Rice	
		Grain yield (kg ha^{-1})	Empty ears (%)	Seed yield (kg ha^{-1})	Empty pods (%)	Grain yield (kg ha^{-1})	Empty grain (%)
001	Opt	8600	4.2	3232	7.3	8207	4.6
	-B	8276	8.6	3147	10.2	7691	8.8
002	Opt	4389*	7.5	3033	8.1	8466	6.5
	-B	3933	20.2	2787	11.7	7850	11.4
003	Opt	7035	4.6	2938	5.2	10500	6.2
	-B	6584	11.7	3010	7.1	10167	9.3
004	Opt	8461*	3.9	–[1]	–	–	–
	-B	7671	12.4	–	–	–	–
005	Opt	7724*	6.1	2889	6.8	–	–
	-B	6917	21.4	2678	10.6	–	–
006	Opt	–	–	–	–	9035	4.3
	-B	–	–	–	–	8227	11.8
007	Opt	–	–	–	–	9168	5.7
	-B	–	–	–	–	8642	10.5

* Significant yield increase ($p<0.05$)
[1] not tested

Table 4. Effects of boron application on sugarbeet yield and quality on 4 soils of Heilongjiang province.

	Soil code							
	001		003		004		007	
	Opt	-B	Opt	-B	Opt	-B	Opt	-B
Yield (kg ha^{-1})	40900	40000	47017*	40596	43730*	38602	41062*	37566
Sugar content (%)	12.9	13.1	12.4	11.9	11.7	11.4	12.6	12.2
Sugar yield (kg ha^{-1})	5276	5240	5783	4831	5116	4401	5174	4583
Rate of disorder Occurrence[1] (%)	2.7	5.7	1.8	11.4	3.1	8.6	3.6	14.2

* significant ($p<0.05$)
[1] Disorder in the table represented the incidence of small crinkled leaves and rotten heart, which are symptoms of B deficiency in sugarbeet.

cal disorders in sugarbeet (crinkled leaves and rotten heart).

Adsorption studies demonstrated that very little applied B was adsorbed by these soils. Thus, relatively small amounts of B (6–9 kg borax ha^{-1}) are needed to satisfy crop requirements for this plant nutrient. These quantities are not likely to cause B toxicity. Boron application should be recommended to farmers at the above rates for the studied crops.

The normal practice for conducting fertilizer experiments in Heilongjiang is to collect soil samples, and analyze their basic fertility without testing for all plant nutrients. The trial then tests the plant nutrients suspected of being deficient, but without regard to other plant nutrients that may be limiting yield responses or to the possible adsorption of the added plant nutrients by the soil. The Systematic Approach has the advantage of eliminating all possible plant nutrients that could limit responses to others under study while estimating the quantities needed to be applied to account for any adsorption phenomena. In other words, more information is available about the soil prior to establishing field trials. This increases the effectiveness and efficiency of soil fertility research in Heilongjiang. The Systematic Approach methodology is very effective in finding yield limiting factors of soil nutrition and is a basis for true balanced fertilization.

The field trials of maize, soybean and rice were not completed in the same year, but the climate was relatively normal and stable in those years (1991 and 1993). Weather might cause some affect on B efficiency to the crops, but it was not significant. The main factor affecting crop yield differently was soil properties.

Acknowledgements

This research was supported in part by the Potash and Phosphate Institute-China Program.

References

Beijing Office of PPIC 1992 Soil Survey on Nutrient Status by the Method of Systematic Approach. Chinese Publishing House of Agricultural Sciences and Technology, Beijing, PRChina.

He N and Meng X 1987 The Principle of Plant Nutrition. The Publishing House of Sciences and Technology, Shanghai, 238–249 p.

Shkolnik M Y 1984 Trace Elements in Plants. Elsevier Science, Amsterdam. 463 p.

Wang X 1982 Summary on the Plot Experiments and Its Demonstration on Micronutrients in the Whole Province. Heilongjiang Agric. Sci. 1, 32. (In Chinese)

Yang R 1986 Review and Prospect on the Experiments of Micronutrients in the Province. Heilongjiang Agric. Sci. 1, 31. (In Chinese)

Zhang, X 1989 Relationship of Soil Available Boron and Wheat Development. Heilongjiang Agric. Sci. 4, 36. (In Chinese)

R.W. Bell and B. Rerkasem (eds.), Boron in Soils and Plants, **00**: 23–27, 1997.
© *1997 Kluwer Academic Publishers.*

Sunflower response to boron as affected by liming

E.C.A. Souza[1], C.A. Rosolem[2] & E.L.M. Coutinho[1]
[1] *Department of Soils and Fertilizers, College of Agronomy, Jaboticabal, and*
[2] *Department of Agriculture, College of Agronomy, Botucatu, Sao Paulo State University, CP 237, 18603-970 Botucatu SP, Brazil*

Key words: aluminum, base saturation, boron, nutrition, pH

Abstract

In Sao Paulo state, Brazil, liming has often caused boron (B) deficiency in sunflowers (*Helianthus annuus*). An experiment was conducted in pots to study the effects of liming on soil and added B, as well as its consequences for plant growth and nutrition. Sunflower plants were grown up to 52 days after emergence in 4.2 l pots receiving 0, 2.1, 4.2 and 6.3 g of dolomitic limestone pot^{-1} and 0, 1.0 and 2.0 mg B kg^{-1}. There was an increase in soil calcium, magnesium, base saturation and pH due to liming. Calcium in the plant tops was also increased but potassium contents were decreased by liming. When the B was applied as fertilizer, there was an increase in B availability due to liming. This was possible because B was applied when most of the Al-hydroxides of the soil were already precipitated, at pH (CaCl$_2$) around 5.2. The highest dry matter yield was observed in the presence of applied lime and B, when the plant top B concentration was 114 mg kg^{-1} dry wt and soil B was around 1.0 mg kg^{-1}.

Introduction

Sunflower (*Helianthus annuus*) is sensitive to aluminium (Al) toxicity and calcium (Ca) deficiency (Foy et al., 1974). In sandy, acid soils, plant development is impaired, plant emergence is uneven, the roots are under-developed and yields are very low (Blamey, 1976).

Sunflower yield responses to lime are well documented. Blamey and Nathanson (1977) related the response to lime to a decrease in Al saturation in the soil. Ungaro et al. (1985) verified a linear response to lime as the pH and base saturation of the soil were increased. Impre and Eulalia (1988) concluded that a pH (KCl) of 6.0 is optimal for sunflower growth.

In Brazilian soils, liming is recommended to raise the soil base saturation to 70%, which corresponds to a pH around 6.0 as determined in water and 5.4 in CaCl$_2$ (Quaggio and Ungaro, 1985). Under these conditions, boron (B) deficiency is often observed and B application is recommended (Ungaro et al., 1985). The availability and plant uptake of added B are decreased at pH (water) higher than 6.3 (Peterson and Newman, 1976).

The severity of B deficiency depends on several factors, including the period of time elapsed following liming. Freshly precipitated Al(OH)$_3$ adsorbs larger quantities of B than soil, even though ageing of Al(OH)$_3$ greatly reduces its B retention capacity (Tisdale et al. 1985). If the added B availability is expected to decrease at pH (water) above 6.3, B deficiencies would not be expected to occur in Brazilian soils when the base saturation of the soil is raised up to 65–70%, with a pH (water) around 6.0. Since ageing reduces the B retention capacity of Al(OH)$_3$, the effect of liming on indigenous and added B availability may be different because B fertilization is usually made some time after liming.

An experiment was conducted to study the effects of liming on indigenous and added B as well as its consequences on B availability to sunflower and plant growth.

Materials and methods

Sunflower plants, triple hybrid C 621, were grown in 4.2 l pots containing an Allic Dark Red Latosol (Haplorthox, 230 g clay kg^{-1}, 750 g sand kg^{-1}). Originally the soil showed a pH ($CaCl_2$) of 4.9, 7 mg of P (resin extractable) kg^{-1}, 0.1 $mmol_c$ K^+ kg^{-1}, 5.0 $mmol_c$ Ca^{2+} kg^{-1}, 3.0 $mmol_c$ Mg^{2+} kg^{-1}, 47 $mmol_c$ H^+ + Al^{3+} kg^{-1}, base saturation of 22%, 23 g organic matter kg^{-1} and 0.5 mg of B (hot water) kg^{-1}.

Lime was applied at 0, 2.1, 4.2 and 6.3 g pot^{-1}. After a wet incubation period of 20 days, B ($Na_2B_4O_7.10H_2O$), was applied at 0, 1 and 2 mg kg^{-1} soil. Nitrogen and P were applied at 200 mg kg^{-1} and K at 150 mg kg^{-1}.

The experimental design was a randomized block with treatments arranged as a 4×3 complete factorial with three replications.

Three sunflower plants were grown per pot for 52 days after plant emergence. At harvest, the tops of the plants were cut at soil level, oven dried and wet digested with a nitric-perchloric acid mixture prior to determination of Ca, potassium (K) and B concentrations. The soil was sampled and analyzed for selected fertility characteristics including B, that was extracted with hot water and analyzed as in Cruz and Ferreira (1984).

Results and discussion

Soil base saturation increased linearly from 22 to 56% ($y=23.6+5.11x$, $R^2=0.99$), corresponding to a linear increase in soil pH ($CaCl_2$) from 4.8 to 5.4 (pH= $4.91+0.87x$, $R^2=0.91$) with the increase in lime application from 0.0 to 6.3 g pot^{-1}. Soil Ca and Mg were raised to 30 $mmol_c$ kg^{-1} and 20 $mmol_c$ kg^{-1}, respectively.

With increasing lime rate, Ca concentration in plant tops was increased from 6.0 to 10.8 g kg^{-1}. These values are low when compared with those obtained by Sfredo et al. (1983) in recently matured leaves at flowering stage (17 to 28 g kg^{-1}). The differences compared to this experiment may be due to plant age and the plant part that was analyzed. The K concentrations in the plant tops were decreased from 46 to 30 g kg^{-1}, which is within the sufficient range defined by Sfredo et al. (1983).

Hot water soluble B levels in the soil were not affected by liming in the pots without B application, but there was an increase in B levels due to liming in

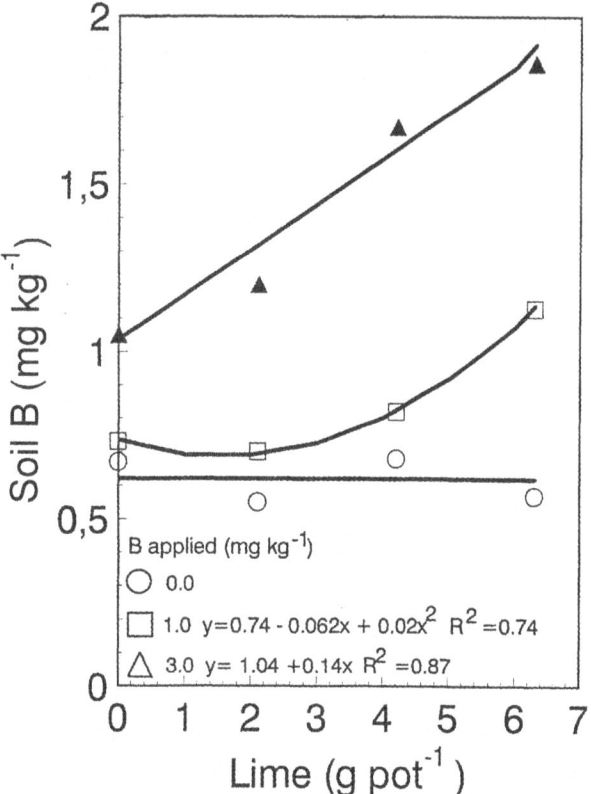

Figure 1. Effects of lime and B application rates on hot water soluble B in an Allic Dark Red Latosol following growth of sunflower for 52 days.

the pots that received B fertilizer (Fig. 1). Available B (hot water) was also positively correlated with pH for acid soils by Berger and Truog (1945). Organic matter mineralization could explain these results, but if this were true, the same trend would be observed in pots without B fertilization (Tisdale et al., 1985), which was not the case in our experiment.

Fresh Al hydroxide gels tend to adsorb large amounts of B, but their B sorption capacity decreases rapidly with age of the gels (Beyrouty et al., 1984). In our experiment, in a soil with 75% of sand, lime was applied 20 days before B fertilization; therefore, when B was applied most of the Al precipitation had probably already occurred. Furthermore, the amount of B adsorbed depends on soil texture and increases with clay content (Mezuman and Keren, 1981).

It is important to note that there was some B adsorption. When we applied 1 mg B kg^{-1}, there was only a slight increase in B availability up to 3 g lime pot^{-1}, which corresponds to 40% of base saturation and a pH ($CaCl_2$) of 5.2. In this pH, there was no effect of lime on B availability. With further increases in pH and base saturation, B availability was increased. For the highest B rate, the increase was linear (Fig. 1). The maximum soil pH (CaCl₂) obtained in the present experiment was 5.4, corresponding roughly to a soil pH (water) of 5.9, which is still below the pH (water) of 6.3 above which Peterson and Newman (1976) reported a substantial decline in B availability. As reported by Raij (1991), at a pH (CaCl₂) around 5.2, Al activity in São Paulo soils is very low and non-toxic, i.e. most of Al will be precipitated. In this way, further increases in pH will precipitate very small amounts of Al and cause little extra B adsorption.

Bloesch et al. (1987) reported a reduction in B sorption in the presence of phosphate, indicating that the two ions compete for the same type of sites. Competition would be expected, since phosphate is known to bond to Al-hydroxide surfaces. In this experiment, we applied a considerable amount of basal P. The increase in B availability with increase in soil pH in this study may be consequential upon the increased availability of P with increase in pH (Siqueira, 1989), which leads to a greater competition of P with B for adsorption sites.

All of these inferences are supported by the results obtained for B concentration in the sunflower tops (Fig. 2). There was no effect of lime on B concentrations in the tops of plants grown without B fertilization; however, when B fertilizer was applied, the B concentration in the plant tops increased linearly with increase in lime application rate. Blamey and Chapman (1979) also found an increase in B uptake by sunflower when the pH (KCl) was raised up to 7.0.

Dry matter yields of sunflower plants increased with applied B (Fig. 3). The increase was linear without lime and quadratic in the presence of lime. At low rates of B and lime, the plants showed symptoms characteristic of B deficiency: they were stunted and some showed death of the terminal bud. Optimum yields were observed with the application of 4.2 g lime pot^{-1} and 1 B mg kg^{-1}, corresponding to a base saturation of 45%, a pH(CaCl₂) of 5.2 and 0.8 mg hot water soluble B kg^{-1} soil.

No visual symptoms of B toxicity were observed at the highest lime rates, despite B concentrations in the plant tops > 110 mg kg^{-1}, the lowest concentration at which Aitken and McCallum (1988) reported the development of B toxicity symptoms in sunflower.

The maximum dry matter yield was observed when the B concentration in plant tops was 114 mg kg^{-1} (Fig. 4). This value, in spite of the differences in organs

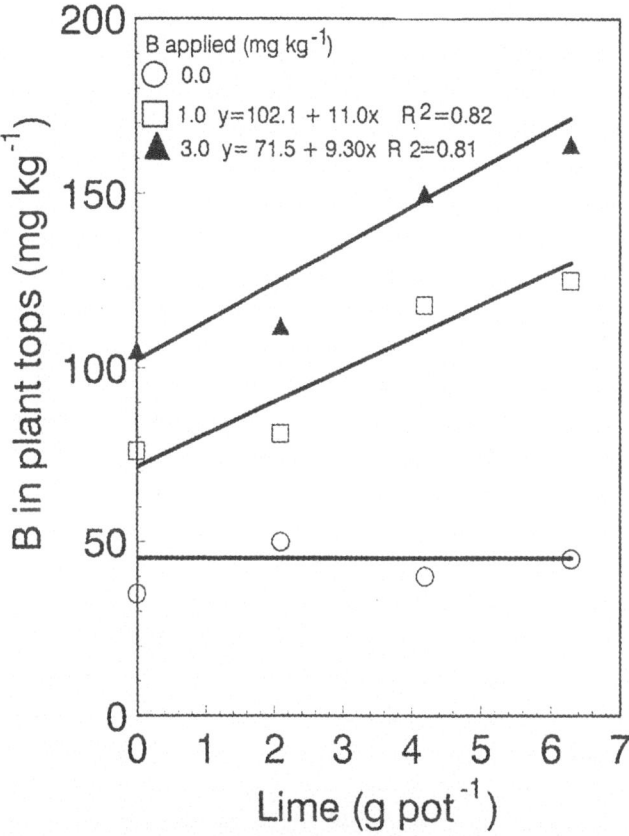

Figure 2. Boron concentration in sunflower plant tops, 52 days after plant emergence, as affected by lime rate and boron fertilization.

and plant age, is within the range reported by Sfredo and Sarruge (1990) as sufficient in sunflower leaves at flowering: 92 to 128 mg kg^{-1}. This B concentration in the plant tops was obtained when the B extractable level in the soil was around 1.0 mg kg^{-1}. Aitken and McCallum (1988) determined that 1.9 mg B l^{-1} in the soil solution was toxic to sunflower.

In summary, the effect of Al-hydroxides precipitation due to liming on B availability to sunflower in this sandy soil depends on when the nutrient is applied. Whereas most of the B added before liming is adsorbed when the hydroxides precipitate, when B is added after most of the lime reaction has already occurred in the soil, the adsorption is low. This occurs because Al precipitation after lime reaction is not as marked as it is at a lower pH.

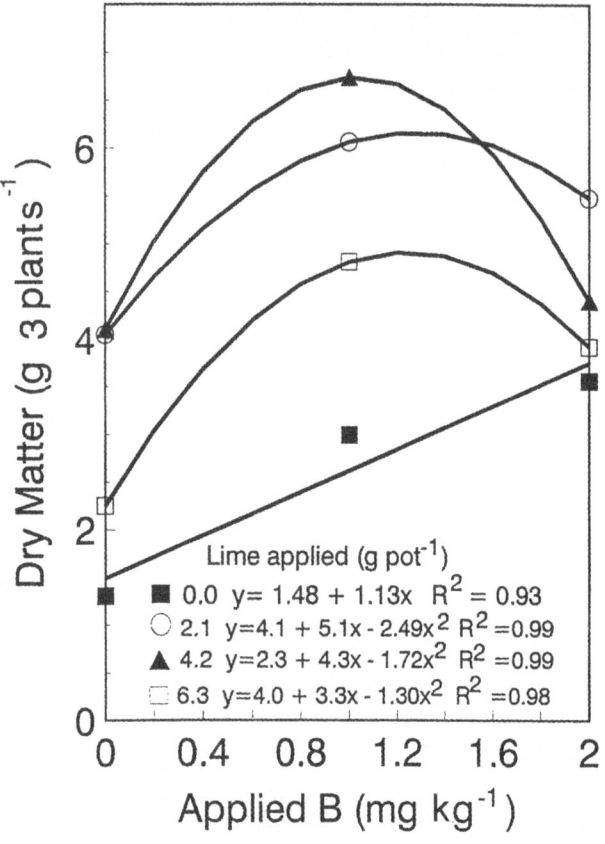
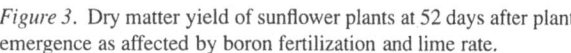

Figure 3. Dry matter yield of sunflower plants at 52 days after plant emergence as affected by boron fertilization and lime rate.

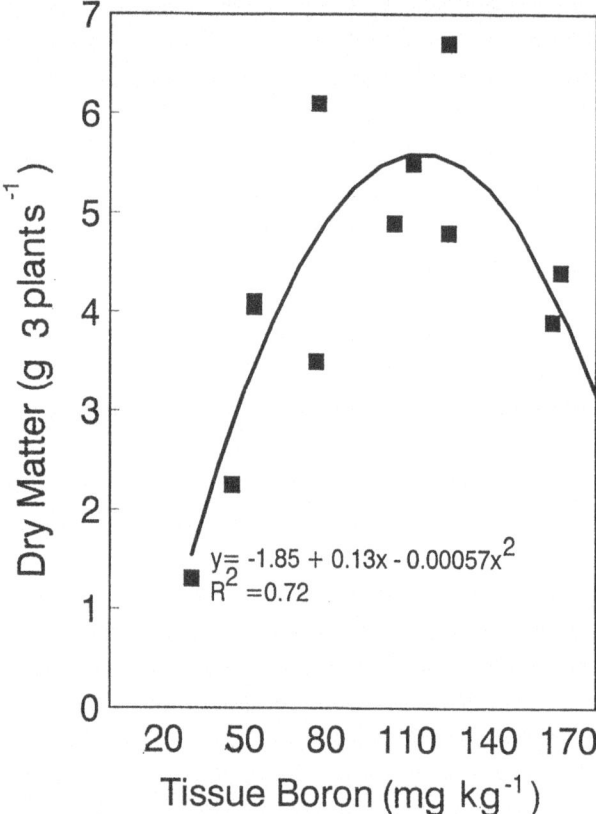

Figure 4. Relationship between boron concentration in sunflower plant tops and dry matter yield of sunflower 52 days after plant emergence.

Acknowledgements

We acknowledge the help of the graduate students I.P. Garcia, V.C. Pipolo and J.S. Assis in conducting the experiment, and the careful and friendly review by an anonymous referee.

References

Aitken R L and McCallum L E 1988 Boron toxicity in soil solution. Aust. J. Soil Res. 26, 605–610.

Berger K C and Truog E 1945 Boron availability in relation to soil reaction and organic matter content. Soil Sci. Soc. Am. Proc. 10, 113–117.

Beyrouty C A, Van Scoyoc G E and Feldkamp J R 1984 Evidence supporting specific adsorption of boron on synthetic aluminum hydroxides. Soil Sci. Soc. Am. J. 48, 284–287.

Blamey F P C 1976 Boron nutrition of sunflowers on an Avalon medium sandy loam. Agrochemophysica 8, 5–10.

Blamey F P C and Chapman J 1979 Soil amelioration effects on boron uptake by sunflowers. Commun. Soil Sci. Plant Analysis 10, 1057–1066.

Blamey F P C and Nathanson K 1977 Relationships between aluminium toxicity and sunflower yields on an Avalon medium sandy loam. Agrochemophysica 9, 59–66.

Bloesch P M, Bell L C and Hughes J D 1987 Adsorption and desorption of boron by goethite. Aust. J. Soil Res. 25, 377–390.

Cruz M C P and Ferreira M E 1984 Selection of methods for the determination of boron availability in soils. Pesq. Agropec. Bras. 19, 1457–1464.

Foy C D Orellana R G, Schwartz J W and Fleming A L 1974 Responses of sunflower genotypes to aluminum in acid soil and nutrient solution. Agron. J. 66, 293–296.

Impre K and Eulalia V 1988 Fertilization and liming of sunflower on acidic sandy soil. Noventermeles 37, 541–547.

Mezuman U and Keren R 1981 Boron adsorption by soils using a phenomenological adsorption equation. Soil Sci. Soc. Am. J. 45, 722–726.

Peterson L A and Newman R C 1976 Influence of soil pH on the availability of added boron. Soil Sci. Soc. Am. J. 40, 280–285.

Quaggio J A and Ungaro M R G 1985 Sunflower. *In* Fertilizer Recommendations for São Paulo. Ed. B van Raij, pp 39 Agronomic Institute of São Paulo, Campinas, Brazil.

Raij B van 1991 Soil Fertility and Fertilization. Ceres Agronomic Publ. Co. and Brasilian Association for Potash and Phosphate Research, São Paulo, Brazil. 343 p.

Sfredo G J and Sarruge J R 1990 Micronutrient concentrations in organs of sunflower plants. Brasilian J. Agric. Research 25, 1727–1732.

Sfredo G J, Sarruge J R and Haag H P 1983 Nutrient uptake by two sunflower cultivars in field conditions: I. Macronutrient concentrations. Annals of the Luiz de Queiroz College of Agronomy 40, 1165–1187.

Siqueira J O F 1989 Crop response to lime and phosphorus in Rio Grande do Sul and Santa Catarina and its economy. *In* Amend-ments to Correct Soil Acidity. Ed. S Volkweiss, J Kaminski and F Becker, pp. 151–176. Federal University of Santa Maria, Rio Grande do Sul, Brazil.

Tisdale S L, Nelson W L and Beaton J D 1985 Soil Fertility and Fertilizers. Macmillan Publ. Co., New York, 754 p.

Ungaro M R G, Quaggio J A, Gallo P B, Dechen S C F, Lombardi F and Castro O M 1985 Sunflower behavior in relation to soil acidity. Bragantia 44, 41–48.

R.W. Bell and B. Rerkasem (eds.), Boron in Soils and Plants, 29–34.
© 1997 *Kluwer Academic Publishers.*

Covering plants at night in the winter increased seed yield of transplanted oilseed rape (*Brassica napus* L. cv. Zheyouyou 2) on a low boron soil

Zhengqian Ye[1,2], R. W. Bell[1], Longbin Huang[1], Yuai Yang[2] & B. Dell[1]
[1] *Division of Sciences, Murdoch University, Perth, 6150, Western Australia*
[2] *Department of Agricultural Chemistry, Zhejiang Agricultural University, Hangzhou, 310029, PR China*

Key words: boron deficiency, borax, canopy cover, low temperature, oilseed rape, seed yield

Abstract

Winter oilseed rape (*Brassica napus* L.) is a major crop in the middle and lower Yangtse river basin, China, where boron (B) deficient soils are widespread. Appearance of B deficiency in oilseed rape often coincides with cold weather during its winter growth. To understand effects of cold weather on plant response to low soil B, a field experiment with oilseed rape cv. Zheyouyou 2 grown in a red soil with low B availability was conducted in Zhejiang province, China. Canopy covers made from transparent plastic sheets were used only at night to modify the microclimate during the early reproductive growth between pre-exposure of flower buds and green bud stage. Canopy treatment (T) increased the minimum air temperature inside the cover by up to 1.5°C when the minimum air temperature in the open was below 0°C in early February. Covering plants for 15 days in early February strongly increased shoot dry weight at all levels of B supply. That covering plants increased shoot dry weight of B deficient plants without increasing their leaf B concentration suggests that internal B requirements were decreased. However, because later plant responses at maturity gave contradictory responses, it was concluded that further study of low temperature × B interactions need to be done in controlled environments.

Introduction

Low temperature in winter and early spring is a major constraint in winter crop production especially during the reproductive phase when it may cause the failure of reproduction (Lardon and Triboi-Blondel, 1994). Low boron (B) supply to seed crops may also result in the failure of seed production (Rerkasem et al., 1993). Moreover, plant response to B may be affected by environmental factors, such as temperature (Morghan and Mascangni, 1991). Reports on the nature of the interaction between B and temperature have been inconclusive. Studies by Mahalakshmi et al. (1995) and by Walker (1969) which were done at 5°C and 12°C, respectively, were of uncertain relevance to the present study because they examined effects on B toxicity in the former and on B adequate plants in the latter case. Whilst increasing root temperature alleviated or even overcame B deficiency in cassava plants responses occurred in the range 19–33°C which is well above the range of interest for winter crops (Forno et al., 1979).

Winter oilseed rape (*Brassica napus*) is reputed to be sensitive to B deficiency (Yang et al., 1993). It is a major winter crop in the middle and lower Yangtse river basin in China, where the weather in winter and early spring is cold (see Fig. 1), and soils with low B level are widespread (Yang et al., 1993). However little is known about the effect of low temperature on B response of winter oilseed rape.

To reduce night time heat loss and protect plants from the cold, covering is often used in winter crop production (Mather, 1974). Therefore this present study, which aimed at understanding if cold weather modifies plant response to B deficiency under field conditions, involved covering treatment plants with a plastic canopy at nights for 15 or 17 consecutive nights in late winter and early spring. Our hypothesis in the present study was that changing the plant's micro environment by imposing a night time canopy cover improves plant

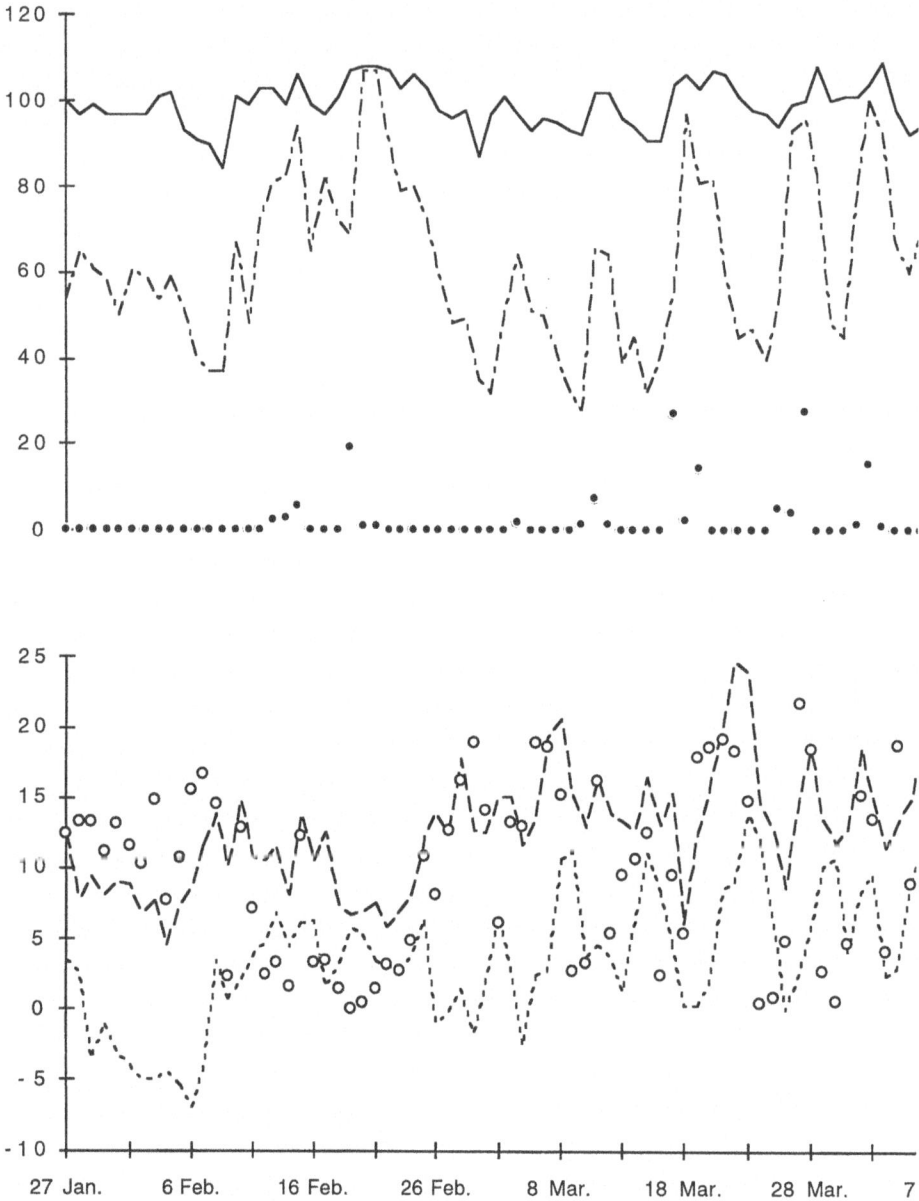

Figure 1. Weather conditions at field site. Key to symbols: ____ max. relative humidity (R.H.) %, – - min. R.H. %, ● Total rainfall (mm day^{-1}); — — max. air temperature (°C), – – min. air temperature (°C), ○ Total solar radiation (MJ day^{-1}).

growth, and decreases plant B internal requirement; or, conversely, increasing B supply to plants improves their resistance to low temperature.

Materials and methods

A field experiment was conducted in Zhejiang province, China in 1994–95, in a paddy soil (red soil)

with: pH(H$_2$O) 5.5; organic matter, 2.24%; hot 0.01 M CaCl$_2$ extractable B, 0.26 mg kg^{-1} soil (Spouncer et al., 1992). Oilseed rape cv Zheyouyou 2 seedlings were transplanted from a seedling bed (seed sown on September 26, 1994) into main fields at the 6 to 7 leaf stage on November 5 at a spacing of 27 × 31 cm. Plants received the following basal fertilisers (kg ha^{-1}): NPK compound fertiliser (15:15:15), 450; urea, 100; Mg SO$_4$.7H$_2$O, 75; CaSO$_4$, 75;

$CuSO_4.5H_2O$, 10; $ZnSO_4.7H_2O$, 15; $MnSO_4.H_2O$, 10; $Na_2MoO_4.2H_2O$, 1.5. Additional fertilisers were applied at seedling stage (kg ha^{-1}): NPK compound (15:15:15), 450; and at stem elongation stage: urea, 150.

Treatments were arranged in a split plot design with main plots (5.4 m × 3.7 m) comprising soil application of 0, 1.5 or 15 kg borax ha^{-1} (B0, B1.5, B15), and sub-plots comprising an area of 3.7 m × 1.8 m which were either uncovered (T0), or covered overnight at T1 or T2 (see details following). Each treatment had four replicates. Canopy covers made from clear plastic sheets were used to cover the plants only during the night (between sunset and the following sunrise) to modify microclimate. The canopy was about 40 cm high. This treatment is referred to as canopy treatment (T) in the following text. It was applied over two periods determined according to plant growth stage and weather conditions: T1, Jan. 27–Feb. 11, 95 (GS 3,1–3,3); T2, Feb. 11–28, 95 (GS 3,3–3,4) (GS: Growth Stage, after Sylvester-Bradley, 1985). The non-canopy treatment was used as the control (T0).

An automatic weather station was used to record air (2.5 m above ground) and soil (10 cm below soil surface) temperatures, rainfall, solar radiation, relative humidity, and wind speed. The soil temperature sensor did not work properly when the temperature was below 0°C. The weather station was located about 50 m away from the experimental site. As only one set of sensors was available, the weather station was only used to record the climate data in the open. Air temperatures in- and outside plastic covers were recorded daily with the use of min./max. thermometers.

Plant samples were collected at seedling stage, and before and after canopy treatments from control and treatment plots. Youngest open leaf blade (YOL, after Huang et al., 1996) samples were collected and dried at 70°C. They were digested in concentrated nitric acid at 130°C and concentration of B was determined by inductively coupled plasma atomic emission spectrometry (Zarcinas et al., 1987). Percentages of plants at specific growth stages were recorded. Plants were harvested on 24 May 95. Seed yields were recorded at maturity.

Results and discussion

Weather conditions in the field

Weather conditions from late winter to spring, including relative humidity, air temperature, total rainfall, and total solar radiation, are shown on Fig. 1. Over the whole growing season, coldest temperatures occurred in February and early March. The spring weather fluctuated markedly. Minimum temperatures <5°C continued into late March and even in mid April. There was a little rainfall in the winter, but the soil contained sufficient water in the winter due to a perched watertable 70 cm below the ground surface. The rainy season started in mid March.

The coldest temperatures occurred during Jan 29 to Feb 7, within the period of T1. The daily minimum air temperature in this period ranged from $-6.9 \sim -1.0$°C. Whilst it was much warmer during the period of T2, minimum air temperatures below 0°C occurred on Feb 26 and 27. The mean minimum and mean maximum air temperatures during the period, T1 and T2, were -2.0 and 9.4°C, and 3.8 and 9.7°C, respectively. Canopy treatment increased the air temperature inside the cover. The minimum air temperature inside the cover increased by up to 1.5°C when the minimum air temperature in the open was below 0°C in early February (Table 1).

Effect of cold weather on plant growth rate with different B levels

Plant growth was severely inhibited and 25% of plants died in low B soil without B application (Table 2). In B0 plots, the plants had YOL B concentrations <10 mg B kg^{-1} which is the minimum B concentration required for adequate leaf function (Huang et al., 1996).

Plant growth slowed down when it became colder. The effect of cold weather on plant growth was more obvious in B0 plot. Without canopy treatment, plants with B0 declined in dry weight from Jan 27 to Feb 1 due to the loss of old leaves; by contrast, plants with B continued growing even in the coldest period. When it warmed up in late February, plants grew faster. The plant biomass increment from Feb 11 to Feb 28 was more than 4 times that from Jan 27 to Feb 11.

32

Table 1. Effect of plastic covers over plants on minimum air temperatures in- and outside covers (°C) and number of days when minimum air temperature was below 0°C outside the cover.

Period	Outside	Inside	Days of air temperature $\leq 0°C$ outside the cover
Jan 27–Feb 11	$-7 \sim 3$	$-5.5 \sim 3.5$	10
Feb 11–Feb 28	$0 \sim 8$	$0.5 \sim 9.5$	2

Table 2. Effect of boron (B) application on youngest open leaf blade (YOL) B concentration (mg B kg^{-1} dry weight), shoot dry weight (g DW plant^{-1}), and seedling survival rate (%). Values are means of four replicates with standard errors in parentheses[a].

B (kg borax ha^{-1})	Shoot dry weight				YOLB (Dec 24, 94)	Survival rate
	(Dec 24,94)	(Jan 27, 95)	(Feb 11)	(Feb 28)		
B0	2.8 (0.2)	4.3 (0.4)	3.9 (0.3)	7.0 (0.3)	6.7 (1.0)	75 (5)
B1.5	6.9 '0.7)	8.8 (1.0)	9.4 (1.1)	13.9 (0.7)	19.2 (1.8)	94 (1)
B15	7.0 (0.4)	8.4 (1.2)	9.4 (1.1)	13.6 (2.4)	29.5 (6.1)	94 (2)

[a] Shoot and leaf samples were from non-canopy treatment at the date of sample collection shown.

Effect of covering plants

Vegetative growth
Covering plants during T1 improved plant vegetative growth (Table 3), while YOL B concentration was not affected by canopy treatment. By contrast, when it became warmer, the canopy treatment did not increase plant growth (T2); neither did it affect leaf B concentrations.

Reproductive growth
After removal of the canopy cover, the canopy treatment continued to influence growth of B fertilised plants (Table 4). In B1.5 and B15 plots, T1 or T2 treated plants set pods earlier than T0 and finally set more seeds. The effect of canopy treatment in B0 plots was not observed as severe B deficiency strongly depressed setting and filling of pods.

Responses of seed and pod numbers per plant suggest that canopy treatment and adequate B supply improved plant reproductive growth (Table 4), and that both low temperature and low B can depress reproductive growth. Whereas covering plants and B application interacted in the effects on seed set and seed yield per plant, canopy treatment had no effect on numbers of pods set. These results suggest that the short term canopy treatments mainly affected seed development, while B influenced both pod and seed development.

Thus canopy treatment reinforced reproductive growth of B1.5 and B15 plants exposed to cold.

Conclusions

From the seed yield results, there was no support for the hypothesis that increasing B supply to plants increased their resistance to low temperature. Whilst it might be argued that the increase in air temperature inside the plastic canopy was relatively small, it was nevertheless effective in stimulating vegetative growth by the end of a 15 day period (T1). And the canopy cover had no effect on vegetative growth at the end of T2 when temperature was the same inside as outside the canopy and temperatures were generally warmer. That shoot dry weight increased at the end of T1 despite no increase in leaf B suggests that increasing night temperatures may have decreased internal B requirements of oilseed rape plants. In contrast, to the shoot dry weight results which represent the short term response of oilseed rape to canopy covering over a 15–17 day period, the responses of pod set, seed set, and seed yield were rather more complex to interpret. Unlike the direct responses of shoot dry weight to B and changed canopy microclimate, seed yield responses were affected by a combination of factors acting on pod set and pod filling subsequent to the canopy treatment period as well as the indirect consequences of the canopy

Table 3. Effects of boron (B) and canopy treatment (T) on shoot dry weight (g DW plant^{-1}) and youngest open leaf blade (YOL) B concentration (mg B kg^{-1} dry weight). Values are means of four replicates with standard errors in parentheses.

B	T[a]	Shoot dry weight		YOL B	
		Feb. 11	Feb. 28	Feb. 11	Feb. 28
B0	T0	3.9 (0.3)	7.0 (0.3)	6.7 (0.4)	9.5 (0.6)
B0	T1/T2	7.0 (1.3)	8.2 (1.7)	5.5 (0.5)	9.1 (0.5)
B1.5	T0	9.4 (1.1)	13.9 (0.7)	15.2 (0.4)	19.8 (0.5)
B1.5	T1/T2	15.3 (1.3)	16.6 (1.3)	16.3 (0.3)	22.1 (0.5)
B15	T0	9.4 (1.1)	13.6 (2.4)	24.3 (1.6)	32.0 (1.4)
B15	T1/T2	11.8 (1.2)	15.1 (2.6)	22.0 (2.3)	31.6 (2.3)
F test[b]					
B		**	**	**	**
T		**	ns	ns	ns
B ×		ns	ns	ns	ns

[a] T: T0, control; T1/T2, plants covered at nights during either 27 Jan ∼ Feb 11 (T1) or Feb 11 ∼ 28 (T2). Date shown was the plant collection time, i.e., at the end of each canopy treatment.
[b] B, T and B × T represent the main effects of B and T, and their interaction, respectively. ns = non-significant; *, ** significant at 5% and 1% probability levels, respectively.

Table 4. Effects of B application (B) and canopy treatment (T) on oilseed rape development and seed yield. Values are means of four replicates with standard errors in parentheses.

B	T	Pods visible[a] (% of plants)	Pods (plant^{-1})	Seeds (plant^{-1})	Seed weight (g plant^{-1})	Seed yield (kg ha^{-1})
B0	T0	0 (0)	58 (33)	519 (307)	2.5 (1.5)	225 (131)
B0	T1	0 (0)	104 (48)	951 (537)	4.9 (2.4)	469 (222)
B0	T2	0 (0)	39 (23)	353 (214)	1.8 (1.1)	169 (99)
B1.5	T0	2.3 (0.5)	258 (55)	2434 (109)	13.0 (0.8)	1445 (91)
B1.5	T1	5.7 (1.7)	298 (35)	2889 (256)	15.6 (1.3)	1620 (102)
B1.5	T2	4.3 (0.9)	267 (17)	2898 (205)	15.0 (1.4)	1725 (170)
B15	T0	3.6 (0.9)	316 (21)	2402 (90)	13.3 (0.2)	1480 (45)
B15	T1	7.5 (2.0)	499 (138)	4635 (803)	25.0 (5.3)	2125 (164)
B15	T2	7.4 (2.8)	346 (57)	3136 (327)	16.8 (2.0)	1905 (152)
F test[b]						
B		**	**	**	**	**
T		*	ns	**	**	**
B×T		ns	ns	*	*	ns

[a] % of plants with visible pods (≥2 cm long) on main stem was assessed on April 1, 1995. The other data were recorded at harvest.
[b] B, T and B×T represent the main effects of B and T, and their interaction, respectively. ns = non significant; *, ** significant at 5% and 1% probability levels, respectively.

treatment period. It is concluded that further investigation of the interaction between low air temperature and B supply needs to be conducted in controlled environments for defined periods of plant growth. Such studies may well focus on stages of plant growth which are known to be sensitive to low B or low temperature.

34

Acknowledgement

This project is supported by an AusAID scholarship to the senior author and by research funds from the Australian Centre for International Agricultural Research (Project 9120), the Zhejiang Provincial government and the Chinese Ministry of Agriculture.

References

Forno D A, Asher C J and Edwards D G 1979 Boron nutrition of cassava, and the boron x temperature interaction. Field Crops Res. 2, 265–279.

Huang L, Ye Z, and Bell R W 1996 The importance of sampling immature leaves for the diagnosis of boron deficiency in oil seed rape (*Brassica napus* cv. Eureka). Plant Soil 183, 187–198.

Lardon A and Triboi-Blondel A M 1994 Freezing injury to ovules, pollen and seeds in winter rape. J. Exp. Bot. 45, 1177–1181.

Mahalakshmi V, Yau S K, Ryan J, and Peacock J M 1995 Boron toxicity in barley (*Hordeum vulgare* L.) seedlings in relation to soil temperature. Plant Soil 177, 151–156.

Mather J R 1974 Climatology: Fundamentals and Applications. McGraw-Hill, Inc. USA. 412 p.

Moraghan J T and Mascangni H J 1991 Environmental and soil factors affecting micronutrient deficiencies and toxicities. *In* Micronutrients in Agriculture. 2nd Edn. Eds. J J Mortvedt, F. R Cox, L. M Shuman and R M Welch, pp. 371–425. Soil Sci. Soc. Amer. Book Series No. 4.

Rerkasem B, Netsangtip R, Lordkaew S and Cheng C 1993 Grain set failure in boron deficient wheat. Plant Soil 155/156, 309–312.

Spouncer L R, Nable R O and Cartwright B 1992 A procedure for the determination of soluble boron in soils ranging widely in boron concentrations, sodicity, and pH. Commun. Soil Sci. Plant Anal. 23, 441–453.

Sylvester-Bradley R 1985 Revision of code for stages of development in oilseed rape (*Brassica napus* L.). Aspects of Applied Biology 10, 395–400.

Walker J M 1969 One-degree increments in soil temperatures affects maize seedling behaviour. Soil Sci. Soc. Amer. Proc. 33, 729–736.

Yang Y, Xue J, Ye Z and Wang K 1993 Responses of rape genotypes to boron application. Plant Soil 155/156, 321–324.

Zarcinas B A, Cartwright B and Spouncer L R 1987 Nitric acid digestion and multi-element analysis of plant material by inductively coupled plasma spectrometry. Commun. Soil Sci. Plant Anal. 18, 131–146.

R.W. Bell and B. Rerkasem (eds.), Boron in Soils and Plants, 35–37.
© 1997 *Kluwer Academic Publishers.*

Effects of boron and iron on yield and yield components of wheat

K. Zada[1] & M. Afzal[2]
[1] *Department of Agronomy, NWFP Agricultural University, Peshawar - 25120, Pakistan*
[2] *Department of Soil Science, NWFP Agricultural University, Peshawar - 25120, Pakistan*

Key words: borax, ferrous sulphate, Pakistan, wheat, yield components, yield

Abstract

Effects of boron (B) and iron (Fe) on yield and yield components of wheat (*Triticum aestivum*) were studied at the Malkandher Farm of the NWFP Agricultural University, Peshawar, during 1992–93, because the soil analysis showed that the soil was low in B and Fe. A factorial combination of three levels of B (0, 2, and 4 kg ha^{-1}) and Fe (0, 10, 20 kg ha^{-1}) were applied. Irrigation was witheld until 28 days after emergence so plants experienced an early water stress. Boron application increased all the yield components and grain yield of wheat as compared to control without the micronutrient application. Iron application increased tillers per m^2, grain weight and grain yield but had no effect on spikelets per spike and grains per spike. However, the combined application of Fe and B did not show additive effects on yield or yield components. Application of B increased grain yield more than the increase in grain yield due to Fe.

Introduction

Despite their importance in crop production, the response of field crops to micronutrients has not been widely studied in Pakistan. There is evidence of possible responses to boron (B) application in wheat (*Triticum aestivum*). The yield and yield components of wheat were affected positively and negatively by B depending up soil status and the doses used. Kausar et al. (1988) reported a 21% increase in yield of wheat with application of 4 kg B ha^{-1}. Boron at the rate of 3 kg ha^{-1} increased grain number and grain weight spike^{-1}, although higher rates depressed these yield parameters (Seth and Singh, 1985). Despite the importance of B in plant nutrition, not enough work has been done to determine its status in the soils of the North West Frontier Province (NWFP) and in plants and according to Sillanpa (1982), this could partly be due to the difficulties in B analysis compared to other trace elements. Keeping in view the essentiality of B and its deficiency in our soils (Table 1), an experiment was designed to explore the possibility of using B and Fe for enhancing growth, and increasing yield of wheat.

Table 1. Physio-chemical properties of soil of the experimental farm of the NWFP Agriculture University, Peshawar.

Characteristics	Units	Values
Clay	g kg^{-1}	312
Silt	g kg^{-1}	530
Sand	g kg^{-1}	158
pH		7.9
Nitrogen	g kg^{-1}	0.89
Organic matter	g kg^{-1}	108
Lime (CaCO$_3$)	g kg^{-1}	114
Electrical conductivity	mSm^{-1}	180
Available P	mg kg^{-1}	22
Available K	mg kg^{-1}	470
Available Fe (EDTA Extractable)	mg kg^{-1}	6.63
B (Hot water soluble)	mg kg^{-1}	0.85
Available Zn (EDTA Extractable)	mg kg^{-1}	0.83

Materials and Methods

To study the effect of B and Fe on yield and yield components of wheat, a field experiment was conducted at the Malkandher Farm, NWFP Agricultural University, Peshawar, during 1992–93, because the soil samples

Table 2. Yield and yield components of wheat as influenced by boron and iron application.

B (kg ha^{-1})	Fe	Tillers m^{-2}	Spiklets Spike^{-1}	Grains Spike^{-1}	1000 Grain wt (g)	Grain Yield (kg ha^{-1})
0	0	60.2b	15.8b	30.1c	26.2b	3012c
2	0	72.5a	17.3ab	33.2b	28.6a	4225ab
4	0	73.1a	19.2a	39.2a	30.2a	4345a
0	10	70.2a	16.5b	31.0c	28.2ab	4210ab
2	10	72.8a	16.2b	31.8c	29.0a	3900bc
4	10	71.1a	17.2b	31.9c	29.0a	4215ab
0	20	71.3a	15.9b	30.2c	28.8a	4300a
2	20	70.9a	16.9b	31.8c	28.1ab	3847bc
4	20	72.5a	18.2a	35.1ab	29.10a	4450a

Means followed by the same letters in vertical columns do not differ significantly ($p < 0.05$)

tested showed low levels of the above micronutrients (Table 1). All possible combinations of three rates of B (0, 2, 4 kg B ha^{-1} as borax) and Fe (0, 10, 20 kg Fe ha^{-1} as ferrous sulfate) were applied in the experiment. A basal dose of N-P at the rate of 100–60 kg ha^{-1} was uniformly applied to the experimental plot. A randomized complete block (RCB) design with 3 replications and plot size of 3 m by 5 m was used. Wheat variety Barani-90 was sown at the rate of 100 kg ha^{-1} on December 05, 1992 and harvested on May 15, 1993. Data were recorded for yield and yield components and statistically analysed as appropriate for a randomized complete block design (Steel and Torrie, 1980).

The wheat growing season was typical of the area. The minimum temperature for December and January ranged from −3 to 1°C and for the months of May it ranged from 15 to 12°C: the respective maximum temperatures ranged from 7 to 23°C and from 29 to 40°C. Rainfall was low and ill distributed, thus wheat was reliant on irrigation from 28 days after emergence when the first irrigation water was applied.

Results and discussion

All the yield components were affected by B and Fe application (Table 2). Application of both the elements increased tillers, B more so than Fe application. Maximum tillers were produced by the plants that received the highest dose of B, about 21.4% more than the control. This was followed by the combination of both the nutrients at both the levels with 20% increase over control.

The physiological response to B was more pronounced than that to Fe in the development of spikelets. Boron at the rate of 4 kg ha^{-1} gave maximum number of spikelets per spike (19.2), significantly more as compared with other treatments and had the maximum% increase (21.6) over the control treatment (Table 2). On the average, 0.5 to 21.6% increases over control were observed in number of spikelets by the application of B and Fe.

Significant increases in numbers of grain per spike were observed with B and Fe applied alone and in combinations. Boron application at the rate of 4 kg ha^{-1} produced maximum numbers of grain per spike followed by its combination with Fe at the higher level, giving 20.2 and 16.7% increase over the control, respectively. Grains per spike were generally reduced by the combined application of B and Fe.

Both B and Fe at low and higher levels increased grain weight. Boron applied at the rate of 4 kg ha^{-1} produced heavier grain (30.2 g 1000 grains^{-1}) representing a 15.1% increase over the control. The grain weight also increased with combined application of B and Fe and the effect was more at the higher level of B and Fe with 11.1% increase over control. Comparatively, B application produced heavier grains than did Fe.

All the yield parameters significantly contributed towards grain yield with corresponding increases from 28 to 47% in grain yield over control treatment. Boron alone at both the rates increased yield as did Fe alone. The increases in grain yield were in close conformity with the corresponding increases in the respective yield components. In the mixed applications, there was no indication of either additive or synergistic effects on grain yield. The combination of both the micro-nutrients at higher levels gave the maximum grain yield of 4450 kg ha^{-1}, having 48% increase over control.

Responses to B in the present study contrast with those of Rerkasem and Loneragan (1994) who found that the main effect of low soil B on wheat was a decrease in grain set. Weight of individual grains often increased at low soil B, probably as a consequence of reduced grain set, whereas tiller number, and spikelets per ear were generally unaffected by B deficiency (Rerkasem et al., 1989). In the central parts of NWFP, temperature in winter months is low and thus early growth of wheat in Peshawar valley is very slow. The differences in the results obtained by Rerkasem and colleagues and our results may be due to low early temperature and early moisture stress. This could also be due to the difference in response of the wheat varieties used as Alkan et al. (1996) reported significant differences among genotypes in response to increased B supply.

Acknowledgements

The senior author is grateful to the Australian Agency for International Development for financial support under its International seminar support Scheme.

References

Alkan A, Torun B, Koleli N, Cakmak I, Kalayci M, Ekiz H and Yalmaz A 1996 Genetic variability in response of different wheat genotypes to boron toxicity in Turkish soil. *In* 5th International Wheat Conference, June 10–14, 1996, Ankara, Turkey, Abstracts. pp 269.

Kausar M A, Tahir M and Sharif M 1988 Wheat response to field application of boron. *In* Proc. Nat. Sem. Micronutrients in Soils and Crops in Pakistan, pp 132–138. Transformation and Integration of Provincial Agricultural Network, Peshawar, Pakistan.

Rerkasem B and Loneragan J F 1994 Boron deficiency in two wheat genotypes in a warm, subtropical region. Agron. J. 86, 887–890.

Rerkasem B, Saunders D A and Dell B 1989 Grain set failure and boron deficiency in wheat in Thailand. J. Agric. (Chiang Mai Univ.) 5, 1–10.

Seth A, and Singh R P 1985 Comparative performance of some macro-micronutrients in relation to growth and yield. Indian J. Agron. 30, 219–23.

Sillanpaa A M 1982 Micronutrients and nutrient status of soils a global study. F.A.O. Soils Bull. 48.

Steel R G D, and Torrie J H 1990 Principles and Procedure of Statistics. McGraw Hill Book Co. New York.

R.W. Bell and B. Rerkasem (eds.), Boron in Soils and Plants, 39–42.
© *1997 Kluwer Academic Publishers.*

Flue gas desulfurization residue: a boron source for alfalfa production

R H Dowdy, J J Sloan, G W Rehm & M S Dolan
USDA-ARS and University of Minnesota, 439 Borlaug Hall, Department of Soil, Water and Climate, 1991 Upper Buford Circle, St. Paul, MN 55108, USA

Key words: alfalfa, boron, coal combustion, extractable boron, fly ash, *Medicago sativa* L.

Abstract

Very little literature documents the fate of flue gas desulfurization (FGD) residues under field conditions when applied at agronomic rates. The objective of this study was to document the availability of FGD residue-borne boron (B) for alfalfa (*Medicago sativa*) uptake when applied to marginally B deficient soils. The FGD residue was applied at rates of 0, 0.46 and 3.74 Mg ha^{-1} (0.038 and 3.1 kg B ha^{-1}, respectively) on a silt loam soil immediately prior to alfalfa seeding. Soil pH increased temporarily, but returned to background levels by the end of the first growing season. Alfalfa yields were unaffected by these rates of residue applications. Shoot B concentrations decreased as the growing season proceeded for the control treatment, but second and third cuttings of alfalfa produced on residue amended soil had increased B levels demonstrating that FGD residue was a readily available B source, particularly later in the growing season when native soil B availability decreased. Residue-borne B did not leach below 0.15 m.

Introduction

Alfalfa is a very important economic crop in United States agriculture as: a high quality animal feed; a deep rooting plant that enhances soil structure; and a crop that can help prevent excess soil nitrate-N from reaching ground water. It also has a high demand for boron (B), exhibiting B deficiency symptoms when grown on many soils. Earlier, Stinson (1953) suggested that soils with <0.5 mg B kg^{-1}, as hot-water-extractable B, appeared to be deficient for alfalfa production. Subsequent research (Su et al., 1994) supports this conclusion.

One source of B is the flue gas desulfurization (FGD) residue produced by limestone scrubbing of sulfur dioxide (SO$_2$) from flue gases of coal-fired electrical generating stations. Quantities of this material have been increasing rapidly, estimated to be 14.1 million Mg in 1994 (Am. Coal Ash Assoc., 1995), for two reasons: an ever-increasing demand for electrical power; and, passage of the 1990 amendments to the Clean Air Act to remove SO$_2$ from flue gases prior to discharge into the atmosphere. Although numerous studies have investigated the B-supplying power of FGD residues for plant growth (Carlson and Adriano, 1993), most studies have been restricted to laboratory and short-term growth chamber/greenhouse investigations. In addition, reported field studies have involved use of FGD residues at very high rates of application for: enhancing soil hydraulic properties (Ghodrati et al., 1995); correcting soil acidity (Korcak, 1980); and/or, disposal near the source site to reduce transportation expenses (Warren, 1992). Although Plank and Martens (1974), and others, have shown that coal fly ash (particulates in flue gas streams minus FGD residues) can serve as a plant B source, Kukier et al. (1994) cautioned that general conclusions concerning the effectiveness of FGD residues as a soil amendment are tenuous because of tremendous variability in their chemical reactivity and differences in the physical and chemical properties of residue amended soils.

The objective of this study was to document the availability of FGD residue-borne B for alfalfa uptake when applied at agronomic rates to a productive soil that was marginally B deficient.

Table 1. Elemental composition of flue gas desulfurization residue.

Element	Concentration[a] (g kg^{-1})	Element	Concentration (mg kg^{-1})	Element	Concentration (mg kg^{-1})
Al	69	B	824	Mo	7.4
Ca	167	Cd	<0.5	Na	5.2
Fe	16.5	Cr	33	P	1700
Mg	17.9	Cu	52	Ni	17
S	56	K	2500	Pb	23
		Mn	514	Zn	34

[a] 110°C Weight basis.

Materials and methods

The FGD residue resulted from the combustion of pulverized, sub-bituminous coal (low S, high ash, medium C) with a calcium carbonate equivalent of 42% (pH=11.8, 1:1 H$_2$O), a hot-water-extractable B content of 48 mg B kg^{-1} (extracted by refluxing 1:40, residue:water for 10 min); and an electrical conductivity of 8.8 dS m^{-1} (saturated paste) (Dahnke, 1988). Elemental composition of FGD residue (Table 1) was determined by microwave acid pressure decomposition of 0.500 g samples digested in 10 ml concentrated HNO$_3$ acid for 10 min at 170 to 180°C (USEPA Method 3051, SW 846). Digests were analyzed by inductively coupled plasma source emission spectroscopy (ICP).

Field plots (3 × 4.5 m) were established on a Renova silt loam soil (Typic Hapludalf, fine-loamy, mixed, mesic) in a complete, randomized block design with four replications. This soil had an initial pH = 6.5, 13 g organic matter kg^{-1}, and a hot-water-extractable soil B content of 0.5 mg B kg^{-1}, considered marginally deficient for optimal alfalfa growth (Su et al., 1994). The FGD residue was broadcast at rates of 0, 0.46, and 3.74 Mg ha^{-1}, (0, 0.38, and 3.1 kg B ha^{-1}, respectively) (110°C basis) and disked into the surface 0.05 m of soil immediately prior to alfalfa seeding (*Medicago sativa* cv. Pioneer 5246) on 11 Aug 1995. The 3.74 Mg ha^{-1} rate of FGD residue application was selected to supply 3 kg B ha^{-1}, which is the recommended rate of B application for alfalfa production on soils testing <1.0 mg B kg^{-1} (Rehm et al., 1994). All treatments received broadcast P (22 kg ha^{-1}) and K (56 kg ha^{-1}) prior to seeding. Three alfalfa cuttings were harvested during the summer of 1996, on 4 June, 5 July, and 14 August.

Soil samples (five randomly selected cores per plot) were collected prior to FGD residue application and 31, 257, and 441 days after treatment (DAT) in three depth increments: 0 to 0.15, 0.15 to 0.30 and 0.30 to 0.45 m. Soil pH was determined in a 1:1, soil:0.01 M CaCl$_2$ slurry, and plant available B was extracted with hot-water, as discussed for FGD residue, and the extract analyzed by ICP. Shoot samples were dried (65°C), ground, and ashed by microwave, acid pressure decomposition of 0.250 g plant material in 0.5 mL of concentrated HNO$_3$ and 2.0 mL of 30% H$_2$O$_2$. The supernatant digestate was analyzed by ICP.

Results and discussion

Despite the very reactive residue, soil pH was unchanged at 31 DAT (Table 2). By the following spring (257 DAT), surface soil pH had increased ($p<0.01$) to 7.1 for the 3.74 Mg ha^{-1} (high) treatment, significantly higher than either the control or low (0.46 Mg ha^{-1}) treatments; but, by October (441 DAT) soil pH was again the same for all treatments. A long-term pH increase in this well-buffered soil was not anticipated because of the relatively low rates of residue application. Similar results were observed by Ransome (1984) for FGD residue applications up to 40 Mg ha^{-1} on a more acid (pH = 4.5) sandy soil.

Alfalfa production (Table 2, 15% moisture basis) was good (approximately 9 Mg ha^{-1}) for a first-year stand in our climate. Yields were not significantly different between treatments. However, the second cutting (5 July) was limited by lack of timely rain. Although alfalfa growth was not enhanced by FGD residue applications, neither were yields reduced as a result of fresh residue amendments, a common occurrence with higher rates of residue applications (Carlson and Adriano, 1993).

Shoot B concentrations, as a function of FGD residue-applied B, are presented in Table 2 for all three cuttings. Based on the literature summary and research

Table 2. Effect of flue gas desulfurization (FGD) residue applications on soil and alfalfa properties.

	FGD Residue (Mg ha^{-1})			Linear correlation
	0	0.46	3.74	
B applied (kg ha^{-1})	0	0.38	3.1	–
Soil pH				
31 DAT	6.40	6.51	6.54	ns
257 DAT	6.80	6.83	7.07	***
441 DAT	6.65	6.56	6.52	ns
Yield (Mg ha^{-1})				
1st cutting	3.79	3.48	3.60	ns
2nd cutting	2.01	1.88	1.92	ns
3rd cutting	3.22	3.25	3.17	ns
B content (mg kg^{-1})				
1st cutting	33.8	32.9	30.3	ns
2nd cutting	31.3	32.8	36.9	*
3rd cutting	24.1	28.4	35.3	***
Extractable B (mg kg^{-1})	0.57	0.67	1.14	***

*, **, *** significant at 0.10, 0.05, and 0.01 levels of probability, respectively.

findings of Kelling and Matocha (1990), all alfalfa B concentrations were in the adequate range (90 to 100% yield potential) of 20 to 50 mg B kg^{-1}. No significant difference ($p>0.10$) in shoot B concentration was noted for the first cutting; but by the third cutting, FGD residue amendments enhanced shoot B concentrations ($p<0.01$), reaching 35 mg B kg^{-1} for the high treatment. Equally important, from a kinetic perspective, is that the unamended soil was not releasing B as fast as alfalfa would assimilate available B. In fact, had water not been limiting plant growth, it is possible that inadequate B availability would have limited growth during the latter part of the growing season. This situation would be magnified during succeeding growing seasons.

Hot-water-extractable B, 441 DAT, was significantly correlated ($p<0.01$) with applied B, reaching a mean concentration of 1.14 mg B kg^{-1} in the surface, 0 to 0.15 m of soil (Table 2). Extractable B in the surface layer of the unamended soil at this time was 0.57 mg B kg^{-1}, very close to the value of 0.5 mg B kg^{-1} considered marginally deficient for alfalfa production (Su et al., 1994). Shoot B concentrations of the third cutting were significantly correlated ($p<0.01$) with extractable B, similar to the correlation observed between shoot B and B applied. Combined, these two observations suggest FGD residue-borne B is a ready source for mid-to late-season B uptake, when availability of native soil B is decreasing. Below the 0.15 m depth, extractable B was not correlated with applied B, suggesting that FGD residue-borne B was not leaching out of the surface soil, which is an environmental concern.

Conclusions

The modest amounts of FGD residue used in this field study resulted in only a temporary pH increase, but pH returned to background levels by the end of the growing season. Alfalfa yields were not affected by residue applications. Yield differences were not expected although extractable soil B levels were in the marginally deficient range for alfalfa growth. Shoot B concentrations decreased as the growing season progressed for alfalfa grown on unamended soil, coinciding with significant increases in shoot B grown on residue amended soil. Hence, FGD residue can serve as a ready B source for alfalfa production, particularly later in the season when availability of native soil B may becoming limiting. Finally, FGD residue-borne B did not leach below 0.15 m, minimizing any environmental threat to ground water quality.

References

American Coal Ash Association 1996 Coal combustion by-product production and use. Am. Coal Ash Assoc. Alexandria, VA, USA.

Carlson C L and Adriano D C 1993 Environmental impacts of coal combustion residues. J. Environ. Qual. 22, 227–247.

Dahnke W C 1988 Recommended chemical soil test procedures for the North Central Region. Agr. Exp. Stn. Bull. 499, North Dakota State University, Fargo. 37 p.

Ghodrati M, Sims J T and Vasilas B L 1995 Evaulation of fly ash as a soil amendment for the Atlantic costal plain: I. Soil hydraulic properties and elemental leaching. Water, Air and Soil Pollution 81, 349–361.

Kelling K A and Matocha J E 1990 Plant analysis as an aid in fertilizing forage crops. In Soil Testing and Plant Analysis, 3rd. ed. Ed R L Westerman, pp 603–643. Soil Science Society America, Madison, Wisconsin.

Korcak R F 1980 Fluidized bed material as a lime substitute and calcium source for apple seedlings. J. Environ. Qual. 9, 147–150.

Kukier U, Sumner M E and Miller W P 1994 Boron release from fly ash and its uptake by corn. J. Environ. Qual. 23, 596–603.

Plank C O and Martens D C 1974 Boron availability as influenced by application of fly ash to soil. Soil Sci. Soc. Amer. Proc. 38, 974–977.

Ransome L S 1984 Scrubber sludge amendment of a coarse sandy soil cropped to soybeans and alfalfa: A field study emphasizing boron distribution. Ph.D. Thesis, Univ. Minnesota, St. Paul. 162 p.

Rehm G, Schmidt M and Munter R 1994 Fertilizer recommendations for agronomic crops in Minnesota. Minn. Extension Service, BU-6240-E, University of Minnesota, St. Paul. 23 p.

Stinson C H 1953 Relation of water-soluble boron in Illinois soils to boron content of alfalfa. Soil Sci. 75, 31–36.

Su C, Evans L J, Bates T E and Spiers G A 1994 Extractable soil boron and alflafa uptake: Calcium carbonate effects on acid soil. Soil Sci. Soc. Am. J. 58, 1445–1450.

Warren C J 1992 Some limitations of sluiced fly ash as a liming material for acid soils. Waste Manag. Res. 10, 317–327.

R.W. Bell and B. Rerkasem (eds.), Boron in Soils and Plants, 43–48.
© 1997 *Kluwer Academic Publishers.*

Long-term field experiment on the application of slow-release boron fertilizer Part 1 Effect of boron on crop growth

Susumu Eguchi & Yoshio Yamada
Agricultural Research and Development Center, Ferro Enamels (Japan) Ltd. 1152 Tsurumi, Beppu, Oita Pref. Japan

Key words: boron uptake, citric acid soluble boron fertilizer, hot water soluble boron, long term field experiment, yield.

Abstract

A slow-release boron (B) fertilizer was applied annually for 15 years to investigate the effect of B on the growth of crops in three soils: a diluvial soil, volcanic ash soil and granitic soil. Cumulative B applications were 20.4, 17.9 and 31.4 kg ha^{-1}, respectively. The relative yield in B treated plots (called FTE plots) was always higher than that in those without the applied B (called original plots). Symptoms of deficiencies in B appeared in original plots after 2–4 years cultivation of crops, and worsened year after year. On the other hand, toxicity symptoms due to excess of B never appeared in any crops in FTE plots; rather, those crops showed healthy growth throughout the 15 years. Similarly the B concentration in crops in FTE plots always remained in the adequate range. The content of hot water soluble B in the soil of original plots, where serious deficiency symptoms appeared in crops, were generally less than 0.3 mg kg^{-1}. The hot water soluble B contents in the soil of FTE plots increased to about 1.0 mg kg^{-1} in the 15th year. Although the granitic soil received the largest B application, hot water soluble B in this soil increased to a lesser extent than in the other soils. The total amount of B taken up by crops was about 10% of the amount of B applied, suggesting that either significant residual B remained in the soil, or significant leaching occurred. These issues are the subject of the following paper.

Introduction

Boron (B) deficiency in susceptible crops is a widespread problem in many countries (Gupta, 1979). Although B is essential for crop growth, the amount required depends on the type of crop, the range of proper application is rather narrow, and its harmful effect can be induced by excessive application (Yamasaki, 1966). There are concerns that such harmful effects will be seen more often in the near future as B application becomes more common.

There are many papers published on B responses in a single year (Yamanouchi, 1979). By contrast very few experiments have studied the successive application of B to the same plots over many years to observe its effect on crop growth, the amount of it absorbed by crops and its accumulation in soils, apart from the paper of Sawaguchi et al. (1992).

Citric acid soluble B has been specified as a slow-release fertilizer since 1960 and is commonly applied on its own, or in compound fertilizers in Japan. The present paper reports on a long-term experiment, involving a slow-release B fertilizer over 15 years, from 1980 to 1995, which was carried out to observe the behavior of B in soils and its effects on crop growth. This paper presents the changes in crop yield, the occurrence of deficiency or excess symptoms, and the results of content analyses of B in crops and hot water soluble B in soils. The following paper examines the fate of applied B in the soil-plant system (Eguchi and Yamada, 1997).

Materials and methods

Experimental conditions and characteristics of soils

This study took place on the experimental field of the Agricultural Research and Development Center of Ferro Enamels (Japan) Ltd., in Oita Pref. Japan (200 masl) and continued for 15 years from 1980 to 1995. Three kinds of soil with a hot water soluble B content of less than 0.5 mg kg^{-1} were selected: Beppu diluvial soil (Hapludand), Shonai volcanic ash soil (Melanudand) and Aki granitic soil (Dystrochrept). Each soil was poured into 9 square meter (3 m × 3 m) plots to a depth of 50 cm. Two types of plot for each soil were prepared: original plot and FTE plot. Plots were surrounded by a concrete wall but were freely drained underneath. The average air temperature was 16.1° C and the average yearly precipitation was 2149 mm from 1981 to 1995. The physical and chemical properties of these soils are shown in Table 1.

There were no replicates of the plots in this experiment as the experiment was carried out under well controlled conditions in the artificially developed 9 m^2 field, in which fertility was confirmed to be consistent among the plots at the commencement of the experiment.

Application rate and method of B fertilizer

The B fertilizer used in the experiments was FTE No.1, (contents of B and Mn are 2.8% and 14.7%, respectively) which is soluble in 2% citric acid but insoluble in water. In the FTE plot of diluvial soil, the total amount of B applied was 20.4 kg ha^{-1} during the 15 years; for volcanic ash soil it was 17.9 kg ha^{-1}; and for granitic soil it was 31.4 kg ha^{-1}, respectively. Urea and ammonium sulphate, calcium superphosphate, and potassium sulphate were applied in both plots. In principle, the yearly amounts of N, P and K were 500, 220 and 330 kg ha^{-1}, respectively. FTE No. 1 was mixed well with the major nutrient fertilizers and applied to the whole surface of each FTE plot, which were then plowed to a depth of 15 cm. After furrowing, seeding or planting were carried out.

Cultivation history of crops

The cultivation history of the crops is shown in Table 2. Nineteen varieties of vegetables were cultivated and 26–29 successive crops in total were grown in each soil.

Observation and analytical method

Growth of plants was observed during the growing season, and yield was determined at harvest. The harvested crops were divided, air dried and pulverized to prepare samples for chemical analysis. Soil was sampled from 0 to 15 cm in depth after harvesting, air dried and passed through a 2 mm sieve. The plant samples used for the analysis of B were dry ashed, extracted with 0.5 N HCl and then determined for B by the curcumin method (Munson and Nelson, 1973). The soil samples to analyze hot water soluble B were extracted with hot water and filtrates were determined by the curcumin method (Gupta, 1967).

Results

Changes in yield ratio and occurrences of nutritional disorders

In the original plot, for all three kinds of soil, the yield ratio was always lower than that in the FTE plot, after the 2nd to 4th cropping. Boron deficiency continued to appear almost every year following 1983 in the diluvial soil, 1985 in the volcanic ash soil and 1984 in the granitic soil, as shown in Figure 1. In FTE plots, yield remained adequately high compared with farmers' yields in this area, and free from deficiencies in B, and the toxic symptoms of B and Mn. The yield ratio of the original plot declined over time with successive cultivation. Whereas in the first 5 years, the yield ratio of the original plots were 73% to 78% in all 3 soils, in the final 5 years, the yield ratio had declined in all soils, but especially in the granitic soil where it had declined to 44%.

Content of hot water soluble B in soil

The hot water soluble B content in each soil in all original plots was lower than that in the FTE plots and showed some fluctuations, decreasing slightly with successive cultivation, as shown in Figure 2. The hot water soluble B content in granitic soil was lower levels than in the other two soils of the FTE plots. The hot water soluble B content of all soils was rather stable during the first 10 years, with minor fluctuations. Thereafter, it increased to 1.1 mg kg^{-1} in diluvial soil and volcanic ash soil. However, that of the granitic soil, to which the largest amount of B was applied, increased to only 0.8 mg kg^{-1} after 15 years cultivation.

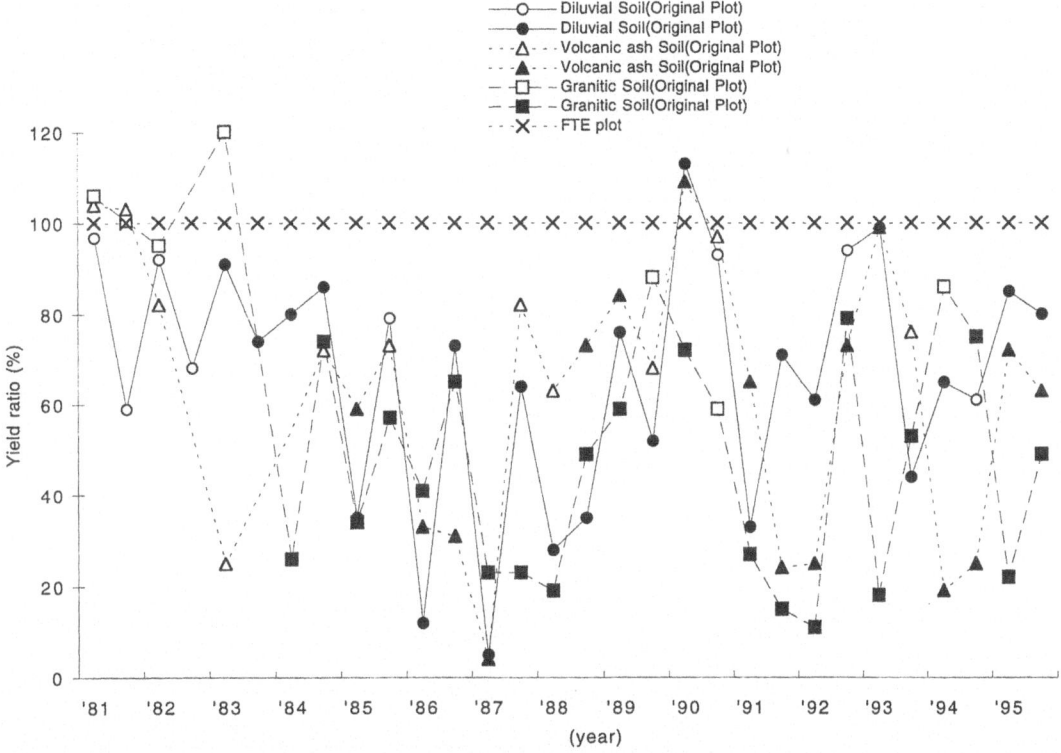

Figure 1. The change of yield ratio and occurrence of boron deficiency symptoms (closed symbols) with successive crops on 3 soils. Yield ratio was calculated as the ratio of yields in the original plot (without B) to those in FTE plots (with added B).

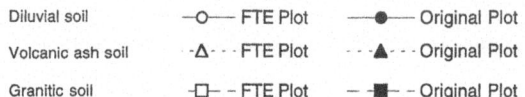

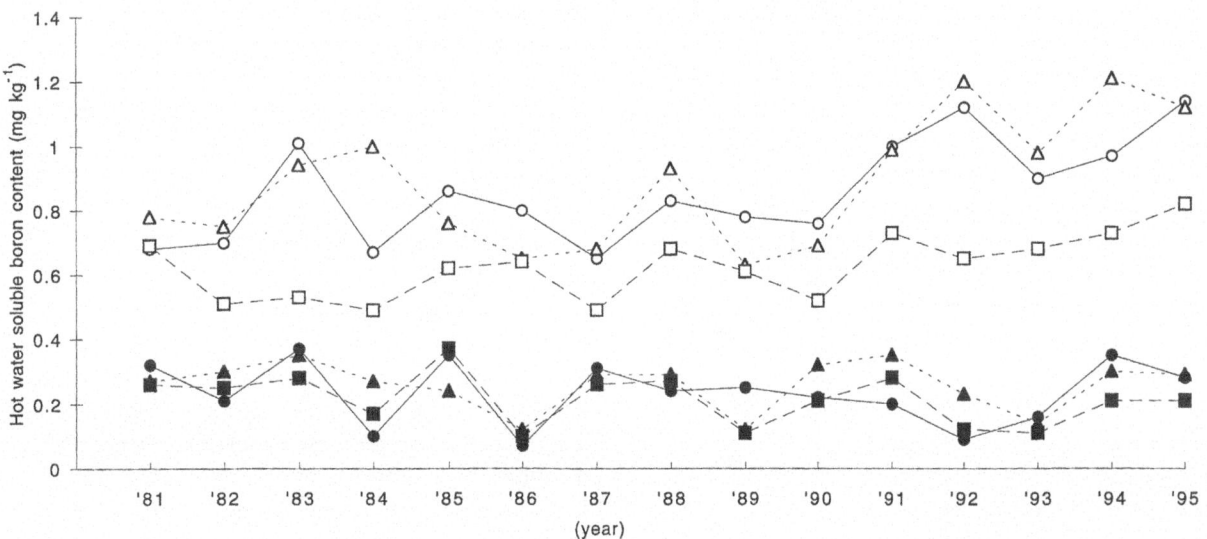

Figure 2. The change of hot water soluble boron in three kinds of soil in response to boron application (FTE plot) or no boron (original plot) over 15 successive years.

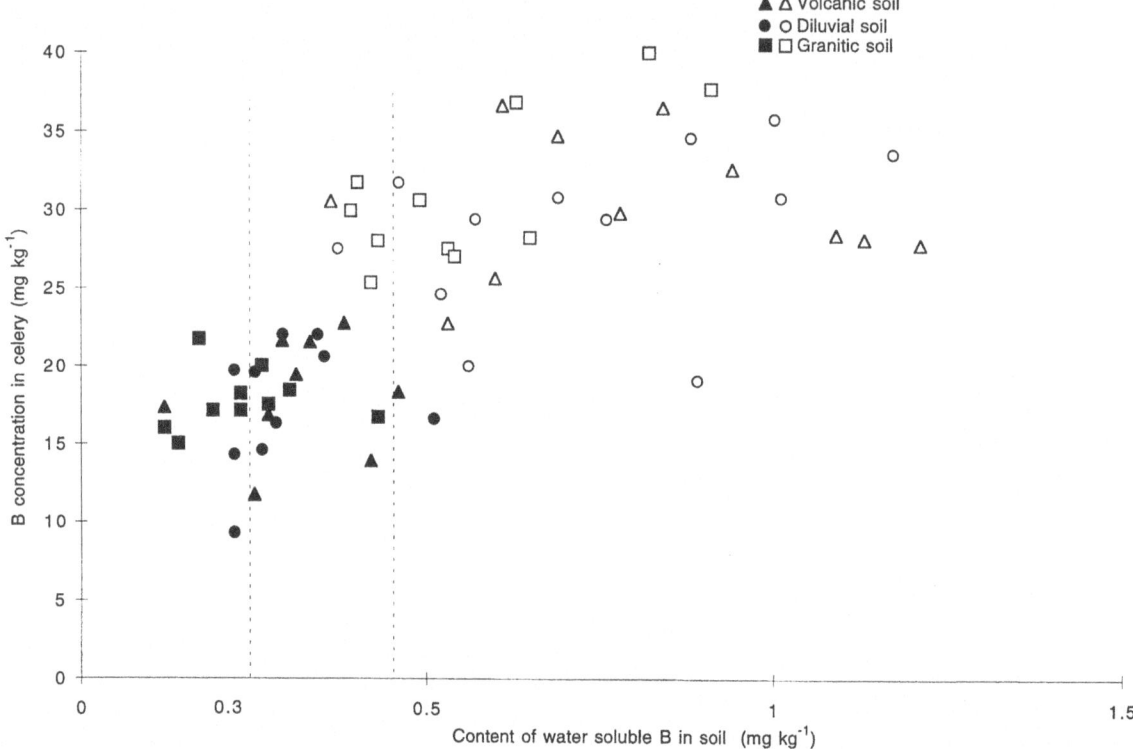

Figure 3. The relation between the content of hot water soluble boron in soil and concentration of boron in celery grown in three soil types. Closed symbols represent crops which exhibited boron deficiency symptoms.

The relation between the B content in plants and the hot water soluble B content in soil

There was a close relationship between the B content in plants and the hot water soluble B content in soil for celery (*Apium graveolens*) and chinese cabbage, (*Brassica campestris pekinensis*) which were often cultivated during the 15 years, as shown in Figure 3. According to the results, B deficiency symptoms appeared whenever the hot water soluble B content in soil was less than 0.3 mg kg^{-1} in celery and less than 0.4 mg kg^{-1} in chinese cabbage. When the hot water soluble B content was between 0.3 and 0.5 mg kg^{-1}, B deficiency appeared in about 50% of crops: no B deficiency symptoms were observed in any of the celery and chinese cabbage crops where the hot water soluble B levels exceeded 0.5 mg kg^{-1} (data of chinese cabbage are not shown).

Boron concentration in celery

Celery was successively cultivated in the granitic soil from 1984, while in the diluvial soil and volcanic ash

soil from 1985. It was assumed that the effect of the successive application of B over a long period on the growth of a crop could be examined by the change in B concentration in crops. The B concentrations in celery were mostly in the range of 25 and 40 mg kg^{-1} in the FTE plots, and in the range of 5 and 20 mg kg^{-1} in the original plots regardless the kind of soil, as shown in Figure 4. Clear changes in the B content in crops could not be found in the FTE plots and original plots during the experiment.

Amount of B taken up by crops

As shown in Table 3, the amount of B taken up by crops in granitic soil was the greatest among the three kinds of soil for the first five years, but drastically decreased later in the original plot. In the case of volcanic ash soil, B uptake also decreased with time. By contrast, in the diluvial soil, it was relatively stable. The amount of B taken up by crops in the FTE plot of granitic soil for the first five years was also the greatest among the three soils, but was rather stable for the next ten years even though it received higher B applications than the other

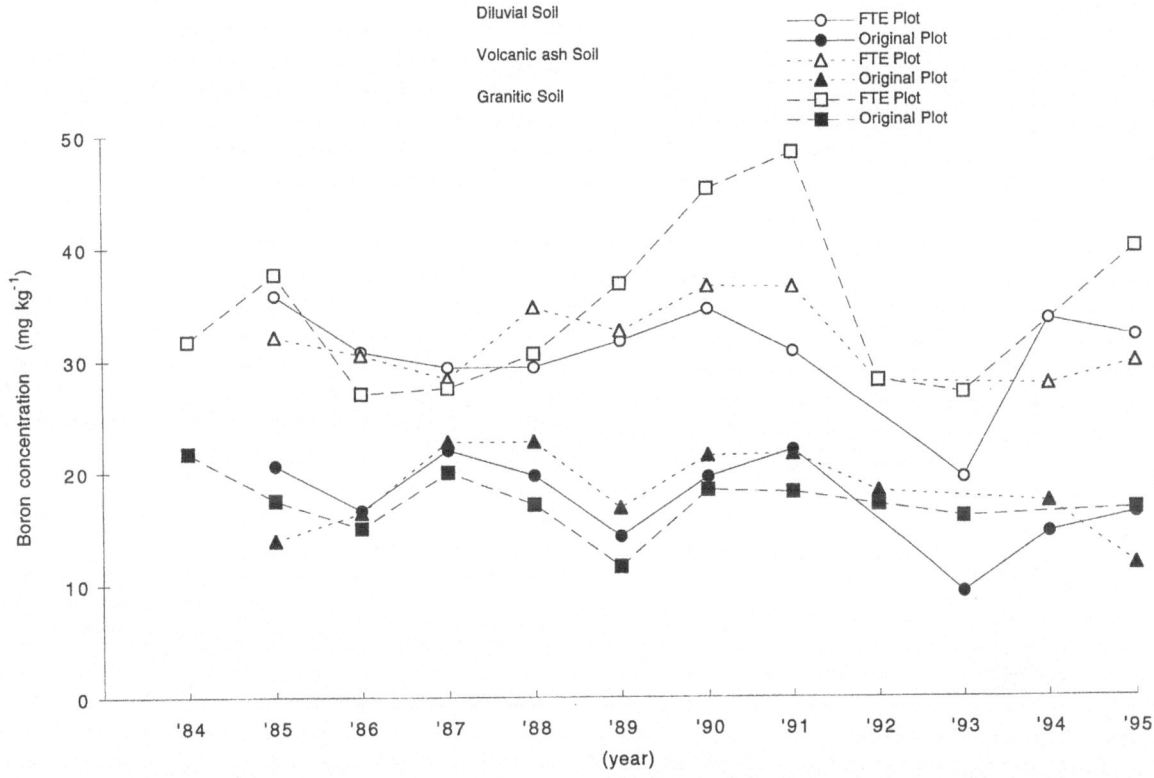

Diluvial Soil
Volcanic ash Soil
Granitic Soil

○--- FTE Plot
●— Original Plot
△··· FTE Plot
▲··· Original Plot
□-- FTE Plot
■-- Original Plot

Figure 4. The effect of boron application (FTE plot) on no boron (original plot) or boron concentrations in celery grown in three soils over 15 successive years.

soils. When the amount of B taken up in the FTE plot by crops in all soils every five years were compared, these amounts were almost the same although there was some fluctuation. The amount never increased with successive cultivation. The total amounts of B taken up by crops in diluvial soil, volcanic ash soil and granitic soil were 6.8%, 10.6% and 7.6% of the B applied, respectively.

Discussion

Successive applications of FTE in the three soils during the 15 years increased the yield of vegetable crops and resulted in healthy growth without any symptoms of B or Mn toxicity. This result was similar to that reported by Sawaguchi et al.(1982). They reported that when repeated application of FTE on volcanic ash soil was made over 12 years, symptoms of excess B and Mn were never observed, even after 11 years, on plants of azuki beans (*Phaseolus angularis*), which is very sensitive to B toxicity. And the yield of azuki beans in

the FTE plot was 9% to 12% higher compared to that in the original plot.

The hot water soluble B content in soil where the crop showed serious B deficiency symptoms was less than 0.3 mg kg^{-1}. While the hot water soluble B content in the original plots fluctuated slightly from year to year and showed a slight decreasing tendency, there was no significant difference in the content between the soils at the beginning and 15 years later.

Small year to year variations in hot water soluble B could be caused by the change of pH, the soil moisture, or the release of B due to the weathering of soil minerals. The increased amount of hot water soluble B in the soils after the 15 years of successive applications of FTE was 0.82 mg kg^{-1} in the diluvial soil, 0.84 mg kg^{-1} in the volcanic ash soil and 0.57 mg kg^{-1} in the granitic soil. These figures are comparable with the amount of 0.90 mg kg^{-1}, which was obtained in the experiment conducted by Sawaguchi et al.(1992).

Hot water soluble B content in the FTE plot of granitic soil was lower than that in the plots of diluvial and volcanic ash soil. The main reason seems to be

48

that soil containing much sand, as seen in Table 1, can increase leaching.

Boron deficiency appeared in 50–70% of celery and chinese cabbage crops when the water soluble B was between 0.3 and 0.5 mg kg^{-1}. Hence these results reconfirmed 0.5 mg kg^{-1} as the marginal content above which deficiency will not appear. Watanabe (1995) recently reported in his paper on B deficiency in garlic (*Allium sativum*) in Thailand that the marginal content of water soluble B in the soil was 0.30 to 0.33 mg kg^{-1}. This marginal content is applicable only to garlic.

In conclusion, the present study showed that the symptoms of B toxicity due to B accumulation never appeared even in soil supplied with B successively over 15 years. In this situation, it is extremely important to study the behavior of 90% of the applied B which was not taken up by crops. These results are reported on in a subsequent paper (Eguchi and Yamada, 1997).

Acknowledgments

We are grateful to Dr. Yoshiaki Ishizuka, Professor Emeritus Hokkaido University, for his precious conception which was shown us at the starting point of this study and for his valuable guidance throughout this experiment. Our thanks are due to Mr. K Motohira, Mr. M Nagano, Mr. M Akitomo, Mr. T Abe and Miss M Ido for their valuable suggestions and helpful cooperation.

References

Eguchi S and Yamada Y 1997 Long-term field experiment on the application of slow-release boron fertilizer – Part 2 – Behavior of boron in the soil. *In* Boron in Soils and Plants. Proceedings. Eds R W Bell and B Rerkasem pp. 49–55. Developments in Plant and Soil Sciences Vol. 76. Kluwer Academic Publ. Dordrecht, The Netherlands.

Gupta U C 1967 A simplified method for determining hot-water-soluble boron in podzol soils. Soil Sci. 103, 424–428.

Gupta U C 1979 Boron nutrition of crops. Adv. Agron. 31, 273–307.

Harada T and Tamai M 1968 Some factors affecting behavior of boron in soil. Soil Sci. Plant Nutr. 14, 215–224.

Munson R D and Nelson W L 1973 Principles and practices in plant analysis. *In* Soil Testing and Plant Analysis. Eds L M Walsh and J D Beaton, pp. 223–248. Soil Science Society of America, Madison Wisconson.

Sawaguchi M, Hasegawa S, Kumagai H and Nakamoto H 1992 Accumulation in soil and crop responses of B and Mn with successive applications of the micronutrients fertilizer. Hokuno 59, 419–423.

Watanabe H 1995 Boron deficiency in garlic (*Allium sativum*) grown on Tropaqualfs in Northern Thailand. Association for International Cooperation of Agriculture and Forestry 16, 22–39.

Yamasaki T 1996 Micro and macro nutrients. Hakuyusha, 240–262.

Yamanouchi M 1979 Boron nutrition in crop plants. Bull. Fac. Agric. Tottori Univ. 31, 37–39.

R.W. Bell and B. Rerkasem (eds.), Boron in Soils and Plants, 49–56.
© 1997 *Kluwer Academic Publishers.*

Long-term field experiment on the application of slow-release boron fertilizer Part 2 Behavior of boron in the soil

Susumu Eguchi & Yoshio Yamada
*Agricultural Research and Development Center, Ferro Enamels (Japan) Ltd. 1152 Tsurumi, Beppu, Oita Pref.
Japan*

Key words: balance sheet of boron, behavior of boron, forms of boron, long term field experiment

Abstract

A long-term field experiment involving repeated boron (B) applications on three kinds of soil was carried out over 15 years to investigate the fate of applied B in these soils.

After the repeated cultivation of 26–29 successive crops without the application of B, the amount of total B and the various forms of B, including hot water soluble, adsorbed and fixed forms, all decreased. This suggests that when crops absorb B from the hot water soluble fraction, adsorbed B replenishes the hot water soluble B pool, and when the adsorbed B decreases, fixed B releases B to the adsorbed B pool. While about 10% of B applied was absorbed by plants, 30–40% of the B was left in the soil and about 40–60% was leached out from the top soil. In the forms of B left in the soil, the fixed form was the greatest, followed by the adsorbed form and hot water soluble form. Hot water soluble B showed a tendency to slightly increase over time with repeated boron applications, but its concentration was far lower than the toxic level of B in soil after 15 years of application.

Introduction

Intensive agriculture has been extended in many places, and the amount of micronutrients removed from the field has been increasing with increased yields. This is a significant cause of the micronutrient deficiencies which have been reported in various parts of the world recently. This tendency is likely to become more prevalent. Boron (B) deficiency is the most common and widely distributed of all micronutrient deficiencies. The establishment of countermeasures to alleviate B deficiency is an urgent matter. As general guidelines for B fertilizer recommendations, Cooke (1982) suggested that when the hot water soluble B in soil was less than 0.5 mg kg^{-1}, deficiency is likely to occur and all crops are to be treated with B; when it is 0.5–1.0 mg kg^{-1}, deficiency may appear and insurance dressings are to be considered; when it is more than 1.0 mg kg^{-1}, deficiency is unlikely and B treatment is not necessary; and when it is 3–5 mg kg^{-1}, crops may be poisoned from excess B. Since there is only a small range between a toxic amount and supplies that are too

little for crops, careful application of boron should be carried out. Accordingly study of the behavior and fate of fertilizer B in soil is very important. While there are many papers on this problem, there are few reports, besides the report of Sawaguchi et al.(1992), on the behavior of B in fields in which boron was applied and crops were cultivated successively for 12 years. Sawaguchi et al. discussed only the hot water soluble B in volcanic ash soil and its response to long-term cultivation and B fertilizer supply.

We carried out the long-term field experiment which was started in 1980, where the application of B to the field continued for 15 years. In a previous paper, the effect of B on the growth of crops was reported (Eguchi and Yamada,1997). In this paper, forms of B in the soil were divided into hot water soluble, adsorbed and fixed forms, and the behavior of these forms of B in response to B fertilizer application and continued cropping in these soils was studied.

Materials and methods

Three kinds of soil, Beppu diluvial soil (Hapludands), Shonai volcanic ash soil (Melanudands) and Aki granitic soil (Dystrochrepts) were filled into respective pairs of 9 m² concrete frames. The depth of soil in each plot was adjusted to 50 cm after settling. Only NPK fertilizer was applied in one of the paired plots (called original plot). Boron fertilizer (FTE No. 1) in addition to NPK fertilizer, was applied in the other plot (called FTE plot). Forty kg of FTE No. 1 ha^{-1} (2.8% B, 14.7% Mn) was applied in principle to diluvial and volcanic ash soils before spring cultivation, and to granitic soil before spring and autumn cultivations. Thus annual B rates were: 1.12 kg, 1.12 kg and 2.24 kg; and the total amounts of boron applied in 15 years were: 20.4 kg, 17.9 kg and 31.4 kg B per ha in the diluvial, volcanic ash and granitic soils, respectively. Boron in NPK fertilizers and also in rainfall was determined to confirm the input of B. Various kinds of crop, mostly vegetables, were grown for 15 years, usually with two crops per year (Eguchi and Yamada, 1997). Boron taken up by crops after 3 and 15 years was calculated from the data of plant analysis. Various forms of B in the soils with and without FTE fertilizer were analyzed before and after 3 and 15 years cultivation using the following methods.

Total B in soil was determined according to the method of Yamada and Hattori (1986). A soil sample was passed through a 2 mm sieve, was pulverized well with an agate mortar and about 200 mg of it was weighed accurately and mixed with 1.5 g of Na_2O_3 in a nickel crucible and fused at 55°C for 20 min. After cooling, the crucible was set in a teflon beaker and the fused material was dissolved with 5 mL water. The solution was then removed in the beaker and the crucible was washed well with water. Next 4 mL of H_2SO_4 (1:2) was gradually added to acidify the solution, after which the residue was completely dissolved by heating. After cooling, all solution was removed into a 50 mL polyethylene bottle with a narrow neck and filled up to about 25 mL with water. Four mL of extract solvent (200 mL of 2 ethyl-1,3-hexanediol dissolved in 1000 mL of chloroform) was added to the solution and shaken for 3 min. After extraction, all liquid in the polyethylene bottle was filtered through a dry paper filter to separate the organic phase from the aqueous phase. One mL of separated organic phase was taken in a dry polyethylene bottle and 1 mL of the curcumin acetic acid solution (1 g of curcumin was dissolved in 500 mL of acetic acid solution at 60–70

°C) and 0.25 mL of sulfuric acid were added. Everything was mixed well. This reaction system was left for 30 min., stirring occasionally. After this, it was dissolved with 95% ethanol and made up to 50 mL. After 30 min, absorbance of this solution was measured at 550 nm.

The content of B in standard soil (NIST SRM 4353 Rocky Flat Soil No. 1) was determined by this method, and the result obtained confirmed that this method can be used for the determination of the total amount of B in the soil.

Determination of hot water soluble B in soil was done by the method mentioned in a previous paper (Eguchi and Yamada, 1997).

Determination of 0.5 N HCl soluble B in soil

Five g of soil sample passed through a 2 mm sieve was added to 100 mL of 0.5 N HCl. After 60 min. shaking, filtration was carried out. Boron in the filtrate was determined according to the curcumin method for hot water soluble B. Fixed and adsorbed forms of B were calculated by the following equations:

$$\text{Fixed form B} = \text{Total B} - 0.5\text{ N HCl soluble B}$$

$$\text{Adsorbed form B} = 0.5\text{ N HCl soluble B} - \text{hot water soluble B}$$

Method of calculation to investigate the behaviour of native and applied B in soil

The numbers in parentheses appearing in Table 1 and 3 are used to show the method of calculation. The amount of B leached out was calculated as (5) = (1) + (3) − (2) − (4) in the original plot, and (9) = (1) + (3) + (7) − (6) − (8) in the FTE plot.

The various forms of applied B in the soil in the FTE plot were calculated under the assumption that the same amounts of native B in the soil at the commencement of the experiment were absorbed by plants and leached out from the top soil, and remained in the top soils of the original plot and the FTE plot, regardless of the B application.

The amount of applied B absorbed by plants was calculated as (10) = (8) − (4), and the amount of applied B left in the soil was (11) = (6) − (2). Accordingly the amount of applied B leached out from the top soil was calculated as (12) = (7) − (10) − (11). Various forms of applied B left in soil of the FTE plot were calculated as the difference between the respective forms of B in the soils of the FTE plot and the original plot. Thus,

Table 1. Balance sheet of boron (kg ha^{-1}) in original plot and boron fertilised plots (called FTE plot).

	Diluvial soil			Volcanic ash soil			Granitic soil		
	1980	1983	1995	1980	1983	1995	1980	1983	1995
Original plot									
(1) Total B in original soil[a]	10.8			11.4			30.4		
(2) Total B in soil		10.4	9.6		11.0	9.8		30.0	29.1
(3) Amount of B added in rainfall and fertilizers		0.49	2.36		0.37	2.33		0.30	2.33
(4) B taken up by plants		0.23	1.00		0.39	1.01		0.39	1.07
(5) B leached from 0–30 cm layer of soil									
(1) + (3) - (2) - (4)		0.60	2.49		0.44	2.96		0.35	2.55
FTE plot									
(6) Total B in soil		13.1	16.7		12.9	16.8		31.6	38.1
(3) Amount of B added in rain fall and fertilizers		0.49	2.36		0.37	2.33		0.30	2.33
(7) Amount of fertiliser B applied		6.82	20.4		3.36	17.9		4.48	31.4
(8) B taken up by plants		0.50	2.79		0.62	2.91		0.50	3.45
(9) B leached out from 0–30 cm layer of soil									
(1) + (3) + (7) - (6) - (8)		4.46	14.0		1.69	8.97		3.09	22.6

[a] Total B in original soil was identical for the original plot and the FTE plot. Total B (kg ha^{-1}) in soil are calculated from the bulk density, depth of top soil (30 cm), content of total B (mg kg^{-1}) and percentage of the soil passed through the 2 mm sieve.

the water soluble form of applied B in the soil was calculated as (15) = (14) − (13), the adsorbed form of applied B was (18) = (17) − (16) − (15) and the fixed form of applied B was (19) = (11) − (15) − (18).

Determination of B in rainfall and fertilizers

Fertilizers were dissolved in water. Boron in rainfall and solution of fertilizers was determined according to the curcumine method.

Results

Input B from rainfall and fertilizers

The amounts of B in rain fall and fertilizers applied were analyzed. Boron was not found in nitrogen fertilizers, but concentrations of B in rainfall, calcium superphosphate and potassium sulfate were 0.002 mg kg^{-1}, 40 mg kg^{-1} and 10 mg kg^{-1}, respectively. Annual input of B was calculated from these data, annual precipitation and amounts of P and K fertilizers applied. These were 43 g, 117 g and 8 g ha^{-1} respectively. Accordingly, total input of B depended on the amount of fertilizers applied, and usually it was 0.17 kg ha^{-1}.

Balance sheet of B in three kinds of soil

Three kinds of soil at the commencement of the experiment and after 3 and 15 years cultivation, were analyzed for total B to create a balance sheet of B after cultivation. The results obtained are shown in Table 1. The data of the total boron absorbed by plants, originally reported by Eguchi and Yamada (1997), was also included for completeness in this table. The amount of B in the granitic soil was greater than in the other soils. The contribution of input B was included in this table, as it was not negligible small. Amounts of B taken up by crops in original plot were relatively small, but those in FTE plot were 2–3 times greater. The amount of B leached out in original plot exceeded the input B. The difference is assumed due to the B leached out frcm soil, which is included in the Table 2. Remarkable amounts of B were leached out in FTE plot.

Behaviour of B in the soils of original plot

As shown in Table 2, the total B in the soil decreased steadily when crop cultivation was done successively in the original plots. The hot water soluble B and adsorbed form of B levels fluctuated slightly but showed the decreasing tendency except in the volcanic ash soil. The fixed form of B decreased in the same manner as the total B. The amount of total B taken up by plants during the 15 years of cultivation was far greater than

Table 2. Forms of boron in the soils of the original plot at the beginning of the experiment (1980), and after 3 and 15 years.

	Total B in soil kg ha^{-1}	0.5N HCl soluble B kg ha^{-1}	Hot water soluble B kg ha^{-1}	Adsorbed B[a] kg ha^{-1}	Fixed B[b] kg ha^{-1}	Leached B kg ha^{-1}	Taken up B by plant kg ha^{-1}
Diluvial soil							
1980	10.8	1.36	0.33	1.03	9.40	0	0
1983	10.4	1.39	0.40	0.99	9.03	0.11	0.23
1995	9.6	1.14	0.29	0.85	8.49	0.13	1.00
Volcanic ash soil							
1980	11.4	1.65	0.29	1.36	9.78	0	0
1983	10.1	1.68	0.40	1.28	9.29	0.07	0.39
1995	9.8	1.69	0.30	1.39	8.10	0.63	1.01
Granitic soil							
1980	30.4	1.85	0.37	1.48	28.6	0	0
1983	30.0	2.21	0.41	1.80	27.8	0.05	0.39
1995	29.1	1.65	0.31	1.34	27.5	0.22	1.07

[a] Adsorbed form B = 0.5N HCl soluble B - hot water soluble B

[b] Fixed form B = Total B - 0.5N HCl soluble B

the amount of hot water soluble B in the original soil. The amount of B leached out in the original plot was very small, but it was somewhat greater in the volcanic ash soil.

Behaviour of applied B in the soils of FTE plot

The behavior of applied B in the soil of FTE plot was calculated by the method described previously, and is shown in Table 1 and 3, and Figure 1. The amount of B taken up by plants was far less compared with the amount of B applied, especially during the first 3 years. The rate of absorption by plants increased over time, but it was 10% of the applied B at most. The rate of B left in the soil was relatively greater during the first 3 years, and decreased during the later period, but 30–40% of the B applied was left in the soil. This was somewhat higher in volcanic ash soil than in the other soils. The rate of B leached out from the 0–15 cm layer was 40–60% of the B applied, and was higher in the granitic soil.

The forms of applied B left in the soil

The forms of the applied B left in soil are shown in Table 3 and Figure 2. The rate of the fixed form of B, which was insoluble in 0.5 N HCl solution, was higher than other forms of B, except in the case of volcanic ash soil after 3 years cultivation. It was increased with repeated B application and cultivation, and reached 54–

80% of the residual B from the B fertilizers applied. The amount of hot water soluble B showed a slight increasing tendency with cultivation using B fertilizer repeatedly, as mentioned in the previous paper (Eguchi and Yamada,1997), but the proportion of B left in the soil decreased to 10–13%. As a proportion of B left in the soil, adsorbed B usually exceeded hot water soluble B, but both showed the same tendency. The amount of adsorbed B increased, but the proportion of it in the B left in the soil decreased to 10–34%. The rate of adsorbed form in the volcanic ash soil was higher than in the other soils.

Discussion

In this paper the balance sheet of B, and the change in the forms of B in the soils after 3 and 15 years cultivation are reported. To do so, estimation of the total amount of B in the soil was very important. Two important considerations were depth of top soil, and the accuracy of the determination of total B in the soil. The depth of top soil, from which samples were taken, was decided at 15 cm because the soil in this layer was mixed well by repeated cultivation, and the majority of crop roots were distributed in this layer and absorbed most of their nutrients from it. Besides, soil samples could not accurately be taken from the lower layer during the study.

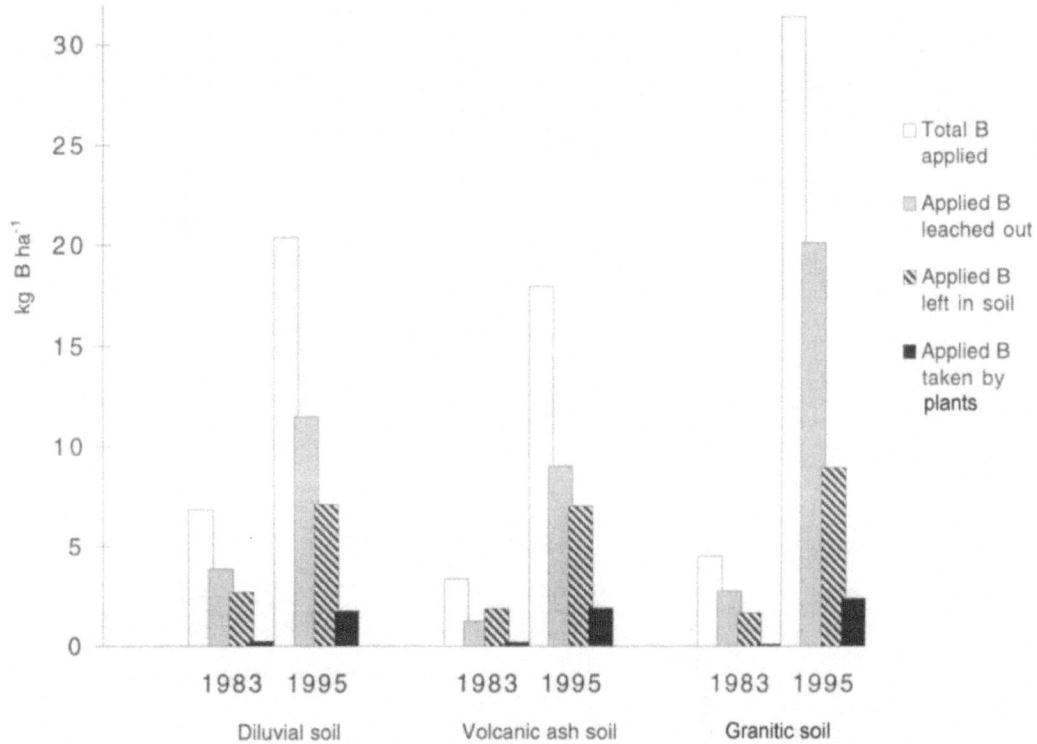

Figure 1. Behaviour of boron applied to the field.

Table 3. Fate of boron applied (kg ha^{-1}) in the soils of the FTE plot. Figures inparetheses are percentages of the applied B.

	Diluvial soil		Volcanic ash soil		Granitic soil	
	1983	1995	1983	1995	1983	1995
(7) Total B applied from 1980	6.8 (100)	20.4 (100)	3.4 (100)	17.9 (100)	4.5 (100)	31.4 (100)
(10) Applied B taken up by plant from 1980 (8)-(4)	0.27 (4)	1.79 (9)	0.23 (7)	1.90 (11)	0.11 (3)	2.38 (8)
(11) Applied B left in the soil from 1980 (6)-(2)	2.69 (39)	7.09 (35)	1.88 (56)	7.01 (39)	1.63 (36)	8.93 (29)
(12) Applied B leached out from 1980 (7)-(10)-(11)	3.86 (57)	11.5 (56)	1.25 (37)	9.01 (50)	2.74 (61)	20.1 (64)
(13) Water soluble B in original plot	0.40	0.29	0.40	0.28	0.41	0.31
(14) Water soluble B in FTE plot	1.06	1.18	0.97	1.14	0.77	1.21
(15) Applied B left in the soil as water soluble form (14)-(13)	0.66 (10)	0.89 (4)	0.57 (17)	0.85 (5)	0.35 (8)	0.90 (3)
(16) 0.5N HCl soluble B in FTE plot	1.39	1.14	1.68	1.69	2.21	1.65
(17) 0.5N HCl soluble B in FTE plot	2.95	3.43	3.11	4.93	2.90	3.46
(18) Applied B left in the soil as adsorbed form (17)-(16)-(15)	0.90 (13)	1.40 (7)	0.86 (26)	2.39 (13)	0.34 (8)	0.91 (3)
(19) Applied B left in the soil as fixed form (11)-(15)-(18)	1.13 (17)	4.80 (24)	0.45 (13)	3.77 (21)	0.94 (21)	7.12 (23)

The determination of total B was the next problem. An accurate method for this was reported by Yamada and Hattori (1986). Total amounts of B in standard soil were analyzed repeatedly by this method and the result obtained confirmed it as a reliable method. The soil sampling method was rather less problematic as the soil in the field was mixed well by the repeated cultivation.

Soil in the 0–15 cm layer was almost homogenized and the data obtained confirmed this because soil B values did not substantially fluctuate. Soil sampling was always carried out carefully.

It is well known that there is a high correlation between the amount of B absorbed by plants and the hot water soluble B in soil, but there is not always

54

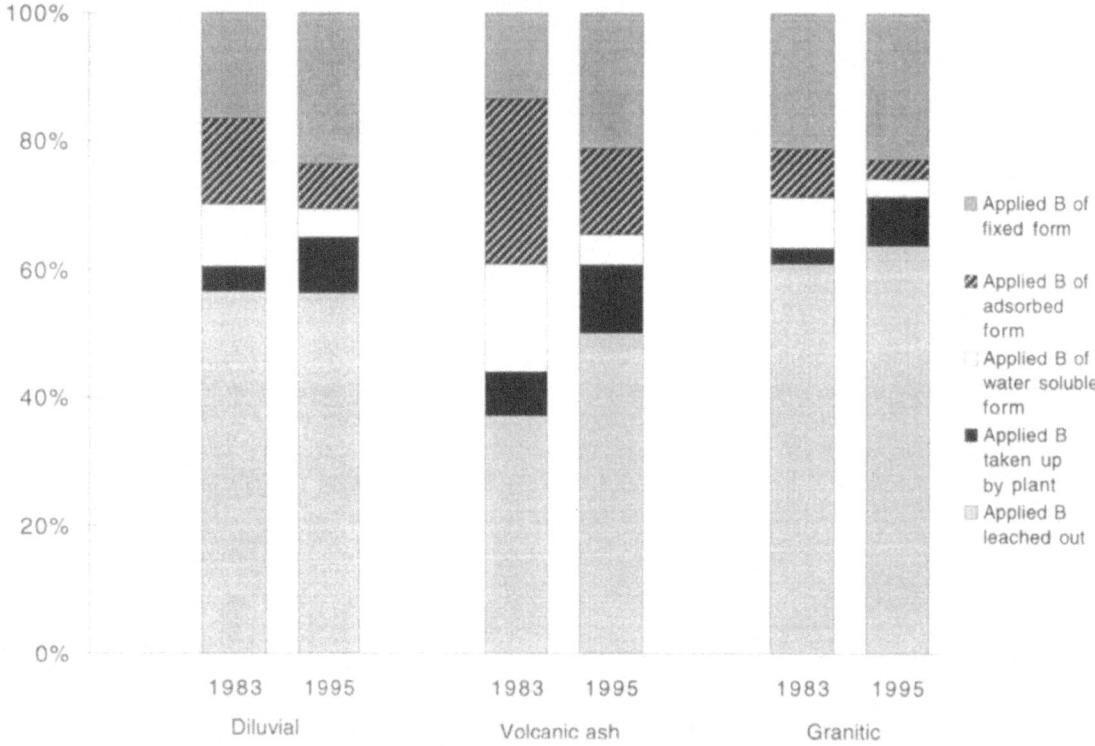

Figure 2. Relative distribution of boron applied to the field.

agreement on the significance of other forms of B in soil (Moraghan and Mascagni, 1992). Recently, Jin et al. (1987, 1988) and Tsadilas et al. (1994) tried to clarify the availability of soil B fractions separated by several chemical extractants. Their results are interesting to examine the availability of each fraction for plant growth, but further studies using more plant species are expected.

For our purpose, a B fraction which included almost all adsorbed forms of B was considered most suitable. Thus, we chose 0.5 N HCl as a moderately powerful extractant. As shown in the Table 2, 0.5 N HCl soluble B was 3–6 times larger than water soluble B. The difference between 0.5 N HCl soluble B and hot water soluble B is assumed to represent adsorbed forms of B as already mentioned, with the potential to become water soluble when the hot water soluble B pool is depleted. Table 4 shows the correlation between the amount of B absorbed by plants and hot water soluble B, 0.5 N HCl soluble B and adsorbed B, using three kinds of soil and five kinds of crop. Positive correlation was found in all cases ($p < 0.01$) although the correlation with the adsorbed form is somewhat lower than with

the hot water soluble B or 0.5 N HCl soluble B. This indicates that plants can access a part of the adsorbed form of B during their growth.

The forms of B insoluble in 0.5 N HCl probably belong to the occluded B in crystal lattices, or the organic B immobilized by soil organisms, as was already reported by Liu et al. (1981). This form is less available for plant uptake during short periods of cultivation, and is tentatively named a fixed form. Hereafter the behavior of various forms of B, such as fixed form, adsorbed form and hot water soluble form, will be discussed.

Balance sheet of B in the plots with and without B application

As already mentioned, input B from rainfall and basal NPK fertilizers was not insignificant and affected the balance sheet of B. However the amount of B leached out was slightly larger than this input. Thus all the input B and some part of B in soil were leached out. Nutritional condition of B became worse year by year in original plot, as the total output of B (4) + (5), was always larger than the input B during the successive cultivation. When the amount of B in fertilizers is less than this level, the situation becomes more severe. In the FTE plot, input B from rainfall and NPK fertilizers was relatively small compared with the B fertilizer applied. In this plot, the amount of B leached out was substantial.

Behaviour of B in the soils of original plot

The total amount of B in the granitic soil was rather high, but its hot water soluble B level was low as mentioned by Liu et al. (1981). The total B in the soils of the original plot was decreased steadily with successive cultivation. The total loss of B, (4) + (5) was far greater than the sum total of hot water soluble B and input B. Accordingly, the extra B must be found in the fixed form of B and adsorbed form of B, which decreased during the cultivation period. It was assumed that among the inorganic and organic components of the fixed form of B, some parts of them were weathered or mineralized gradually and changed to adsorbed and water soluble forms during the long period of cultivation, as suggested by Gupta et al. (1985). Also a part of the adsorbed form must have changed to the water soluble form, supplemented by the fixed form.

Behaviour of B in the soils of FTE plot

The behaviour of the B applied is an important key to the rational application of B. It was demonstrated by this long-term experiment in B application that the rate of absorption by plants was at most 10% of the B applied. By contrast, the rate of B leached out of the 0–15 cm layer reached 40–60%, and was higher in granitic soil. Slow-release B fertilizer used in this experiment was FTE, which was citric acid soluble and water insoluble; however a high proportion of B was still leached out. As mentioned in the previous paper, the average precipitation in this area is more than 2000 mm, and granitic soil is a typical sandy soil (Eguchi and Yamada, 1997). These factors may have promoted leaching. Adjustment of the pH of the soil in the field was always carried out before cultivation to within pH 6.5–7.0. Accordingly the effect of pH on the leaching or adsorption of B may not be great. The rate of B left in the soil was higher after 3 years cultivation (40–55%) than after 15 years cultivation (30–40%), and higher in volcanic ash soil.

Certainly some plant roots distribute in the 15–50 cm layer and can absorb B from this layer. Accord-

ingly leached B to this layer is not totally lost, but the significance of B in this layer is probably very small. Anyhow, such a high rate of leaching creates a concern in considering the application of B in the future.

Among the forms of B left in the soil, fixed form was the greatest, followed by the adsorbed and then the hot water soluble form. This was true except in the volcanic ash soil after 3 years cultivation. The rate of fixed B increased with the cultivation successively. This indicates that chemical and biological fixation of B was proceeded during the period of cultivation. However it is difficult to justify such a higher rate of the fixed form of B by the formation of inorganic and organic matter which is occluded in crystal lattices or immobilized in biomass. Further study is necessary. The amount of the adsorbed form of B increased with cultivation, but the proportion of this form decreased and was comparatively high in volcanic ash soil. This shows that the adsorption capacity might be larger in the earlier period, and also larger in volcanic ash soil. It was a reasonable result because the absorption coefficient of phosphate and organic matter were high in this soil as mentioned by Bingham et al. (1971) and Harada and Tamai (1968). The higher rates of the adsorbed form and hot water soluble form of B in volcanic ash soil may bring the higher absorption of B by plants than in the other soils. The amount of water soluble B showed a slight increasing tendency, but far lower concentration than the toxic level of B in soil, in spite of the continuous application of B over 15 years.

Acknowledgments

We are grateful to Dr. Yoshiaki Ishizuka, Professor Emeritus of Hokkaido University, for his precious guidance during this study. Our thanks are due to Mr. K Motohira, Mr. M Nagano, Mr. M Akitomo, Mr. T Abe and Miss M Ido for their valuable suggestion and helpful cooperation.

References

Bingham F T, Page A L, Coleman N T and Flach K 1971 Boron adsorption characteristics of selected amorphous soils from Mexico and Hawaii. Soil Sci. Soc. Am. J. 35, 546–552.

Cooke G W 1982 Fertilizing for Maximum Yield. Granada, UK, 465 p.

Eguchi S and Yamada Y 1997 Long-term field experiment on the application of slow-release boron fertilizer – Part 1 – Effect of boron on crop growth. *In* Boron in Soils and Plants. Proceedings.

56

Eds R W Bell and B Rerkasem. pp. 43–48. Kluwer Academic Publishers, Dordrecht, the Netherlands.

Gupta U C, Jame Y M, Campbell C A, Leyshon A J and Nicholaichuk W 1985 Boron toxicity and deficiency. A review. Can. J. Soil Sci. 65, 381–409.

Harada T and Tamai M 1968 Some factors affecting the behavior of boron in soil. Soil Sci. Plant Nutr. 14, 215–224.

Jin-yun J, Martens D C and Zelazny L W 1987 Distribution and plant availability of soil boron fractions. Soil Sci. Soc. Am. J. 51, 1228–1231.

Jin-yun J, Martens D C and Zelazny L W 1988 Plant availability of applied and native boron in soils with diverse properties. Plant and Soil 105, 127–132.

Liu Z, Zhu Q Q and Tang L H 1981 Boron deficient soils and their distribution in China. Soil Research Report (Institute of Soil Science Academia Sinica) 5, 1–13.

Liu Z, Zhu Q Q and Tang L H 1983 Micronutrients in the main soils of China. Soil Sci. 135, 40–46.

Moraghan J T and Mascagni H J Jr 1991 Environmental and soil factors affecting micronutrient deficiencies and toxicities. *In* Micronutrients in Agriculture, 2nd ed. J J Mortvedt, F R Cox, L N Shuman and Z M Welch, pp. 371–426. Soil Science Society America Book Series No 4.

Sawaguchi M, Hasegawa S, Kumagai H and Nakamoto H 1992 Accumulation in soil and crop responses of B and Mn with successive application of the micronutrient fertilizers. Hokuno 59, 419–423.

Tsadilas C D, Yassoglou N, Kosmas C S and Kallianou Ch 1994 The availability of soil boron fractions to olive trees and barley and their relationships to soil properties. Plant and Soil 162, 211–217.

Yamada H and Hattori T 1986 Determination of total boron in soil by the curcumine-acetic acid method after extraction with 2-ethyl-1,3 hexanediol. Soil Sci. Plant Nutr. 32, 135–140.

R.W. Bell and B. Rerkasem (eds.), Boron in Soils and Plants, 57–61.
© 1997 *Kluwer Academic Publishers.*

The influence of boron fertilizer on distribution of extractable boron in soil profiles in rape-rice rotations in southeast China

K Wang[1], JM Xu[2], Y Z Wei[3], YA Yang[3] & RW Bell[4]
[1] *Institute of Remote Sensing Applications in Agriculture*
[2] *Department of Soil Science*
[3] *Department of Agricultural Chemistry, Zhejiang Agricultural University, Hangzhou, PR China*
[4] *School of Environmental Science, Murdoch University, Murdoch WA 6150, Australia*

Key words: borax, crop rotation, depth, leaching, residual boron, soil boron

Abstract

Soil application of borax is recommended but not widely adopted to correct boron (B) deficiency of oilseed rape (*Brassica napus* L.) in southeast China. In order to estimate the residual value of applied B fertilizers field experiments in three important soils of southeast China were conducted to study the effect of B fertilizer on the depthwise distribution of soil extractable B over a 3 year period in rape-rice (*Oryza sativa* L.) rotations. A single application of up to 30 kg borax ha^{-1} increased soil extractable B concentrations significantly not only in the 0–20 cm layer but also in the 20–40 cm layer. Repeated B fertilizer application increased soil extractable B concentrations with each successive application and to greater soil depths. Applications of 15 and 30 kg borax ha^{-1}, while increasing B levels in the profile in the year of addition, had no residual effect on soil extractable B in the next year in the blue purple paddy soil, but did in the 0–20 cm layer of alluvial and red soils in the second year. There was limited B leaching from borax applications up to 30 kg ha^{-1} except in the red soil. At 15 kg borax ha^{-1}, soil B levels were maintained over 3 successive years under the rape-rice rotation at levels greater than control soils.

Introduction

Boron (B) is normally present in the soil solution as a nonionized molecule throughout most of the pH range suitable for plant growth, making it susceptible to leaching. Leaching occurs most readily from the surface layers of soils especially sandy textured soils with neutral to acidic pH but much less so in heavy clay soils (Pinyerd et al., 1984; Saarela, 1985).

Boron deficiency is widespread in southeast China and foliar B fertilizer is commonly applied to correct it especially in oilseed rape (*Brassica napus* L.) (Yang et al., 1993). Given the prevalence of B deficiency and the lesser effectiveness of foliar B application, soil application of B fertilizer is recommended (Yang, pers comm). In Zhejiang province, soil applied B fertilizer had residual effects on oilseed rape yields 2–3 years after application, especially in heavy textured soils with high rates and repeated applications (Yang,

pers comm). However in Zhejiang province, the mean annual rainfall of 1500–1700 mm is mostly concentrated in the months March to October, making the applied B highly susceptible to leaching. On the other hand, roots of many crops may go beyond the surface layer to draw part of their nutrients from the subsurface layer. Therefore, information about the depthwise distribution of available B in soils following B fertilizer application is valuable for improving B soil tests, fertilizer applications, and avoiding B toxicity in the sensitive crops (Keren and Bingham, 1985).

In southeast China, two rice (*Oryza sativa* L.) crops are generally grown in rotation with oilseed rape. Changes in soil chemistry that occur in paddy soils are reported to have no specific effect on B forms or availability (Ponnamperuma, 1972). However, the effects of alternating upland field crops with paddy rice on the forms and availability of B do not appear to have been examined. Two possible consequences arise for

Table 1. Location, soil type, and soil properties (0–10 cm) in the three experiments.

Soil properties	Jianshan Agricultural Research station	Jiangshan, Shangyu	Jiashan, Fengnan
Soil type (Chinese Classification	Alluvial soil	Red soil	Blue purple paddy soil
Soil taxonomy Sub-order	Udifluvent	Hapludult	Aquent
pH (KCl)	6.5	5.1	6.4
Organic matter (g kg^{-1})	27	26	43
[1] Clay (g kg^{-1})	200	260	220
[2] ECEC (cmol(+)kg^{-1})	13.1	7.1	22.2
[3] B (mg kg^{-1})	0.35	0.33	0.75

[1] fraction < 0.001 mm

[2] Effective cation exchange capacity

[3] hot 0.01 M CaCl$_2$ extractable boron

the plant availability of B in oilseed rape-rice rotations from the soil chemical changes that occur after the drying of flooded soils. Firstly, B availability to oilseed rape may be depressed by reaction of B with Fe and Al oxyhydroxides (Jin et al., 1988). Secondly, increased sorption of B on oxyhydroxide surfaces may decrease the amount of B which is susceptible to leaching.

The objective of the present study was to examine: the depthwise distribution of extractable B in soils after harvesting of oilseed rape supplied with varied B fertilizer levels; the effects of repeated B fertilizer and of rice-rape rotations on the depthwise distribution of extractable B in soils.

Materials and methods

Field experiments were conducted on alluvial, red and blue purple paddy soils (Tables 1 and 2) to study the residual effects of B fertilizer over three successive years and the risks of inducing B toxicity from repeated annual applications of B fertilizer to oilseed rape (Wang et al., 1997). The present paper is concerned only with the effect of B fertilizer on the depthwise distribution of soil extractable B. The B fertilizer treatments which were replicated 4 times and arranged in a randomised block design, and are listed in Table 3. In each experiment, a triple cropping rotation was followed with each cycle commencing with oilseed rape. Double low oilseed rape (*cv.* Zheyouyou 2) was sown in seed beds in September and seedlings were transplanted from their seed beds at 6 to 7 leaf stage in late October to mid November.

Table 2. Borax application rates (kg ha^{-1}) for the three experiments.

Treatment No. (code)	Year 1	Year 2	Year 3
1 (0+0+0)	0	0	0
2 (15+0+0)	15	0	0
3 (30+0+0)	30	0	0
4 (30+30+30)	30	30	30 (0)[1]

[1] No B application in experiment of Jiashan Fengnan in year 3.

One hundred and twenty plants were transplanted into each plot of 272 cm width and 375 cm length. The B fertilizer for each plot was dissolved in water, and every plant in the plot received the same aliquot of B fertilizer placed 5 cm beside seedlings at transplanting to ensure equal supply to each plant in the plot. The distance between the adjacent fertilization holes was 25 cm. All the plots received the same amount of basal nutrients and supplementary fertilizer. Soil profile samples were collected from 0–10, 10–20, 20–40, 40–60, and 60–100 cm depth at each site before experiments commenced. The same depths were sampled each year from all the treatments after the harvest of each oilseed rape and the second rice crop. About 10 separate samples were collected in each plot and mixed completely before analysis. Soil available B was extracted by boiling in hot 0.01 M CaCl$_2$ for 35 minutes (Spouncer et al., 1992). Boron concentration in extractants was determined with azomethine-H (Zarcinas, 1995).

Table 3. Effect of soil sampling above (A) and between (B) the borax fertilization sites at bolting and at harvest in plots fertilized with increasing rates of borax on two soils. Values are means of four replicates. The LSD test is at $p<0.05$.

Borax Treatment kg ha^{-1}	Sampling Location	Alluvial Soil		Red Soil	
		Bolting stage	Harvest	Bolting stage	Harvest
0	A	0.59	0.63	0.55	0.63
	B	0.63	0.65	0.56	0.60
	LSD	ns	ns	ns	ns
15	A	0.85	0.72	1.49	0.89
	B	0.75	0.69	0.85	0.83
	LSD	0.09	ns	0.11	ns
30	A	1.25	0.82	2.04	0.92
	B	0.85	0.70	0.83	0.85
	LSD	0.18	ns	0.50	ns

Results and discussion

Rainfall, soil properties and soil sampling

Rainfall during the oilseed rape growth season at Jiangshan varied from 1000–1600 mm, whereas rainfall at Jiashan was 700–800 mm. However, rainfall at both sites was considered high enough to be conducive to the leaching of B fertilizer applied to the oilseed rape.

In the 0–10 cm layer, soils varied markedly in pH, organic matter, ECEC and extractable B but not in clay content (Table 1). However, the clay content of the 20–60 cm layer in the red soil was substantially higher than in the alluvial soil (data not shown).

A preliminary investigation of soil sampling methods showed that the position of soil sampling in B fertilized plots after harvesting the first oilseed rape crop had no significant effect on extractable B levels in the 0–10 cm layer (Table 3). It was therefore assumed that the following results on depthwise distribution of B and its response to B fertilizer rates was not an artifact of the B fertilizer application method which placed B fertilizer solution below individual seedlings. Residual B values in soils was unlikely to have been affected by the soil sampling procedure since thorough soil cultivation to 20 cm depth was carried out for each of the two rice crops which followed oilseed rape.

Effect of B fertilizer on extractable B concentration

In all three experiments, the soil extractable B concentrations of the 0–20 cm layer at harvest of oilseed rape in year 1 were significantly increased by B fertilizer especially at the highest rate (30 kg ha^{-1}) (Tables 4, 5 and 6). By contrast, extractable B in the 20–40 cm layer was also significantly increased by B fertilizer only in the red soil (Table 6). Increased soil extractable B concentrations in the subsurface layer of the red soil immediately after harvesting oilseed rape suggests redistribution of B fertilizer after transplanting. There were no significant difference in extractable B in the 0–20 and 20–40 cm layers between applications of 15 kg borax ha^{-1} and 30 kg borax ha^{-1} except in the red soil (Tables 4, 5 and 6).

Repeated B fertilizer application increased soil extractable B concentration with each successive application, and the effects reached a greater soil depth (Tables 4, 5 and 6). Repeated applications of borax increased extractable B to 40–60 cm in the blue purple and red soils, and below 60 cm in the alluvial soil (Table 5 and 6). The result suggested that B fertilizer was more easily leached in the light textured alluvial soil[1] than in the red soil, which is heavier textured in the subsoil.

The residual B from the B fertilizer applied at 15 or 30 kg borax ha^{-1} in the first year in the blue purple paddy soil had no significant effect on the extractable B in the profile in years 2 and 3, but increased the extractable B in the surface 0–20 cm layer in alluvial and red soils. The effect did not persist into the second year after the B fertilizer application (Tables 5 and 6). By contrast on the blue purple paddy soil which contained the highest original extractable B levels, repeated B fertilizer for two successive years increased the available B in the surface layer of the blue purple paddy soil significantly compared to the treatment of no

Table 4. Effect of B fertilizer on the depthwise distribution of extractable B (mg kg^{-1}) in blue purple paddy soil. Apart from the pre-plant values, soil B values were obtained from samples taken after the harvest of oilseed rape. Values are means of four replicates. Values in each row followed by the same letter are not significantly different according to Duncan's test at $p<0.05$.

Depth (cm)	Season	Pre plant	0+0+0[1]	15+0+0	30+0+0	30+30+0
0–10	92–93	0.94	0.91b	1.06ab	1.20a	1.27a
	93–94		0.96b	0.97b	1.21b	1.99a
	94–95		1.02b	0.95b	1.12b	1.20a
10–20	92–93	0.53	0.69b	1.01a	1.10a	1.26a
	93–94		0.51b	0.51b	0.66b	0.96a
	94–95		0.80b	0.88ab	0.95ab	1.06a
20–40	92–93	0.40	nd[2]	nd	nd	nd
	93–94		0.37b	0.34b	0.38b	0.56a
	94–95		0.43a	0.50a	0.39a	0.48a
40–60	92–93	0.31	nd	nd	nd	nd
	93–94		0.32b	0.32b	0.34b	0.39a
	94–95		0.37a	0.35a	0.43a	0.37a
>60	94-95		0.45a	0.39a	0.39a	0.45a

[1] See Table 2 for treatment details
[2] Not sampled.

Table 5. Effect of B fertilizer on the depthwise distribution of extractable B (mg kg^{-1}) in alluvial soil. Apart from pre-plant values, soil B values were obtained from samples taken after the harvest of oilseed rape. Values in each row followed by the same letter are not significantly different according to Duncan's test at $p<0.05$.

Depth (cm)	Season	Pre plant	0+0+0[1]	15+0+0	30+0+0	30+30+30
0–10	92–93	0.45	0.65b	0.70ab	0.70a	0.84a
	93–94		0.45b	0.62a	0.62a	0.65a
	94–95		0.49b	0.54b	0.53b	0.88a
10–20	92–93	0.33	0.43b	0.47ab	0.56ab	0.63a
	93–94		0.41b	0.53ab	0.61a	0.61a
	94–95		0.47b	0.50b	0.51b	0.84a
20–40	92–93	0.18	nd[2]	nd	nd	nd
	93–94		0.44a	0.55a	0.55a	0.52a
	94–95		0.36b	0.38b	0.49b	0.78a
40–60	92–93	0.14	nd	nd	nd	nd
	93–94		nd	nd	nd	nd
	94–95		0.20b	0.19b	0.26b	0.37a
>60	94-95		0.19b	0.20b	0.22b	0.29a

[1] nd – not sampled.

B application in three years (Table 4). This indicated that soil extractable B accumulated from repeated B application and its residual effect became significant.

The extractable B concentrations of treatments with high initial rates of B fertilizer were still significantly higher than those of treatments without B fertilizer after planting two rice crops (data not shown). This suggests that waterlogging of two rice crops did not enhance the leaching of boron significantly, and this may be one of the reasons that the applied B fertilizer still had residual effects on oilseed rape yield in the year after a complete rape-rice-rice rotation (Yang, pers comm).

We conclude that, in the three soils which represent those which cover a large part of southeast China (National Soil Survey Office, 1993), there was limited downwards movement of B in the soil profiles even

Table 6. Effect of B fertilizer on the depthwise distribution of extractable B (mg kg^{-1}) in red soil. Apart from pre-plant values, soil B values were obtained from samples taken after the harvest of oilseed rape. Values in each row followed by the same letter are not significantly different according to Duncan's test at $p<0.05$.

Depth (cm)	Season	Pre plant	0+0+0[1]	15+0+0	30+0+0	30+30+30
0–10	92–93	0.35	0.60c	0.83b	0.85ab	0.93a
	93–94		0.49c	0.52bc	0.58b	0.91a
	94–95		0.50b	0.48b	0.53b	0.76a
10–20	92–93	0.29	0.39c	0.57b	0.75a	0.70a
	93–94		0.40c	0.57b	0.51b	0.77a
	94–95		0.46b	0.46b	0.52b	0.80a
20–40	92–93	0.26	0.39c	0.57b	0.71a	0.72a
	93–94		0.21c	0.23bc	0.22bc	0.43a
	94–95		0.25b	0.35b	0.35b	0.56a
40–60	92–93	0.18	nd[1]	nd	nd	nd
	93–94		0.19a	0.17a	0.18a	0.22a
	94–95		0.19b	0.18b	0.21b	0.25a
>60	94–95		0.17a	0.16a	0.16a	0.19a

[1] nd – not sampled.

with applications of 30 kg borax ha^{-1}. Only on the red soil with repeated annual applications of 30 kg borax ha^{-1} was there significant accumlation of extractable B at >60 cm after 3 years.

A single application of 15 kg borax ha^{-1} maintained extractable soil B levels for 3 years of continuous cropping at values exceeding those in the unfertilized soil. Thus there was no suggestion of significant leaching losses of B from the three soils tested. Further studies on the residual value of B fertilizer in soils of southeast China should concentrate on soil reactions of applied B, and on the amounts removed in harvested crop products.

Acknowledgements

This research was supported by the Australian Centre for International Agricultural Research (Project 9120), the Zhejiang Provincial Government and the Chinese Ministry of Agriculture. The senior author is grateful to the Australian Agency for International Development for financial assistance under its International Seminar Support Scheme.

References

Jin J Y, Martens D C and Zelancy L W 1988 Plant availability of applied and native boron in soils with diverse properties. Plant and Soil 105, 127–132.

Keren R and Bingham F T 1985 Boron in water, soils and plants. Adv. Soil Sci. 1, 229–276.

National Soil Survey Office 1993 Soil Types of China. Vol 1. pp 560–870. Agricultural Press, Beijing.

Pinyerd C A, Odom J W, Long F L and Dane J H 1984 Boron movement in a Norfolk loamy sand. Soil Sci. 137, 428–433.

Ponnamperuma F N 1972 The chemistry of submerged soils. Adv. Agron. 24, 29–96.

Saarela I 1985 Plant-available boron in soils and the boron requirement of spring oilseed rapes. Ann. Agric. Fenn. 24, 183–265.

Spouncer L R, Nable R O and Cartwright B 1992 A procedure for the determination of soluble boron in soils ranging widely in boron concentrations, sodicity and pH. Commun. Soil Sci. Plant Anal. 23, 441–453.

Wang K, Yang Y, Xue J M, Ye Z Q, We Y Z and Bell R W 1997 Low risks of boron toxicity from boron fertilizer use in oilseed rape-rice rotations in southeast China. Nutrient Cycling in Agroecosystems. (FRES 2078 under review).

Yang Y, Xue J M, Ye Z Q and Wang K 1993 Responses of rape genotypes to boron application. Plant Soil 155/156, 321–419.

Zarcinas B A 1995 Suppression of iron interference in the determination of boron using the azomethine-H procedure. Commun. Soil Sci. Plant Anal. 26, 713–729.

R.W. Bell and B. Rerkasem (eds.), Boron in Soils and Plants, 63–67.
© *1997 Kluwer Academic Publishers.*

63

Boron analysis at different stages of the cell cycle in cultured tobacco cells

H.Iikura[1], T.Kataoka[1], M.Tamada[1], T. M.Nakanishi[1] & C.Yonezawa[2]

[1] *Graduate School of Agricultural and Life Sciences, The Univ. of Tokyo, 1-1-1 Yayoi, Bunkyo-ku, Tokyo, Japan 113*

[2] *Dept of Chemistry and Fuel Research, Japan Atomic Energy Research Institute, Tokai-mura, Ibaraki-ken 319-11, Japan*

Key words: cell cycle, flow cytometry, prompt gamma-ray analysis, synchronization, tobacco BY-2

Abstract

The paper presents the data suggesting that boron (B) content in the cultured tobacco (*Nicotiana tabacum*) cells change during the stage of cell cycle. Cultured cell line, tobacco BY-2, was synchronized by treatments with aphidicolin and propyzamide. When the cells started to grow, cells were periodically collected at each stage of the cell cycle. To measure the B content, cells were dried and prepared as tablets and prompt gamma-ray analysis (PGA) with a cold neutron beam was performed. Flow cytometry (FCM) was carried out on the same cell fraction to analyze the cell cycle in more detail. There was a clear difference in B content with respect to the cell cycle stage. Boron content in the cells at G_2+M phase was found to be about half of that at G_0/G_1.

Introduction

Boron (B) is reported to be localized in the cell wall and associated with pectin (Matoh, 1993; Hu and Brown, 1994). Boron deficiency was also reported to increase the membrane permeability to K^+. However the primary function of the B in plants has not been clarified yet.

Recently several reports of prompt gamma-ray analysis (PGA) with thermal neutrons have been reported to determine B content in biological samples (Anderson et al., 1990, 1994). The method allows B determination in biological materials without the need for chemical preparation. The same samples can be run several times. The method has the advantage that the residual radioactivity in the sample is negligible because the lifetime of prompt gamma-rays emitted from the excited state of nuclei is within 10^{-14}s. Prompt gamma-ray is suitable to detect light elements, especially B. However, when H is abundant in the sample, the background level in gamma-ray measurement is high. To reduce the background, a cold neutron source is more preferable than that of thermal neutron. Recently, an atomic reactor installed at Japan Atom-

ic Energy Inst. (JAEI) has been remodeled and PGA with cold neutrons has been performed (Yonezawa, 1993). Therefore, we used the facilities to determine B in cultured cells. The sensitivity to detect B with this method was as high as those by inductively coupled plasma atomic emission spectrometry (ICP-AES), ICP-mass spectrometry (ICP-MS) or spectrophotometric measurement (Yonezawa, 1996).

In the present study, we present B content in cultured cells collected from different stages of the cell cycle by PGA.

Materials and methods

Cells and Culture Condition

Suspension cell line of tobacco (*Nicotiana tabacum*) cv. Bright Yellow 2 (BY-2) was cultured in a modified LS medium (Linsmaier and Skoog medium with 3% sucrose, 200 mg KH_2PO_4 L^{-1}, 0.2 mg 2,4-dichlorophenoxy acetic acid (2,4-D)L^{-1}, 0.1 g inositol L^{-1} and 1 mg thiamin hydrochloride L^{-1}) at pH 5.6 according to the previously described method (Nagata

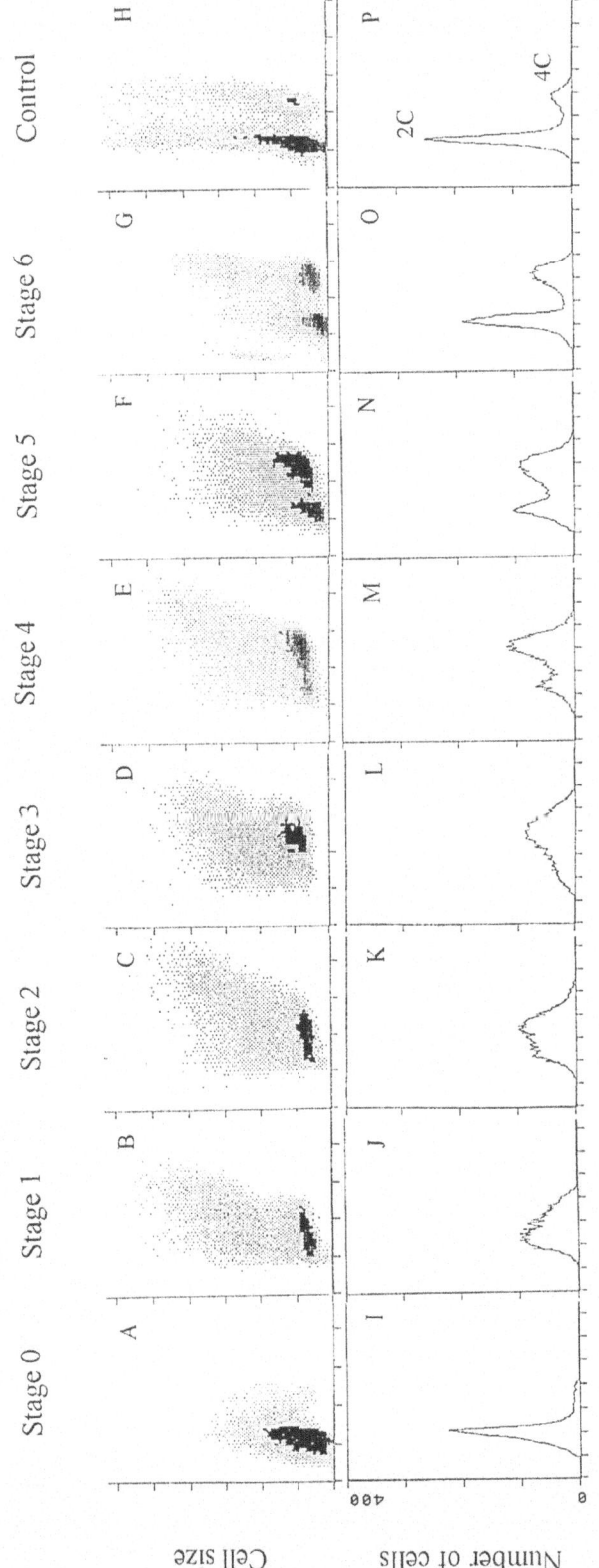

Figure 1. Flow cytometric analysis of the cells stained with PI. (A.I) Stage 0; (B.J) 1; (C.K) 2; (D.L) 3; (E.M) 4; (F.N) 5; (G.O) 6; (H.P) control.

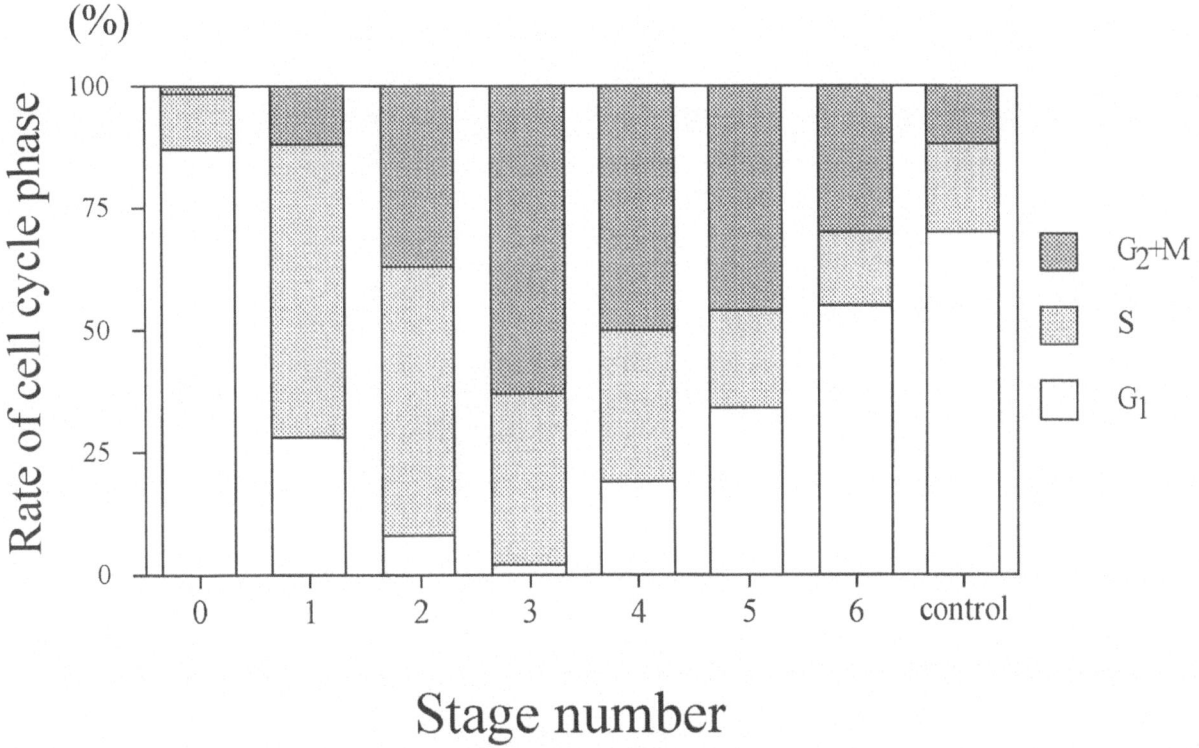

Figure 2. Percentage of cell cycle phase at each synchronized stage.

et al., 1982). The concentration of B in the medium was 1 mg B L^{-1} as boric acid. Cells were grown, at 26° C, in 100 mL of liquid medium in 300-mL Erlenmeyer flasks and shaken at 135 rpm on a rotary shaker in the dark. Subcultures were performed every 7 days with a 5-mL inoculum.

Synchronization of the Cells

To synchronize the cells, 20 mL of 7-day-old suspension culture was transferred to 80 mL of modified LS medium containing 5 mg aphidicolin L^{-1} (Hasezawa and Nagata, 1991). After 24 hr of the treatment, cells were collected and washed 3 times with 0.4 M mannitol. Then the cells were resuspended in a fresh medium and cultured. During the cell growth, cells were collected every hour and washed with 0.4 M mannitol to get the cells at different cell cycle stages. On the other hand, treatment with propyzamide was also performed to synchronize the cells again at G$_2$/M phase. After 4 hr, when the cells began to grow, cells were treated with 3 mM propyzamide and resuspended in a fresh medium. Then the cells were collected and washed as described above. Each cell fraction was divided into

two samples, one for flow cytometry (FCM) and another for B analysis. Stock solutions of aphidicolin and propyzamide were 15 mM and 6 mM in DMSO, respectively, and stored at 0–4° C.

Prompt gramma-ray analysis

Cells were washed 3 times with 0.4 M mannitol and dried at 60° C. The dried samples were milled in a mortar and 300 mg of the powder was pressed to form a small disk (1 cm diameter) by a tablet maker and hydraulic pump. Each tablet was prepared and then analysed for B by PGA as described by Nakanishi et al. (1997).

Flow Cytometry

When the cells started to grow after the synchronizing treatment, cells were collected every one hour and protoplasts were prepared. One hundred mg fresh weight samples of cells were washed twice with 0.4 M mannitol and treated with 1% (w/v) Cellulase Onozuka RS, 0.2% Macerozyme R10 (Yakult Pharma. Co., Tokyo) and 0.01% Pectoliase Y23 (Seishin Pharma.

B content at stage of cell cycle

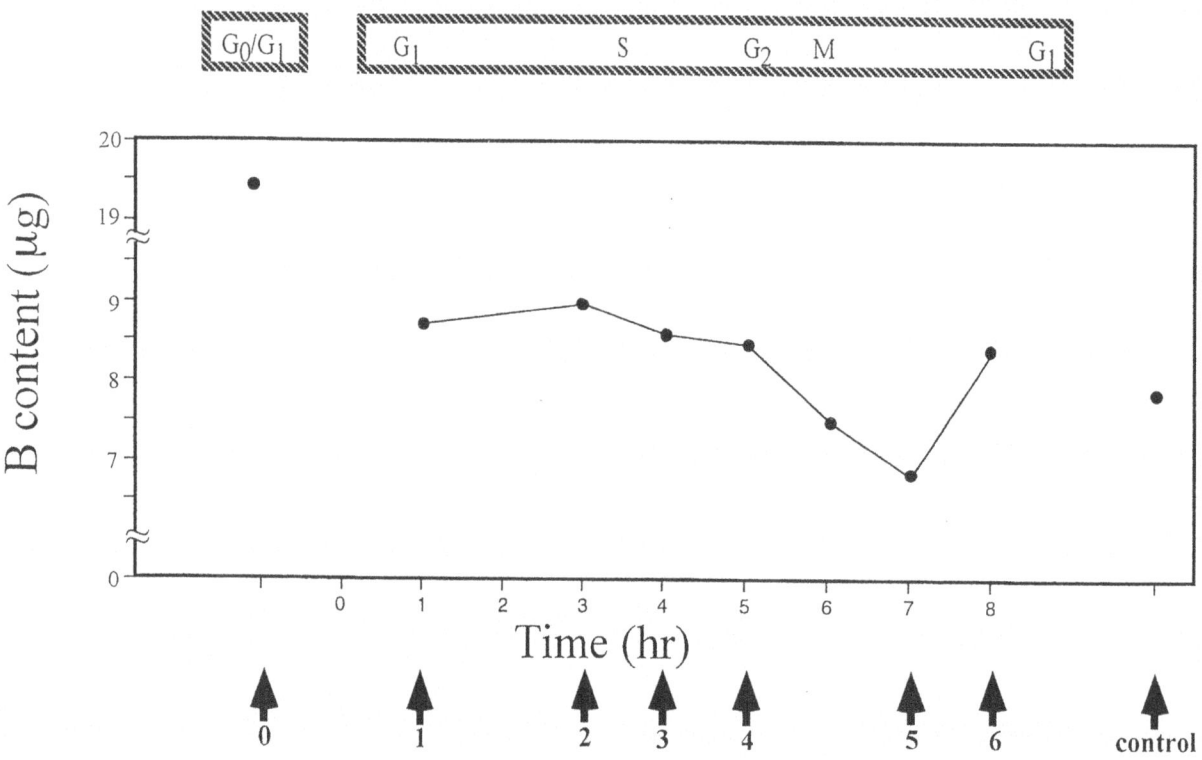

Figure 3. The relation between cell cycle and B content in 1 g cells (dry wt). The number under the arrow shows each growth stage number.

Co., Tokyo) in 0.7 M mannitol at 30° C for 6 h, with gentle shaking. The protoplasts were filtrated through the nylon mesh (50 μm), and washed twice with 0.4 M mannitol. The same quantity of 0.4 M mannitol was added to the protoplast suspension. Then PI (propidium iodide)-staining solution (50 mg propidium iodide L^{-1}, 1 g CH_3COONa L^{-1}, 0.58 g NaCl L^{-1}, 5 mL Nonidet P40 (Sigma Co.) L^{-1} and 5 mg RNase L^{-1}) was added to the 1 mL of the cell suspension. The flow cytometry was carried out right after DNA staining by an Epics Elite ver. 4.02 equipped with argon ion laser (488-nm). Data were analyzed using the software, Multicycle. Twenty thousand protoplasts were analyzed in each cell fraction.

Results and discussion

When the cells are treated with aphidicolin, DNA synthesis is inhibited and the cells are synchronized at

G_1/S phase. With the treatment of propyzamide, cells are synchronized at G_2/M phase because of the inhibition of microtubule formation. By synchronizing the cells with the treatment by aphidicolin and propyzamide, the cells at each cell cycle stage (Stage 1–6) were able to be collected (Figure 1 B–G, J–O). When the cells started to grow again after removal of aphidicolin, the cells were at G_1/S phase for 1 hr (Stage 1), and proceeded to S/G_2 phase in 3 hr (Stage 2). After 4 hr, more than 60% of the cells was found to be a G_2+M phase (Stage 3). Five hr after the treatment with aphidicolin, the stage of the cells was the same as those when propyzamide was removed, i.e. 50% and 20% of the cells were at G_2+M and G_1 phase, respectively (Stage 4). In 2 hr after removal of propyzamide, about 30% of the cells went through the cell cycle again to G_1 phase (Stage 5), and more than 50% of the cells proceeded to G_1 phase after 3 hr (Stage 6). Two-day-old and 8-day-old cells in suspension culture without any treatments were used as control (Figure 1 H. P) and G_0/G_1 phase

(Stage 0; Figure 1 A. I) samples, respectively. Figure 2 shows the proportion of cell cycle phase at each stage of cell harvest.

As is shown in Figure 3, B content decreased gradually as the cells proceeded through the stages of the cell cycle. However, when cell division was finished, B content increased dramatically. It might be correlated with a change in density of the cell wall where most tissue B is localized (Hu and Brown, 1994). Indeed B content at Stage 0 (almost G_0/G_1 phase) was more than twice that in the proliferating phase (Stage 1–6). At present we cannot explain why the cells at G_0/G_1 phase contained higher amount of B.

The improvement of PGA with cold neutrons made it possible to measure trace amounts of B in cells. We hope that the kinetics of B in cells during the cell cycle will be clarified and study of B in plant will become more active in future.

References

Anderson D L, Cunningham W C and Lindstrom T R 1994 Concentrations and intakes of H, B, S, K, Na, Cl, and NaCl in foods. J. Food Comp. Anal. 7, 59–82.

Anderson D L, Cunningham W C and Mackey E A 1990 Neutron capture prompt-gamma activation analysis of foods. Biol. Trace Elem. Res. 26/27, 613–622.

Hasezawa S and Nagata T 1991 Dynamic organization of plant microtubules at the three distinct transition points during the cell cycle progression of synchronized tobacco BY-2 cells. Bot. Acta 104, 206–211.

Hu H and Brown P H 1994 Localization of boron in cell walls of squash and tobacco and its association with pectin. Evidence for a structural role of boron in the cell wall. Plant Physiol. 105, 681–689.

Matoh T, Ishigaki K I, Ohno K and Azuma J I 1993 Isolation and characterization of a boron-polysaccharide complex from radish roots. Plant Cell Physiol. 34, 639–642.

Nagata T, Okada K and Takebe I 1982 Mitotic protoplasts and their infection with tobacco mosaic virus RNA encapsulated in liposomes. Plant Cell Reports 1, 250–252.

Nakanishi T M, Iikura H, Kataoka T, Tamada M, Furukawa J and Yonezawa C 1997 Analysis by prompt gamma-ray method with cold neutrons of boron and other elements in soybean. In Boron in Soils and Plants. Proceedings. Eds R W Bell and B Rerkasem, pp 69–72. Kluwer Academic Publishers, Dordrecht, the Netherlands.

Sakai Y, Yonezawa C, Magara M, Sawahata H and Ito Y 1994 Measurement and analysis of the line shape of prompt gamma-rays from recoiling ^{7*}Li produced in the ^{10}B (n, alpha) ^{7}Li reaction. Nucl. Instr. and Meth. in Phys. Res. 353, 699–701.

Yonezawa C 1993 Prompt gamma-ray analysis of elements using cold and thermal reactor guided neutron beams. Anal. Sci. 9, 185–193.

Yonezawa C 1996 Multi-element determination by a cold neutron-induced prompt gamma-ray analysis. Anal. Sci. 12, 605–613.

R.W. Bell and B. Rerkasem (eds.), Boron in Soils and Plants, 69–72.
© *1997 Kluwer Academic Publishers.*

Analysis by Prompt Gamma-Ray Method with Cold Neutrons of Boron and Other Elements in Soybean

T. M. Nakanishi[1], H. Iikura[1], T. Kataoka[1], M. Tamada[1], J. Furukawa[1] & C. Yonezawa[2]

[1] *Graduate School of Agricultural and Life Sciences, The Univ. of Tokyo, 1-1-1 Yayoi, Bumkyo-ku, Tokyo, Japan 113*

[2] *Dept of Chemistry and Fuel Research, Japan Atomic Energy Research Institute, Takaimura, Ibaraki-ken 319-11, Japan*

Key words: boron, element distribution, prompt gamma-ray analysis, neutron activation analysis, soybean

Abstract

We have employed prompt gamma-ray analysis (PGA) with cold neutrons for the analysis of boron (B) in soybean (*Glycine max*). Primary, first and second leaves, as well as root samples were dried, prepared as thin disks and irradiated with cold neutrons. Prompt gamma-rays emitted from the sample was measured by Ge and bismuth germanate (BGO) detectors. Boron content in root was less than 40% of that in leaves. The same sample was used to determine Al, K, Ca, and Zn content by neutron activation analysis. When two soybean cultivars, Al susceptible (Chief) and Al tolerant (Perry) were compared, the similar B distribution was observed, except in primary leaf, where B concentration in former plant was about two times higher than that of the latter.

Introduction

Generally, inductively coupled plasma atomic emission spectrometry (ICP-AES), ICP mass spectrometry (ICP-MS) and spectrophotometric measurement have been used for boron (B) determination in plant tissues (James et al., 1995; Zarcinas, 1995; Stringari et al., 1996). However, sample preparation to digest sample materials into solution involves time and contibutes to errors (Stringari et al., 1996). Since a suitable radioisotope for B does not exist, the radio-tracer work cannot be carried out. The prompt-gamma analysis (PGA) using thermal neutrons has been used for B determination in biological materials by several workers (Anderson et al., 1990, 1994). However, thermal neutrons induce a high background in gamma-ray measurement, especially in the presence of H. Since the conventional PGA system has suffered from poor beam quality and low counter sensitivity, the method has lower sensitivity than ICP-AES or ICP-MS. Recently, an atomic reactor, JRR-3M, at the Japan Atomic Energy Research Institute has been remodeled and a cold neutron source with stable beam flux and a revised

detection system for PGA were equipped (Yonezawa et al., 1993). Using the facilities, it was possible to determine 10^{-9} g of B (Yonezawa, 1996). The method detects gamma-rays emitted from the sample while it is irradiated with cold neutrons. The neutron converts ^{10}B to ^7Li and an alpha particle and gamma-rays (478 keV) are released. The gamma-ray enables determination of B content in the sample. With cold neutrons, the sensitivity for detection of B was reported to be more than ten times higher than that with thermal neutrons (Yonezawa, 1996). It was also reported that the method has the highest sensitivity in B compared to the other elements, except for H. The advantage of this method is the same as conventional PGA with thermal neutrons, that is non-destructive, multi-element analysis, suitable to determine light elements and negligible induction of radioactivity in the sample. The disadvantage of PGA is that it needs an atomic reactor with a special beam guide for the analysis.

We present here the determination of B in soybean plant by PGA with cold neutrons. Since the method is non-destructive, the same sample was used to perform thermal neutron activation analysis to determine other

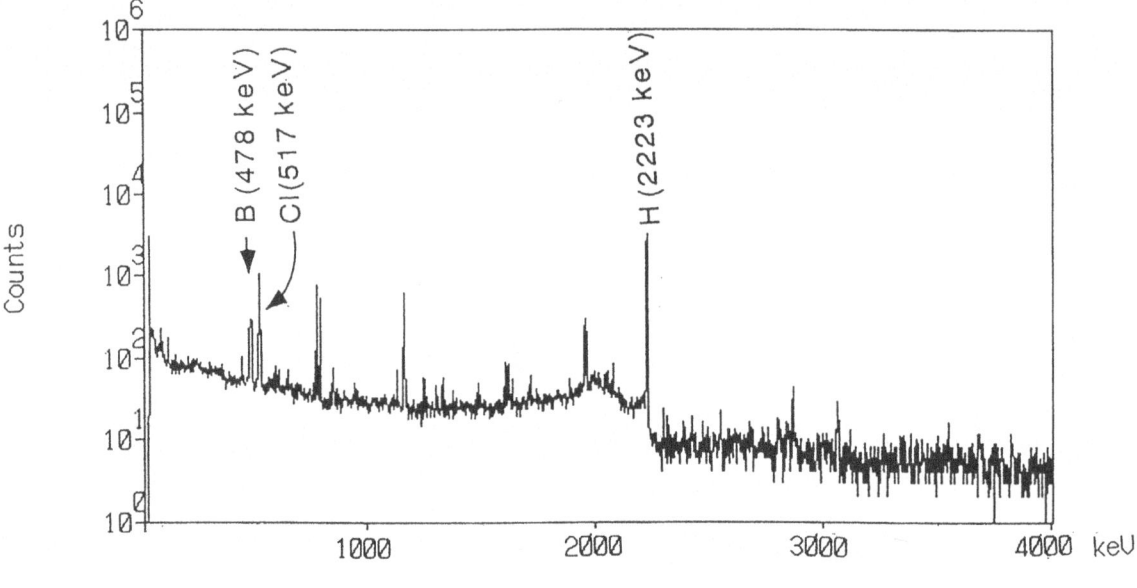

Figure 1a.

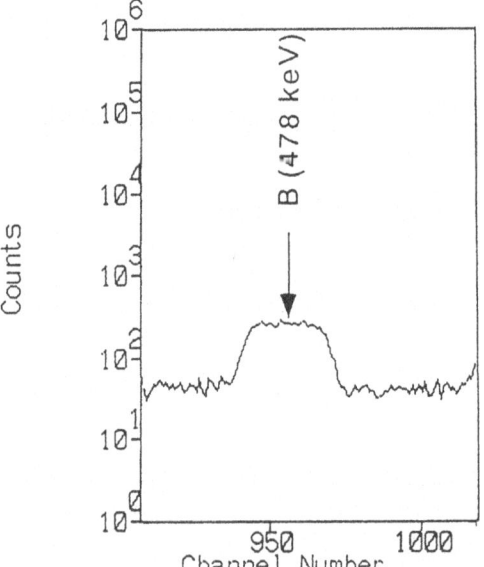

Figure 1. (a) Prompt gamma-ray spectrum of Perry root irradiated with cold neutrons. Measurement was performed for 566 seconds. Besides B (478 keV), H(2223 keV) and Cl (517 keV) peaks were also observed. (b) Magnification of B peak in Figure 1 (a). The broadening of the peak is due to the recoil energy of gamma-ray from ^{7}Li in the sample.

Materials and Methods

The study used soybean plants, (*Glycine max*) cv. Perry and Chief, which were aluminum tolerant and sensitive, respectively. The plants were germinated and grown in water culture. After 30 days of culture, more than ten plants of both cultivars were harvested to collect root, primary leaf, first trifoliate and second trifoliate leaves. All samples were dried at 60° C overnight.

Prompt gamma-ray analysis

The dried root and leaves of Perry and Chief were milled and 100 mg of each sample was pressed to make thin disks, 1 cm in diameter. Then the sample pellets were sealed in fluorinated ethylene propylene (FEP, 25 μm in thickness, 15 mm × 15 mm) bags. The standard samples were prepared by dispersing boric acid solution onto a filter paper. When the solution was dried, the sample was sealed in a FEP bag of about the same size as the plant sample. The FEP bag was fixed in the middle of the sample holder to be set within the beam area during irradiation. The material of the sample holder was polytetrafluoroethylene (PTFE), to minimize the gamma-ray background of the measurement. The atomic reactor used was JRR-3M, installed at the Japan Atomic Energy Research Institute. The beam size of cold neutrons was about 2 cm in diameter. The sample was set at 45 degree to the

elements. The elements detected through the neutron activation analysis were also presented along with B.

Table 1. Boron, aluminium, zinc, potassium and calcium concentrations in soybean cvv. Perry and Chief plant material (dry wt).

	B (mg kg^{-1})	Al (mg kg^{-1})	Zn (mg kg^{-1})	K (g kg^{-1})	Ca (g kg^{-1})
Perry					
root	13.4 ± 0.1	302 ± 6.0	8.53 ± 0.41	69.2 ± 2.5	2.07 ± 0.31
primary leaf	37.1 ± 0.2	62 ± 1.5	0.49 ± 0.06	29.4 ± 2.7	7.19 ± 0.24
first leaf	$56 \quad \pm 0.3$	105 ± 2.0	0.77 ± 0.12	30.1 ± 3.2	$140 \quad \pm 0.40$
second leaf	40.2 ± 0.2	98 ± 2.3	1.29 ± 0.15	39.9 ± 3.1	9.94 ± 0.38
Chief					
root	19.2 ± 0.1	253 ± 5.0	2.76 ± 0.21	47.4 ± 2.6	2.55 ± 0.27
primary leaf	69.7 ± 0.3	141 ± 3.0	1.08 ± 0.11	25.7 ± 2.7	$110 \quad \pm 0.40$
first leaf	50.9 ± 0.3	88 ± 2.2	0.61 ± 0.09	31.2 ± 3.5	7.81 ± 0.35
second leaf	43.4 ± 0.2	75 ± 2.3	$1.4 \quad \pm 0.15$	36.5 ± 3.0	8.47 ± 0.43

Errors: standard deviations calculated from gamma-ray counting.

neutron beam and gamma-rays emitted from the sample were detected by a pure Ge counter, shielded with Bismuth germanate counter, BGO ($Bi_4Ge_3O_{12}$). The neutron flux irradiated was 1.1×10^8 n cm^{-2} s^{-1}. The irradiation was performed from 200 to 500 seconds, to accumulate sufficient counts of the gamma-ray peak. The gamma-ray energy used to analyze B was 478 keV.

Neutron activation analysis

Neutron activation analysis was performed on the same samples used for PGA. The neutron irradiation was carried out by a Triga Mark II reactor, at Rikkyo University. The neutron flux was 1.3×10^{12} n cm^{-2}. First, the samples were irradiated for 30 seconds to determine Al, K and Ca. Then the samples were irradiated again for 24 hours to analyze Zn. The gamma ray energy used to determine ^{28}Al, ^{42}K, ^{49}Ca and ^{65}Zn were 1779, 1525, 3084 and 1116 keV, respectively. All these nuclides were produced from corresponding stable isotopes by (n, gamma) reaction. Standard samples, 100 mg of Orchard Leaves (NIST) and 20 mg of JB3 (Geological Survey of Japan) were prepared in the same way as plant samples and were irradiated at the same time.

Results and Discussion

Figure 1 shows an example of a prompt gamma-ray spectrum. Figure 1 (a) is the whole spectrum of Perry root and (b) is the magnification of the B peak. As is shown in Figure 1 (b), the Na peak (473 keV) was not detected. When high amounts of Na exist, its sharp

peak is superposed on the B peak (data not shown). Background levels lower than the H peak were higher due to compton scattering of gamma-rays from H. When the sample was not dried, the background level was much higher due to H in water (data not shown). As is indicated in the figure, H (2223 keV) and Cl (517 keV) peaks were also observed beside that of B (478 keV). Since the gamma-rays to determine B came from recoiling gamma-rays of ^{7}Li, only the B peak was broader than those of the other elements (Sakai, 1994). Sakai also reported that from the shape of the B spectrum, the atomic density around B could be calculated. However, there was no difference in B peak between root and leaves (data not shown). The gamma-ray measurement was performed for about 200 to 600 seconds for leaves and roots, respectively, which gave gamma-ray counts more than 5000. The background measurement of the filter paper without any sample showed no peak in the energy regions of B, Cl and H. The background counts in the same area as B was 13 in 290 seconds. Boron concentration in root and leaf disks of Perry and Chief are shown in Table 1. In both plants, the B content in root was less than 40% of that in leaf. Boron content in both plant cultivars showed the similar concentrations except in primary leaf, where B concentration in Chief was much higher.

By short duration neutron activation analysis, ^{28}Al, ^{42}K and ^{49}Ca were measured. Since the half lives of these nuclides are short, the same samples were irradiated for 24 hours further to determine their Zn content. The element concentration in plant samples by thermal neutron activation analysis were also shown in Table 1. Boron concentration in leaf was two to three times higher than that in root in both cultivars. When element concentrations in primary leaf in both plants are compared, Al and Zn concentrations in Chief showed higher concentration than those in Perry. The error in the table was calculated from the standard deviation (S.D.) of gamma-ray counting. Therefore, in the case of B, when longer measurement is performed, the error becomes smaller. To determine B amount with an error less than 10%, about 200 counts are needed. We carried out the measurement from 200 to 500 seconds to get more than 5000 counts from 100 mg of the sample. Therefore, 1 hour is needed to measure B content in 1 mg of the sample with 10% of accuracy. Though error in counting depends on the background level and the length of the measurement, the detection limit of B by this method is, therefore, estimated to be 1000 times less than the amount we measured in the sample.

We presented here the prompt gamma-ray analysis with cold neutrons, which seems to be promising in analyzing the trace amount of B in plant material. Since the method to analyze the trace amount of B is still one of the largest problems in studying the role of B in plants, further development of the method is expected.

Acknowledgements

The authors express their sincere thanks to Dr. C. D. Foy, at USDA, USA, for providing the soybean seeds, cvv. Perry and Chief.

References

Anderson D L, Cunningham W C and Lindstrom T R 1994 Concentrations and intakes of H, B, S, K, Na, Cl, and NaCl in foods. J. Food Comp. Anal. 7, 59–82.

Anderson D L, Cunningham W C and Mackey E A 1990 Neutron capture prompt-gamma activation analysis of foods. Biol. Trace Element Res. 26/27, 613–622.

James D W, Hurst C J and Tindall T A 1995 Alfalfa cultivar response to phosphorous and potassium deficiency: Elemental composition of the herbage. J. Plant Nutr. 18, 2447–2464.

Sakai Y 1994 Measurement and analysis of the line shape of prompt gamma-rays from recoiling $^{7}*$Li produced in the ^{10}B (n, gamma) ^{7}Li reaction. Nuclear Instruments and Methods in Physics Res. A 353, 699–701.

Stringari G, Moeller F, Ceschini A and Failla O 1996 Are the differences among samples from agricultural trials analyzed by routine procedures experimental or only analytical? Commun. Soil Sci. Plant Anal. 27, 1403–1416.

Yonezawa C 1996 Multi-element determination by a cold neutron-induced prompt gamma-ray analysis. Anal. Sci. 12, 605–613.

Yonezawa C, Wood A K H, Hoshi M, Ito Y and Tachikawa E 1993 The characteristics of the prompt gamma-ray analyzing system at the neutron beam guides of JRR-3M. Nuclear Instruments and Methods in Physics Res. A 329, 207–216.

Zarcinas B A 1995 Suppression of iron interference in the determination of boron using the azomethine-H procedure. Commun. Soil Sci. Plant Anal. 26, 713–729.

R.W. Bell and B. Rerkasem (eds.), Boron in Soils and Plants, 73–75.
© 1997 *Kluwer Academic Publishers.*

Total boron determination in silicate material with azomethine-H, after extraction with boron specific resin

Rui R.Vale

Universidade de Trás-os-Montes e Alta Doura (UTAD), Secçao de Química, 5000 Vila Real Portugal

Key words: Amberlite IRA-743, azomethine-H, boron determination, boron specific resin

Abstract

A simple method of total boron (B) determination in silicate material, which virtually eliminates interferences from the matrix, is proposed. After sample fusion with sodium carbonate, the melt is dissolved with water and B is extracted from the aqueous phase with a B specific resin, Amberlite IRA-743. Boron is then eluted from the resin with hydrochloric acid and colorimetrically determined with azomethine-H. The preliminary results, using three reference rock standards, indicate a very good reproducibility but the measured mean values are somewhat higher than the reported working values for the standards. However, in two cases no significant differences ($p < 0.05$) were encountered between the reported and measured values.

Introduction

The accurate determination of boron (B) in silicate samples, especially silicate rocks and soils, can play an important role in geological studies, assessment of the B status on toxic mine spoils, and in predicting the ability of soils to supply B in the long term for plant growth.

Nevertheless, most silicate samples have B concentrations below 200 mg kg^{-1} (Walsh, 1985) and B determination at the trace levels is not an easy analytical task due to the complexity of the matrices involved.

The classical method of B determination in silicate samples is by destillation as methylborate (Jeffery, 1975), but this is a time consuming and complex procedure. Many other methods have been proposed for determination of total B, using emission spectrography, atomic absorption spectroscopy (AAS), colorimetry, ion selective electrodes, and inductively coupled plasma-atomic emission spectroscopy (ICP-AES). However, none of the proposed methods have had all the attributes necessary for universal acceptance, such as precision, sensitivity, accuracy, rapidity or convenience for routine application.

Boron has been determined directly in sodium carbonate digests, by ICP-AES, but several problems are associated with this procedure. Iron is a major interfering element due to spectral interference with boron, and the high concentration of salts present in the solution may clog low solid nebulizers or sample introduction tips of the torch (Self, 1993). An alternative procedure using a preliminary phosphate fusion followed by a second fusion with potassium hydroxide has been proposed (Din, 1984). The azomethine-H method for B determination is sensitive and applicable to aqueous media. However, this method cannot be directly applied to mineral sample digests due to the high salt content of the digest solutions and potentially interfering cations such as iron, aluminum, copper, etc..

Hence the procedure of Bingham (1982) requires the addition of ethanol and adjustment of the pH to near neutrality to precipitate excess salts or iron and aluminum. This procedure though precise becomes expensive due to the high amounts of ethanol involved in each determination.

At the Universidade de Trás-os-Montes e Alto Douro (UTAD), Portugal, investigation of a simple, reliable, and economic method for the determination of total B in silicate samples involves B separation from aqueous sodium carbonate digests, with the B specific resin, Amberlite IRA-743. After solubilization of the fused melt with water, B is extracted from the solution

Table 1. Analytical results of boron in reference samples.

	Reported values (mg B kg^{-1})				Measured values (mg B kg^{-1})			
	N[a]	WV[b]	SD[c]	± CLs[d]	N[a]	MV[e]	SD[c]	± CLs[d]
IWG-GIT Granite MA-N	11	17	5.44	3	10	33.2	1.96	1.4
IWG-GIT Granite AC-E	8	21	3.2	2.6	10	23.4	1.35	1.0
ANRT Serpentine UB-N	9	140	17.3	12	10	153	2.47	1.7

[a] Number of results
[b] Working value
[c] Standard deviation
[d] 95% confidence limits calculated with the formule Sd.t/$\sqrt{N}$
[e] Measured mean value

with a small amount of resin by overnight shaking. The B adsorbed to the resin is then eluted with hydrochloric acid and determined with azomethine-H.

Boron specific resins have been used by Carlson and Paul (1968; 1969) in the potentiometric determination of this element in waters and agricultural samples. This article describes the proposed method and reports some preliminary results on its suitability for the determination of total B in silicate samples.

Materials and methods

Chemicals and reagents

- Amberlite IRA-743 resin, 50 to 100 mesh
- Ammonium hydroxide (NH_4OH), 3M
- Azomethine-H reagent: Dissolve 0.9 g of azomethine-H in 100 mL of 2% L-ascorbic acid solution
- Boron standards: Aliquot 0, 1, 2, 3, 4, 5 mL of a B stock solution, containing 15 mg B L^{-1} into a 100 mL polyethylene wide mouth flask; add 3 g of sodium carbonate and extract B as described in the procedure for the samples
- Buffer masking solution as described by Gaines and Mitchell (1979)
- Deionized distilled water, subsequently passed through a column of Amberlite IRA-743
- Hydrochloric acid (HCl), 3M
- Sealed bags of synthetic fiber, containing 0.5 g of resin
- Sodium carbonate, (Na_2CO_3)

Table 2. Concentrations of some elements in the analyte solution.

Element	Concentration (mg kg^{-1})		
	Granite MA-N	*Granite AC-E*	*Serpentine UB-N*
Aluminium	<100	<75	<50
Iron	<10	<50	<60
Copper	<2	<0.6	<0.4
Manganese	<3	<4	<10
Silicon	<160	<160	<200

Reference rock standards

Two reference samples from the International Working Group 'Analytical Standards of Minerals, Ores, and Rocks' (IWG-GIT Granite MA-N and IWG-Granite AC-E) and one from the Association Nationale de la Recherche Technique (ANRT Serpentine UB-N) were used.

Procedure

Fuse 0.500 g of fine-powered sample with 3.000 g of sodium carbonate over a Mecker burner as described by Lim and Jackson (1982). Cool and add 10 mL of water to the crucible. Allow to stand until dissolution of the melt is complete and transfer into a 100 mL wide mouth polyethylene flask. Repeat this operation four more times to completely digest and transfer the melt. Insert a resin bag in the flask and shake the aqueous mixture overnight in the end over end shaker at 100 rpm. Withdraw the bag into a capped tube, wash with two 10 mL portions of water and add 5 mL of 3M NH_4OH in the tube. Shake for five minutes to remove interfering anions from the resin and discard the solution. Wash with two 10 mL portions of water. Add 5 mL of $3M$ HCl into the tube and shake for five minutes. Remove the resin bag and wash it with 10 mL of water. Add the washing to the tube containing the solution. To regenerate the resin, wash with 10 mL of water, add 5 mL of $3M$ NH$_4$OH, shake for five minutes, and wash twice with 10 mL of water. Determine boron in the HCl extracts by pipetting a 2 mL aliquot into a capped plastic tube. Add 4 mL of the buffer-masking solution and 2 mL of azomethine-H solution and mix well. Allow color to develop for one hour at room temperature. Measure absorbance in a spectrophotometer at 420 nm with a 1 cm cell.

Results and discussion

The results of the preliminary studies using the proposed method are presented in Table 1 together with the reported values (Govindaraju, 1995) for the rock standards.

The values were highly reproducible with relative standard deviations of 5.9% (Granite MA-N), 5.8% (Granite AC-E), and 1.6% (Serpentine UB-N). The B concentrations obtained for Granite AC-E and Serpentine UB-N were not statistically different ($p<0.05$) from the reported values. Thus, the B specific resin, Amberlite IRA-743, which was used due to its capacity to adsorb B from neutral and alkaline solutions, ease of handling, and simplicity of the method (Carlson and Paul, 1968) appeared to be effective for the reproducible extraction and determination of B in silicate rocks.

As the element is isolated from the major components of the sample, the method must be quite free from interferences. Table 2 reports the concentrations of some potentially interfering elements in the analyte solution, determined by AAS. The presence of several cation species in the eluate is possibly due to entrappment of small particles of solids in the resin bag during shaking of the suspension which are then solubilized by the acid. However, the possibility that the cations could be bound by the resin cannot be discounted. One way or another, the concentrations determined do not seem to pose a serious analytical problem to the azomethine-H method as indicated by John et al. (1975) and Porter et al. (1981). The buffer-masking solution described by John et al. (1975) and adopted for this proposed procedure may not be required in its entirety, since it was developed for B determinations in soil and plant

ash extracts. If the present method is proved totally successful for B determination in silicate material, a simplified composition of the buffer masking solution may be applicable.

Conclusion

The proposed method of total B determination in silicate material was precise and does not require either hazardous chemicals or expensive equipment. However, some disparity between reported and measured B values for one of the reference standards suggests that further evaluation and adjustments are required, especially if the procedure is to be applied to low-B samples. Further studies are already under way.

References

Bingham F T 1982 Boron. In Methods of Soil Analysis, Part 2, Chemical and Microbiological Properties. Ed AL Page, pp 431–447. American Society of Agronomy, Madison, USA.

Carlson R M and Paul J L 1968 Potentiometric determination of boron as tetrafluoroborate. Anal. Chem. 40, 1292–1295.

Carlson R M and Paul J L 1969 Potentiometric determination of boron in agricultural samples. Soil Sci. 108, 266–272.

Din V K 1984 The preparation of iron-free solutions from geological materials for the determination of boron (and other elements) by inductively-coupled plasma emission spectroscopy. Anal. Chim. Acta 159, 387–391.

Gaines T P and Mitchell G A 1979 Boron determination in plant tissue by the azomethine H method. Commun. Soil Sci. Plant Anal. 10, 1099–1108.

Govindaraju K 1995 Working values with confidence limits for twenty six CRPG, ANRT and IWG-GIT geostandards. Geostand. Newsl. 19, 1–32.

Jeffery P G 1975 Boron. In Chemical Methods of Rock Analysis, pp 151–162. Pergamon Press Inc., New York, USA.

John M K, Chuah H H and Neufeld J H 1975 Application of improved azomethine H method to the determination of boron in soils and plants. Anal. Lett. 8, 559–568.

Lim C H and Jackson M L 1982 Dissolution for total element analysis. In Methods of Soil Analysis, Part 2, Chemical and Microbiological Properties. Ed A L Page, pp 1–11. American Society of Agronomy, Madison, USA.

Porter S R, Spindler S C and Widdowson A E 1981 An improved automated colorimetric method for the determination of boron in extracts of soils, soil-less peat-based composts, plant materials and hydroponic solutions with Azomethine H. Commun. Soil Sci. Plant Anal. 12, 461–473.

Self R 1993 Soil testing for boron on alkaline soils. In Boron and its Role in Crop Production, Ed. U C Gupta, pp 125–135. CRC Press Inc., Boca Raton, USA.

Walsh J N 1985 Determination of boron at trace levels in rocks by inductively coupled plasma spectrometry. Analyst 110, 959–962.

R.W. Bell and B. Rerkasem (eds.), Boron in Soils and Plants, 77–81.
© 1997 *Kluwer Academic Publishers.*

An improved procedure for the extraction of plant available soil boron

Youzhang Wei[1] & B. A. Zarcinas[2]
[1] *Department of Land Science and Applied Chemistry, Zhejiang Agricultural University, Hangzhou, 310029,*
P.R.China
[2] *CSIRO Land and Water, Glen Osmond, SA 5064, Australia*

Key words: boron, constant temperature shaking machine, oilseed rape, sunflower, soil

Abstract

In this study, a constant temperature orbital shaking machine was designed and adopted to standardize conditions for the extraction of hot 0.01 M $CaCl_2$ soluble soil boron(B). The oven temperature for extraction was 100° C, the orbital shaking speed was 250 revolutions per minute (rpm) and the extraction time was 30 minutes. Control of the extraction temperature and time with minimal evaporative loss of water achieved adequate precision (C.V. = 0.8–2.0%) of extractable soil B. Using the sum of a series of sequential fractionation extractions, the theoretical basis of using hot 0.01 M $CaCl_2$ soluble soil B as an effective indicator of plant available B is postulated. Glasshouse pot experiments resulted in a high correlation between hot 0.01 M $CaCl_2$ soluble soil B and oilseed rape leaf B (r=0.999) and sunflower leaf B (r=0.997). This indicates the proposed procedure can effectively determine the plant available soil B fraction.

Introduction

The hot water soluble (hws) soil B extraction procedure of Berger and Truog (1939) has been extensively adopted as an indicator of plant available soil B (Agricultural Chemistry Committee of China, 1983). However, difficulties with the procedure including variable boiling time, differences in extraction temperature due to differences in soil matrix salts, and the loss of water during extraction result in low accuracy and precision of extraction of soil B between and within laboratories. Jin et al. (1987) and Hou et al. (1994) proposed serial extraction procedures in order to determine B associated with different soil fractions. Each extractant attacked different soil components and released the nutrients associated with them with each fraction solubilizing increasingly more difficult to extract forms of soil B. The authors indicated a decreasing plant availability of B with increasing extraction ability.

The purpose of this project was to investigate the correlation of plant available B with the sum of a sequential fractionation scheme and with our proposed hot 0.01 M $CaCl_2$ soluble soil B extraction proce-dure. Our proposed procedure for extracting plant available hot 0.01 M $CaCl_2$ soluble soil B was confirmed in a glasshouse study where good correlation with plant uptake B was obtained using low B Chinese soils with oilseed rape (*Brassica napus*) and sunflower (*Helianthus annuus*). Also, good precision and accuracy of extraction of hot 0.01 M $CaCl_2$ soluble soil B was achieved with analysis by inductively coupled plasma spectrometry (ICPS).

Materials and methods

Inductively coupled plasma spectrometer

A Plasma-Spec direct reading Echelle ICP (Leeman Labs Inc., USA) was used for B determinations. The operating parameters were: forward power 1.0 kw; coolant 12 L min^{-1}; auxiliary 10 L min^{-1}; nubuliser pressure 240 kPa producing a flow 0.35 L min^{-1}; sample uptake rate of 0.9 mL min^{-1} into a modified Scott spray chamber using a Hildebrand grid nebuliser.

Boron was determined at 249.678 nm with Fe interference corrected arithmetically.

Constant temperature orbital shaker

The shaker was 330 mm long × 330 mm wide × 90 mm high and mounted inside a thermostatically controlled constant temperature oven. The shaker tray could hold 25–50 ml tubes (30 mm × 155 mm). The temperature within the compartment could be adjusted from 25 to 200° C and rotated at 250 orbits per min with a maximum displacement of 40 mm.

Glasshouse soil characterisation

The soils used were a loamy siliceous thermic Aeric Endoaquept in glasshouse experiment 1 and a fine-clayey kaolinitic thermic Typic Plinthudult in glasshouse experiment 2. The soils were collected from Tonglu and Fuyang counties, Zhejiang Province, P.R. China, respectively. The respective properties of the surface soils were: pH 5.2, 6.5 (1:2 soil:1 M KCl); organic carbon 18.7, 31.2 g kg^{-1} (Allison et al. 1965); clay content (<0.001 mm) 112, 180 g kg^{-1} (by the pipette method) and CEC 4.8, 12.9 cmol(+) kg^{-1} soil. The extraction techniques are standard procedures of the Agricultural Chemistry Committee of China (1983).

Reagents

The water was distilled and further purified using B-specific resin (IRA-743, Sigma Chemical Co.) and was used for all washing, extraction and glasshouse watering procedures. All glass and plastic ware was cleaned by soaking overnight in 2 N HCl, rinsed with water and oven dried.

Fractionation of boron

The fractionation procedure chosen for this study was a combination of extractants described by Hou et al. (1994) and Jin et al. (1987) and chosen to suit our soil properties. The water soluble and non specifically adsorbed B were extracted by cold 0.02 M CaCl$_2$, the specifically absorbed B by 0.02 M mannitol, the Mn oxyhydroxide bound B by 0.1 M NH$_2$OH.HCl and the organic matter bound B by 0.02 M HNO$_3$ + 30% H$_2$O$_2$. The B concentrations in the extraction solutions were determined by ICPS.

Hot 0.01 M CaCl$_2$ soluble soil B extraction procedure

The procedure of Spouncer et al. (1992) was modified to extract hot 0.01 M CaCl$_2$ soluble soil B using the constant temperature shaker at 100° C for 30 min. The procedure was to weigh 10.0 g of < 2 mm air dried (35° C forced air draft) soil into a 50 mL plastic screw capped test tube and add 0.4 g activated charcoal. The charcoal was added since it is required for the azomethine-H colourimetric procedure of Gaines and Mitchell (1979). Twenty mL of 0.01 M CaCl$_2$ was added, lids replaced and the tubes were given an initial hand shake to move the soil from the bottom. The tubes were placed into the constant temperature shaker at 25° C, heated to 100° C without shaking and then shaken for 30 min at 250 rpm. After extraction the samples were immediately filtered into plastic test tubes through Whatman No. 42 filter papers for B determination by ICPS.

Precision of hot 0.01 M CaCl$_2$ soluble soil boron procedure

Four soils of varying hot 0.01 M CaCl$_2$ soluble soil B were sampled from Jiangshan, Hangzhou, Jiaxian and Jiaojiang counties. The hot 0.01 M CaCl$_2$ soluble soil B values were determined in quadruplicate on each soil by 5 operators on 5 occasions over a 1 month period to assess between operator precision and long term accuracy of the proposed procedure.

Glasshouse experiments

Two glasshouse experiments were established using five B treatments of 0; 0.006; 0.01; 0.02 and 0.04 g borax kg^{-1} of soil. Each plastic pot had 3 kg of soil in experiment 1 and 10 kg of soil in experiment 2. Amendments applied to the soils were 0.29 g urea, 0.20 g potassium chloride, 0.67 g calcium superphosphate and 0.60 g N:P:K (15:15:15) fertiliser kg^{-1} soil, respectively. The amended soils were mixed dry, placed in pots and arranged in a randomised complete block design with four replications. In experiment 1, the test crop was oilseed rape with two seedlings per pot while experiment 2 used sunflower with one seedling per pot. B-free deionised water was added to pots daily to maintain approximately −33 kPa tension. The youngest open leaves were sampled at the seedling stage.

Table 1. Effect of borax application on plant B, sum of sequential fractionation of soil B, hot 0.01 M CaCl$_2$ soluble and hot water soluble soil B. (Values are means in mg B kg^{-1} [soil or plant material] $\pm$ s.d., n=4)

Treatment g borax kg^{-1} soil	B$_1$	B$_2$	B$_3$	B$_4$
0	0.210	0.210 $\pm$ 0.023	0.205 $\pm$ 0.008	14.3 $\pm$ 0.4
0.006	0.354	0.323 $\pm$0.041	0.353 $\pm$0.021	42.4 $\pm$ 3.2
0.01	0.396	0.351 $\pm$ 0.037	0.391 $\pm$ 0.018	53.9 $\pm$ 3.6
0.02	0.574	0.538 $\pm$ 0.017	0.583 $\pm$0.012	79.3 $\pm$ 4.7
0.04	1.23	1.14 $\pm$ 0.18	1.24 $\pm$ 0.14	184 $\pm$ 9

B$_1$: sum of sequential fractionations; B$_2$: determined by the Berger and Truog (1939) method; B$_3$: determined by our method; B$_4$: youngest open leaf B. R$_{ab}$ is the regression coefficient (** p<0.01, * p<0.05). r$_{1,4}$ = 0.999**, r$_{2,4}$ = 0.997**, r$_{3,4}$ = 0.999**.

Statistical analysis

Simple correlation analyses were performed to evaluate the relationships between the dependent variable, B concentrations in the plant tissue, and the independent variables, hot 0.01 M CaCl$_2$ soluble soil B and the sum of the sequential soil B fractions. No transformations were necessary. Student's t-test was used to assess differences in extractability of soil B between our proposed procedure and that of Berger and Troug (1939).

Results and discussion

Relationship between hot 0.01 M CaCl$_2$ soluble soil boron and the sum of the sequential fractionations

It has been generally assumed that the procedure of Berger and Truog (1939) extracts B from soluble, absorbed and organic phases of the soil (Aitken and McCallum, 1988; Berger and Truog, 1939; Gupta et al., 1985; Haddad and Kaldor, 1984). In recent years, Jin et al. (1987) and Hou et al. (1994) proposed the concept of B fractionation in soil to provide a soil chemical explanation of soil B extractions as an indicator of plant available B. Using these procedures, the sum of the sequential soil B fractions and plant B data of experiment 1 are presented in Table 1. Jin et al. (1987) have shown some of these individual fractions to be plant available. B$_1$, the summation of these individual (data not presented) sequential fractions was not significantly different, by Student's t-test at p<0.01, to the values obtained using the method of Berger and Truog (1939), ie. B$_2$.

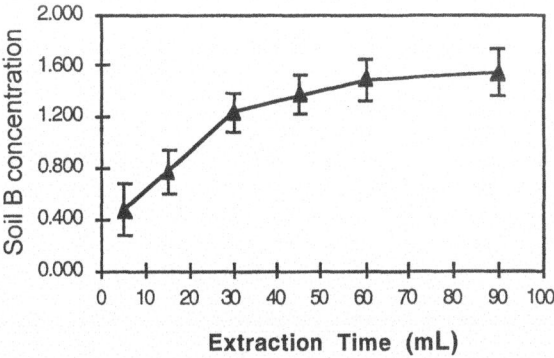

Figure 1. Effect of extraction time on hot 0.01 M CaCl$_2$ soluble soil B (mg B kg^{-1} soil). Vertical bars denote standard errors.

Extraction parameters

Although the Berger and Truog (1939) method for extraction of hws B from soils has been extensively applied, some difficulties using this procedure have been identified. Actual extraction temperature (due to variable soil salt concentrations), agitation during boiling (due to variable soil particle size distributions) and filtering times very much greater than 5 min result in a variable contact time between the soil and hot water (due to clogging of the filter paper by fine soil clay particles) producing low precision of extractable soil B values using this method. The constant temperature shaker designed in this study provided a constant temperature compartment for the extraction of soil B at an extraction temperature of 91 $\pm$ 2° C across all 25 test tubes with solution agitation effectively maintained. The soil to water ratio of 10 g: 20 mL with a shaking speed of 250 rpm resulted in even and complete mixing of the soil suspension. Figure 1 indicates that the

80

Table 2. Soil B concentrations of 4 soils determined by 5 operators on 5 occasions. (Values are means in mg B kg^{-1} soil ± s.d., $n = 4$.)

Operator	Date	Jiangshan	Hangzhou	Jiaxin	Jiaojiang
A	Aug. 1	0.283 ± 0.009	0.440 ± 0.023	0.611 ± 0.021	1.43 ± 0.13
B	Aug. 2	0.297 ± 0.017	0.454 ± 0.019	0.625 ± 0.026	1.52 ± 0.15
C	Aug. 8	0.297 ± 0.021	0.454 ± 0.012	0.625 ± 0.022	1.43 ± 0.17
D	Aug. 16	0.283 ± 0.019	0.444 ± 0.026	0.639 ± 0.031	1.42 ± 0.13
E	Aug. 31	0.283 ± 0.011	0.468 ± 0.029	0.611 ± 0.026	1.50 ± 0.16
C.V. (%)		0.82	2.1	1.4	2.0

C.V. is the coefficient of variation.

Table 3. Soil B extracted by the improved procedure and plant B concentrations from glasshouse experiments 1 and 2. (Values are means in mg B kg^{-1} [soil or plant material] ± s.d., $n = 4$.)

Treatment	Experiment 1 (Oilseed rape)		Experiment 2 (Sunflower)	
g borax kg^{-1} soil	^{a}Soil B	^{b}Leaf B	^{c}Soil B	^{d}Leaf B
0	0.205 ± 0.008	14.3 ± 0.4	0.271 ± 0.013	25.3 ± 0.22
0.006	0.353 ± 0.021	42.4 ± 3.2	0.329 ± 0.028	32.8 ± 1.9
0.01	0.391 ± 0.018	53.9 ± 3.6	0.469 ± 0.037	37.3 ± 3.0
0.02	0.583 ± 0.012	79.9 ± 4.7	1.18 ± 0.11	64.2 ± 3.1
0.04	1.24 ± 0.14	183 ± 9	2.54 ± 0.32	105 ± 7

$r_{x,y}$ is the regression coefficient between soil B and plant B (** $p<0.01$, * $p<0.05$).
$r_{a,b} = 0.999$**, $r_{c,d} = 0.997$**

amount of soil B extracted gradually increased with extraction time and tended to plateau after about 30 min. Table 1 and Figure 1 indicate that the extraction parameters detailed above extracted a concentration of soil B (B3) from glasshouse experiment 1 which was highly correlated with the uptake B by plants (B4) and was not significantly different (Student's t-test, $p<0.01$) from the values obtained using the method of Berger and Truog (1939). Hence, an extraction time of 30 min, oven temperature of 100° C and shaking speed of 250 rpm was chosen.

Extraction precision

The analytical variables which can lead to poor precision when the Berger and Truog (1939) procedure is used to extract soil B have been satisfactorily controlled as indicated in Table 2. The coefficients of variation were 0.82–2.1%, indicating adequate precision was achieved. Moreover, the plastic test tubes were fitted with lids which greatly reduced the loss of water from the tubes during the 30 min extraction period. Also, the use of 0.01 M CaCl$_2$ as the extractant, instead of water, flocculated the soil clay particles which settled to the bottom of the test tubes on cessation of agitation.

Minimal soil clay was therefore available to clog the filter paper hence minimising the contact time between the hot 0.01 M CaCl$_2$ extractant and the soil in the filter paper.

Plant availability of the soil boron

The soil B in the different treatments of glasshouse experiments 1 and 2 was extracted using the recommended procedure and determined by ICPS. The B concentration in the leaves of oilseed rape and sunflower were determined using the method of Gaines and Mitchell (1979). The results presented in Table 3 indicate that the soil B extracted using our procedure was highly correlated with the plant uptake B and is therefore recommended as a rapid, precise and efficient procedure for the extraction of plant available B.

Acknowledgments

This research was supported by Australian Centre for International Agricultural Research (Project 9120); the State Ministry of Agriculture and the Provincial Natural Science Foundation of Zhejiang, P.R.China.

References

Agricultural Chemistry Committee of China 1983 Conventional Methods of Soil and Agricultural Chemistry Analysis. Science Press, Beijing, P. R. China.

Allison L E 1965 Organic carbon. *In* Methods of Soil Analysis, Part 2 Chemical and Microbiological Properties. Ed C A Black, pp 1367–1378. American Society of Agronomy Monograph No. 9, Madison, Wisconsin, USA.

Aitken R L and McCallum L E 1988 Boron toxicity in soil solution. Aust. J. Soil Res. 26, 605–610.

Berger K C and Troug E 1939 Boron determination in soil and plants using quinalizarin reaction. Ind. Eng. Chem. 11, 540–545.

Gaines T P and Mitchell G A 1979 Boron determination in plant tissue by the azomethine-H method. Commun. Soil Sci. Plant Anal. 10, 1099–1108.

Gupta U C, James Y W, Campbell C A, Leyshon A J and Nicholaichuk W 1985 Boron toxicity and deficiency: A review. Can. J. Soil Sci. 65, 381–409.

Haddad K S and Kaldor C J 1984 Boron supplying power, boron absorption capacity and productivity of some acidic soils from the central Tableland of New South Wales. Aust. J. Exp. Agric. Anim. Husb. 24, 120–125.

Hou J, Evans L J and Spiers G A 1994 Boron fractionation in soils. Commun. Soil Sci. Plant Anal. 25, 1841–1853.

Jin Ji-yun, Martens D C and Zelazny L W 1987 Distribution and plant availability of soil boron fractions. Soil Sci. Soc. Am. J. 51, 1228–1231.

Spouncer L R, Nable R O and Cartwright B 1992 A procedure for the determination of soluble boron in soils ranging widely in boron concentrations, sodicity and pH. Commun. Soil Sci. Plant Anal. 23, 441–453.

R.W. Bell and B. Rerkasem (eds.), Boron in Soils and Plants, 83–88.
© *1997 Kluwer Academic Publishers.*

Boron nutrition of radiata pine plantations in Australia

Marcia J. Lambert[1], John Turner[1] & Jim Knott[2]
[1] *FORSCI Pty Ltd, Unit 14, 124 Rowe St, Eastwood, NSW 2122, Australia*
[2] *Forest Consultant, 60 O'Grady St, Albert Park, Victoria 3206, Australia*

Key words: boron, deficiency, radiata pine, plantations

Abstract

Extensive areas of radiata pine (*Pinus radiata*) plantations in Australia, New Zealand and Chile are low in boron (B) and deficiency is recognised as the most common micronutrient limitation in these forest plantations. Knowledge of the relationships between the incidence of B deficiency and site, stand and climatic factors, together with results from fertilizer trials, enables the formulation of pre-emptive B fertilizer treatment for plantations identified to be at risk. The primary objective of applying B fertilizer is to prevent a marked falldown in the expected levels of plantation timber yield and value. Both the severity and extent of B deficiency symptoms are related to periods of low water availability (low rainfall). The actual availability of soil water cannot be managed directly, however, the probability of periods of low rainfall can be determined from historical records and the level of risk of B deficiency estimated. This process is a component of systems to determine requirements for early or later age B fertilizer treatments in plantations.

Introduction

Radiata pine (*Pinus radiata*) is an extensively planted pine species in temperate areas of the Southern Hemisphere. Between New Zealand, Australia, Chile and southern Africa, approximately 3 million hectares have been planted. Extensive areas (at least 20%) of these plantations have been diagnosed, through symptoms or foliage analysis, as boron (B) deficient (Stone and Will, 1965; Snowdon, 1972; Hill and Lambert, 1981; Hopmans and Flinn, 1984; Schlatter and Gerding, 1985a; 1985b). Boron deficiency is recognised as the most common micronutrient limitation in forest plantations and is highly variable both spatially and temporally within even a small plantation area (Lambert and Ryan, 1990), making diagnosis and planning difficult for managing this deficiency. This paper discusses the problems of B deficiency in radiata pine plantations, together with diagnostic systems including risk analysis and appropriate amelioration treatments.

Boron Deficiency in Radiata Pine

Visual symptoms of B deficiency in radiata pine include yellowing of young needle tips, production of resin droplets, needle necrosis and leader dieback; repeated dieback resulting in multiple leaders, lack of apical dominance, and a tree with bushy appearance (Stone and Will, 1965; Proctor, 1967; Snowdon, 1972; Lambert and Turner, 1977; Will, 1978; 1985; Turner et al., 1979; Escobar et al., 1984; Hopmans and Clerehan, 1991). Shoot death is usually followed by rapid production of secondary shoots from fascicle buds below the dead leader (Will, 1985). Affected leaders usually contain a brown pith. Snowdon (1972) indicated damage to the vascular system leading to reduced water flow to the shoot and this represents part of the reason for tip dieback in periods of water stress. Root production is reduced in both size and number (Lambert and Turner, 1977) and roots are also subject to tip death with subsequent forking (Stone, 1990). In a given year, tip dieback and mortality in trees might be confined to seasonally waterlogged drainage lines in isolated plantations whilst in the next year there may be widespread damage. Repeated occurrences of the problem result in unmerchantable stunted 'bushes' and overall low and very patchy stocking. The stem damage includes deformation where leader dieback has occurred, usually in the first few years of the rotation, and as a result of effects on wood quality, the most significant losses are incurred in the valuable butt logs of otherwise actively growing trees. Even a single dieback episode leaves persistent stem defects. The results are a loss in volume of merchantable timber (yield) and a reduction in the value (m^{-3}) of that timber. Thus the economic consequences of B deficiency are far greater than the volume losses. Losses in productivity can be traced to marked B deficiency 'events' in the early years of the rotation and this contrasts with the progressive decline in productivity attributed to P deficiency (Lambert and Ryan, 1990).

Plantations are established in Australia in summer-, uniform- and winter-dominated rainfall areas leading to water stress (depending on area) in spring, early summer and into late summer. Deficiency symptoms, such as tip dieback, generally occur in young trees during periods of water stress in late spring and early summer, that is periods of active growth coinciding with rapid depletion of soil water and/or intense weed competition (Turner et al., 1979). Lambert and Turner (1977) reported that the herbaceous weeds associated with these radiata pine plantations have greater B concentrations as well as greater biomass than the trees and so compete strongly for limited B supplies. Where available B levels are low, higher growth rates can also exacerbate the problem by increasing demand for the limited pool of available B. As the silvicultural trend in forest management is towards higher rates of production via genetic improvement and more intensive site preparation and fertilization with N and P, care must be taken not to drive the stand into B deficiency.

Severe B deficiency symptoms are normally linked with foliage B concentrations of 8 mg kg^{-1} dry wt or less and marginal deficiency in the range 9–12 mg B kg^{-1} (Lambert and Turner, 1977, Will, 1978). Snowdon (1972) found in field surveys that chronic B deficiency symptoms usually occurred with foliage B concentrations in the range of 3–7 mg B kg^{-1}. The concentration of B leading to symptoms has been shown to be related to soil water conditions. Where there is high rainfall, even low foliage concentrations of B (for example, 7 mg B kg^{-1}) do not lead to symptoms, but in dry areas or periods of drought B concentrations in excess of 12 mg B kg^{-1} will lead to dieback (Hill and Lambert, 1981).

Forest Site, Stand and Climatic Effects

Soil Type and Prior Land Use

The spasmodic and sporadic incidence of B deficiency symptoms in plantations suggests that B nutritional status is dependent on site, stand and climatic factors. Soil parent material can be defined in terms of Parent Rock Code (PRC), the primary attribute of the 'Technical Soil Classification for Pinus Plantations in Australia' (Turvey, 1987). This classification system is routinely used in forest management as it focuses on parameters which are observed in the field and have most bearing on forest productivity and nutrition (Turner et al., 1990). Definition of PRC involves grouping of soils according to their geological origins, and twelve primary groupings have been developed which can be sub-classified into consolidated or unconsolidated (Table 1).

Another major site factor contributing to overall productivity and nutrition is that of land use prior to plantation establishment (Turner and Lambert, 1986). At the most basic level, prior land use can be classified as native vegetation (native forest), plantation or grazing/agricultural land (mainly pastures). Soils from native vegetation are essentially un-modified prior to plantation establishment, whilst both plantation and agricultural sites have been subject to considerable soil chemical and physical changes including, application of fertilizer (primarily N and P), erosion, and the use of legumes to raise soil N levels. Previous pasture sites are often characterised by higher levels of weed competition (particularly grass species), modified (often increased) levels of P and N (especially mineral-N), and increased levels of surface soil erosion and compaction than previous native forest or second rotation sites. Until 10 years ago, a significant proportion of radiata pine plantings was on previous native forest sites but the pattern now is for planting on previous pasture or previous plantation sites and an increased diversity of soil parent materials. There is a much greater incidence of B deficiency occurring on previous pasture sites than on previous forest areas.

For any given parent material on previous pasture sites, the B concentrations in foliage were lower than for those previous native forest sites (Table 2). The occurrence of low foliar B concentrations (<12 mg B kg^{-1}) on Parent Rock Codes 2, 3 and 5 is usually associated with severe P deficiency and the fact that B deficiency symptoms are rarely observed on these sites is probably due to the slow growth rate. After

Table 1. Summary of Parent Rock Codes associated with the Technical Soil Classification (Turvey, 1987).

Description	Parent Rock Code	Consolidation	Examples
Carbonaceous Group:			
Dominated by carbon	PRC011	Consolidated	Coal, lignite
	PRC012	Unconsolidated	Peat
Quartzose Group			
Dominated by quartz or	PRC021	Consolidated	Quartzose sandstone
secondary silica	PRC022	Unconsolidated	Quartz sands, gravels
Sesquioxide Group			
Dominated by iron and	PRC031	Consolidated	Ferruginous sandstone, massive laterite
aluminium	PRC032	Unconsolidated	Gravelly laterite
Calcareous Group			
Dominated by secondary	PRC041	Consolidated	Marble, limestone
calcium compounds	PRC042	Unconsolidated	Marl, shelly sand
Argillaceous Group			
Dominated by clay or silt	PRC051	Consolidated	Slate, shale, mudstone
	PRC052	Unconsolidated	Clay alluvium
Micacous-Chloritic Group			
Dominated by micas and/or	PRC061	Consolidated	Phyllites, schist
chlorite	PRC062	Unconsolidated	Highly micaceous sand
Feldspathic-Quartzose Group A			
Medium-coarse grained	PRC071	Consolidated	Granite, feldspathic sandstone
rocks dominated by			
feldspar and quartz	PRC072	Unconsolidated	Colluvium from granite
Feldspathic-Quartzose Group B			
Fine to medium grained	PRC081	Consolidated	Rhyolite
rocks dominated by	PRC082	Unconsolidated	Pumice
feldspar and quartz			
Feldspathic-Micaceous Group			
Medium/coarse grained	PRC091	Consolidated	Granodiorite
rocks dominated by	PRC092	Unconsolidated	Colluvium from granodiorite
feldspar and mica mica			
Feldspathic Group			
Fine to medium grained	PRC101	Consolidated	Trachyte, dacite
rocks dominated by alkali	PRC102	Unconsolidated	Trachyte ash
feldspar			
Ferro-magnesian Group			
Rocks with dark silicate	PRC111	Consolidated	Basalt, dolerite, gabbro
minerals, especially	PRC112	Unconsolidated	Basic volcanic ash
amphibole, pyroxene and olivine			
Magnesium-silicate Group			
Rocks dominated by	PRC112	Consolidated	Serpentinite
magnesium silicates	PRC122	Unconsolidated	Talc

correction of P deficiency on these sites, foliar B concentrations have been found to increase in response to improved root development and water uptake.

Climatic Effects

Rainfall has been found to have a major effect on the concentration of B in the foliage of radiata pine. Dry

Table 2. Mean foliage B concentrations associated with prior land use and parent rock codes.

Soil Parent Material	Previous Land use	B Deficiency Hazard	Foliage B (mg B kg⁻¹) Age (yrs)		% samples (age 2–5 yrs) in specified foliar B range (mg B kg⁻¹)				Primary/site stand factors contributing to low foliar B concentrations
			2–5	>10	0–4	5–8	9–12	>12	
PRC2	Native forest	Low	12.8	23.1	2	32	15	51	P deficiency
	Pasture	Low	–	17.2	–	–	–	–	P deficiency
PRC3	Native forest	Low	–	27.6	–	–	–	–	P deficiency
	Pasture	Low	–	–	–	–	–	–	P deficiency
PRC5	Native forest	Low	11.4	18.4	0	33	35	32	P deficiency
	Pasture	Low	10.8	24.4	3	36	31	30	P deficiency
PRC7	Native forest	High	9.4	21.5	8	45	29	18	Low soil B, coarse texture, P deficiency
	Pasture	High	9.3	–	0	50	38	12	Low soil B, coarse texture, P deficiency, weed comp.
PRC8	Native forest	Mod	10.9	23.3	0	57	26	17	Low soil B, P deficiency
	Pasture	High	9.3	15.5	2	40	40	18	Low soil B, P deficiency weed comp.
PRC9	Native forest	Mod	108	21.8	1	29	44	26	Low soil B, coarse texture, P deficiency
	Pasture	High	7.8	8.8	6	62	27	3	Low soil B, coarse text, P deficiency, weed comp.
PRC10	Native forest	Mod	10.9	21.4	0	27	43	30	Low soil B
	Pasture	Mod	10.3	–	0	27	60	13	Low soil B, weed comp.
PRC11	Native forest	Mod	11.0	18.5	0	28	38	34	Low soil B
	Pasture	High	9.2	–	0	38	54	8	Low soil B, weed comp.

seasons lead to reduced uptake of B both because of reduced availability of B in the soil and reduced transpiration (mass flow). In Australia, there is a further factor in that most plantations are distant from the coast and there are minimal B inputs in rainfall. Hill and Lambert (1981) reported a negative correlation between the threshold foliage B concentration at which widespread deficiency symptoms will occur and the preceding season's rainfall. This relationship showed that under conditions of water stress, trees exhibited a higher deficiency threshold (12 mg B kg⁻¹) but under higher rainfall conditions, foliage concentrations could fall as low as 7 mg B kg⁻¹ before the onset of symptoms. This implies greater susceptibility to B deficiency symptoms under conditions of drought. Hopmans and Clerehan (1991) found that foliage B concentrations assessed each winter varied between 4 and 25 mg B kg⁻¹ over a 6-year period and concluded that an assessment of B deficiency hazard based on foliage B concentrations alone could be misleading unless seasonal rainfall is taken into account.

In soils which are low in B, the main store of soil B is in the surface organic matter which if eroded will leave the trees susceptible to B deficiency. Assessment of B availability in soils, particularly agricultural soils, is usually carried out using hot water soluble

Table 3. Fractionation* of B in surface soils from different parent materials.

Soil Parent Material	Soil B (mg B kg^{-1})						Foliage B (mg kg^{-1})
	Hot Water Soluble	Organic			Mineral	Total	
		Humic	Fulvic	Total			
Rhyolite	0.01	0.7	15.0	15.7	9.8	25.5	13
Granite	0.25	0.0	23.8	23.8	48.5	72.6	25
Basalt	1.62	1.1	49.4	50.5	6.0	58.1	29
Siliceous sandstones	0.53	0.1	43.0	43.1	19.1	62.7	23
Shale	0.34	0.1	60.7	60.8	62.8	123.9	37

* Hot water soluble B was extracted according to Lambert (1989). Organic B was extracted with 0.5 M NaOH; the humic fraction was precipitated with dilute acid and the B content of the supernatant was considered to be an estimate of B contained in the fulvic material. The B remaining in the sample after extraction with NaOH was considered to be mineral B. Boron was estimated in the above extracts according to Lambert (1989).

soil B as an index, however, this has been found to be insensitive in plantation forest soils. Differences were found between different soil parent materials when B in surface soils from radiata pine plantations was fractionated (Table 3). Soils derived from rhyolite parent material had the lowest total B concentrations, while those derived from the fine-grained sedimentary materials were the highest. The B concentrations in the humic fraction appeared to be insignificant, while of the other fractions (fulvic, mineral, hot water soluble), the fulvic acid fraction gave the highest correlation with foliar B and appeared to be the major source of B for tree growth. Hence a forest stand would be dependent upon the mineralization of organic matter to provide B. This helps to explain why there are limited quantities of available soil B during dry seasons.

Relationship with Stand Age and Rainfall

The problems involved in predicting the onset of B deficiency symptoms and defining deficiency (and toxicity) thresholds in radiata pine plantations based solely on foliar B analyses are compounded by the diversity of site, stand and climatic conditions which affect both B foliage concentrations and the development of B deficiency symptoms in stands. Foliar B concentrations for radiata pine plantations in NSW are presented in Table 2 for the various Parent Rock Codes (PRC) strata and prior land use (pre-crown closure at age 2–5 years and in older stands >10 years), together with a B deficiency hazard rating based on the observance of B deficiency symptoms. Relationships between foliar B, stand age, rainfall in the preceeding growth season (September to April) and soil factors have been

developed using data obtained from fertilizer trials and intensive surveys covering more than 2,000 sites in NSW plantations (Lambert et al., 1997). Details of the methodology are provided in Lambert et al. (1997). The data set represented radiata pine plantations over a number of Parent Rock Codes, over a 25-year period for which rainfall was known, and for trees which ranged from 2 to 30 years of age. The lowest foliage B concentrations, and hence most of the B deficiencies were identified as occurring in the younger stands. Foliar B nutritional status was positively correlated with stand age and growth season rainfall. Individual regression equations for each PRC land use strata enabled predictions to be made for 3-year-old plantations under a range of growth season rainfall conditions. The predictions were consistent with B deficiency hazard ratings.

Nutrition Management Strategies

Plantation mangement will rely on application of fertilizers where ever B deficiency is likely because even a single episode of dieback has such adverse effects on stem quality. The aim of prognosis and treatment is to prevent occurrences rather than improving growth of already affected stems.

Based on a knowledge of location, climate and its level of variability, soil parent material, previous land use, and stand conditions, a strategy for stand management including B fertilizer amendment can be developed. This recognises that there are interactions with other nutrients, particularly N and P. For young trees, where B is required, there needs to be an application of 5–8 g B per tree. In many situations in the past, this has been applied as sodium borate, however, the use of ground and prilled colemanite or ulexite has been found to give more beneficial and long term effects. This is usually applied in a soil slit near the tree after planting. On soils with relatively high mineral N (for example, on previous pasture), this may be a B-only treatment or with P as well. On other sites it may be as a NPB application. In older stands (age 2–5 years) after foliage analysis, aerial applications of B may be applied, primarily as prilled ulexite or colemanite at an application rate of 8 kg B ha^{-1}. Treatments of B are not common in much older stands, as the merchantable log length usually has been set and there is little additional economic return.

A system for selecting the optimum treatment for a site has been developed (Knott et al., 1997). Financial losses attributable to B deficiency involve yield forgone at subsequent harvesting operations; lower log prices due to product degradation; forced modification of silvicultural regimes such as shortening rotation lengths; and reduced marketing and plantation management options. Emphasis has been placed on the preventative nature of the treatment which is a form of insurance against drought-induced B deficiency symptoms on high risk sites.

References

Escobar R R, Gonzalez G V, Millan J H and Gonzalez C O 1984 Effectos de la fertilizacion boratada al establecimiento de plantaciones de pino insigne (*Pinus radiata* D. Don). Proc. Fourth National Soil Sci. Symp. Valdivia, Chile. Comm. 2, Sess. 5.

Hill J and Lambert M J 1981 Physiology and management of micronutrients in forest trees in Australia. *In* Proc. Aust. Forest Tree Nutrition Workshop, Productivity in Perpetuity. Ed P J Ryan, pp 93–103. CSIRO, Melbourne.

Hopmans P and Flinn D W 1984 Boron deficiency in *Pinus radiata* D. Don and the effects of applied boron on height growth and nutrient uptake. Plant and Soil 79, 295–298.

Hopmans P and Clerehan S 1991 Growth and uptake of N, P, K and B by *Pinus radiata* D. Don in response to applications of borax. Plant and Soil 131, 115–127.

Knott J, Turner J and Lambert M J 1997 Boron nutrition in *Pinus radiata* plantations: II. An exercise in risk management. For. Ecol. Manage. (under review).

Lambert M J 1989 Methods for chemical analysis of forestry materials including foliage, soil, wood, water and other forest products. State Forests of NSW Technical Paper No. 25. Third edition, 187 pp.

Lambert M J and Ryan P J 1990 Boron nutrition of *Pinus radiata* in relation to soil development and management. For. Ecol. Manage. 30, 45–53.

Lambert M J and Turner J 1977 Dieback in high site quality *Pinus radiata* stands – the role of sulphur and boron deficiencies. N.Z. J. For. Sci. 7, 333–348.

Lambert M J, Knott J and Turner J 1997 Boron nutrition in *Pinus radiata* plantations: 1. Site, stand and climatic effects. For. Ecol. Manage. (under review).

Proctor J 1967 A nutritional disorder of pine. Comm. For. Rev. 46, 145–154.

Schlatter J and Gerding V 1985a Deficiencia de boro en plantaciones de *Pinus radiata* D. Don en Chile. 1. Distribution y origen. Bosque 6, 24–31.

Schlatter J and Gerding V 1985b Deficiencia de boro en plantaciones de *Pinus radiata* D. Don en Chile. II. Principales causes y correccion. Bosque 6, 32–43.

Snowdon P 1972 Observations on boron deficiency in *Pinus radiata*. *In* Proc. of The Australian Forest Tree Nutrition Conference. Ed R Boardman, pp 191–206. Forestry and Timber Bureau, Canberra.

Stone E L 1990 Boron deficiency and excess in forest trees: a review. For. Ecol. Manage. 37, 49–75.

Stone E L and Will G M 1965 Boron deficiency in *Pinus radiata* and *P. pinaster*. For. Sci. 11, 425–433.

Turner J and Lambert M J 1986 Nutrition and nutritional relationships of *Pinus radiata*. Ann. Rev. Ecol. Syst. 17, 325–350.

Turner J, Lambert M J and Edwards D W 1979 A guide to identifying nutritional and pathologic disorders of *Pinus radiata*. For. Comm. N.S.W. Res. Note No. 36, 16 pp.

Turner J, Thompson C H, Turvey N D, Hopmans P and Ryan P J 1990 A soil technical classification system for *Pinus radiata* (D. Don) plantations. I. Development. Aust. J. Soil Res. 28, 797–811.

Turvey N D 1987 A technical classification for soils of *Pinus* plantations: field manual. Bulletin No. 6, School of Forestry, University of Melbourne.

Will G M 1978 Nutrient deficiencies in *Pinus radiata* in New Zealand. N.Z. J. For. Sci. 8, 4–14.

Will G M 1985 Nutrient deficiencies and fertilizer use in New Zealand exotic forests. N.Z. For. Service FRI Bull. No.97, 53 pp.

R.W. Bell and B. Rerkasem (eds.), Boron in Soils and Plants, 89–93.
© 1997 *Kluwer Academic Publishers.*

Diagnosis and prognosis of boron deficiency in black gram (*Vigna mungo* L. Hepper) in the field by using plant analysis

R N Noppakoonwong[1], B Rerkasem[1], R W Bell[2], B Dell[2] & J F Loneragan[2]
[1] *Multiple Cropping Centre, Faculty of Agriculture, Chiang Mai University, 50200 Thailand*
[2] *School of Biological and Environmental Sciences, Murdoch University, Perth, 6150 Western Australia*

Key words: critical boron range, diagnosis, prognosis, soil water, *Vigna mungo*

Abstract

From previous glasshouse experiments, 12–18 mg B kg^{-1} dry wt in the youngest fully expanded leaf blade (YFEL) represents the minimum leaf boron (B) requirement for unrestricted leaf blade elongation (LBE) in black gram (*Vigna mungo*). A field experiment at Chiang Mai, Thailand was undertaken to test the validity of this B concentration range for diagnosis of B deficiency and to examine the possibility of using B concentration in the YFEL during growth for prognosis of seed yield at maturity. Six levels of B fertilizer (0, 0.5, 1, 2, 4 and 8 kg borax ha^{-1}) and basal fertilizers were applied to 6 replicates. The critical B concentration for diagnosis of pod number in the YFEL at the first mature pod was 13–17 mg B kg^{-1} dry wt being within the range established from the glasshouse experiments. Thus, the functional B requirement for LBE, 12–18 mg B kg^{-1} dry wt in the YFEL, can be recommended for diagnosis of B deficiency in black gram. The critical B ranges in the YFEL for prognosis of seed yield were 31 ± 3, 30 ± 4, 15 ± 1, and 9 ± 1 mg B kg^{-1} dry wt at the first inflorescence, first pod, first full pod and first mature pod, respectively. The unusually low value of 9 ± 1 mg B kg^{-1} dry wt is attributed to the recovery of B deficient plants as the result of an increase in B uptake following rain after a drought period. Thus, the use of prognostic values of B deficiency can be limited by the probability of rain after sampling. Despite this limitation, critical values for the prognosis of B deficiency can be used as a guide to the possibility of B deficiency depressing final yield.

Introduction

From previous glasshouse experiments, we suggested that 12–18 mg B kg^{-1} dry wt in the youngest fully expanded leaf blade (YFEL), the minimum boron (B) requirement for unrestricted leaf blade elongation (LBE), may be used as a critical range for diagnosis of B deficiency in black gram (*Vigna mungo*) from vegetative to reproductive growth in black gram (Noppakoonwong, 1991). The minimum B requirement for unrestricted LBE is the concentration of B in an expanding leaf blade associated with 10% depression in its elongation rate.

The present experiment was undertaken to test the validity of this critical range of B concentrations for diagnosis of B deficiency in field-grown black gram crops and to examine the possible value of using leaf B

concentration for the prognosis of seed yield at maturity. Changes in soil water content during the experiment provided information on the relationships of soil water content to B uptake and B response, and were used to assess the effect of soil water on the prognosis of B deficiency in black gram.

Materials and methods

A field experiment was conducted at Chiang Mai, Thailand, on a Typic Tropaqalf with a silty loam texture, pH of 5.5–6.5 and hot water soluble B level of 0.08 mg kg^{-1} soil (Dible et al., 1952). Experimental plots of 7.3 m × 4.5 m were set out in a randomised complete block design with 6 replicates and 6 B levels: 0, 0.5, 1, 2, 4 and 8 kg borax ha^{-1} which were designated as B0, B0.5, B1, B2, B4 and B8, respectively. Basal fertilizers were applied to give rates of 22 kg P ha^{-1} as Ca(H$_2$PO$_4$)$_2$, 40 kg K ha^{-1} as K$_2$SO$_4$, 10 kg Mg ha^{-1} as MgSO$_4$.7H$_2$O, 4.6 kg Zn ha^{-1} as ZnSO$_4$.7H$_2$O, 2.5 kg Cu ha^{-1} as CuSO$_4$.5H$_2$O and 0.12 kg Mo ha^{-1} as Na$_2$MoO$_4$.2H$_2$O. Micronutrient fertilizers and B treatments were applied in solution to 1800 g aliquots of air dry soils in plastic bags, dried and mixed thoroughly before being applied to designated experimental plots. *Bradyrhizobium* strain THA 301 from the BNF Centre Bangkok was inoculated onto black gram seeds (cv. Regur) immediately before sowing. Nine days after sowing (D9), plants were thinned to 50 plants m^{-2}. Twenty YFEL from each plot were sampled at the first inflorescence (D35), first pod (D42), first full pod (D58), first mature pod (D70) and maturity (D90). On D70 and D90, 10 plants from each plot were harvested. Seed yield was obtained from a 2 m^2 area of each plot on D92. Leaf blades were oven dried at 80° C for 48 hours, weighed, dry ashed and their B concentrations were determined using the azomethine-H method (Bell et al., 1989). Soil samples were collected at 0–25 cm on D42, D50, D57 and D92 for gravimetric soil water determination. Daily rain fall was recorded (Figure 1).

Results and discussion

Effect of boron treatment on boron deficiency

Some plants of all treatments except B8, had short and brittle petioles, chlorosis in their leaf blades and finally developed a rosetted appearance at shoot apices. In addition, these plants shed more flowers and pods than normal plants.

The number of pod bearing nodes and pods, the weight of pods and seeds and the B concentrations in the YFEL increased progressively with increasing B supply (Table 1, Figure 2). However, seed yield and B concentration varied widely within each B treatment (Figure 2) resulting in inconsistent responses with B treatments in some replicates. By contrast with the variation of above responses of seed yield and leaf blade B concentrations, there was a close relationship between leaf B concentration and seed yield (Figures 3, 4).

Effect of soil water on boron uptake

The high variation in leaf B concentrations within each treatment may be related to difference among the

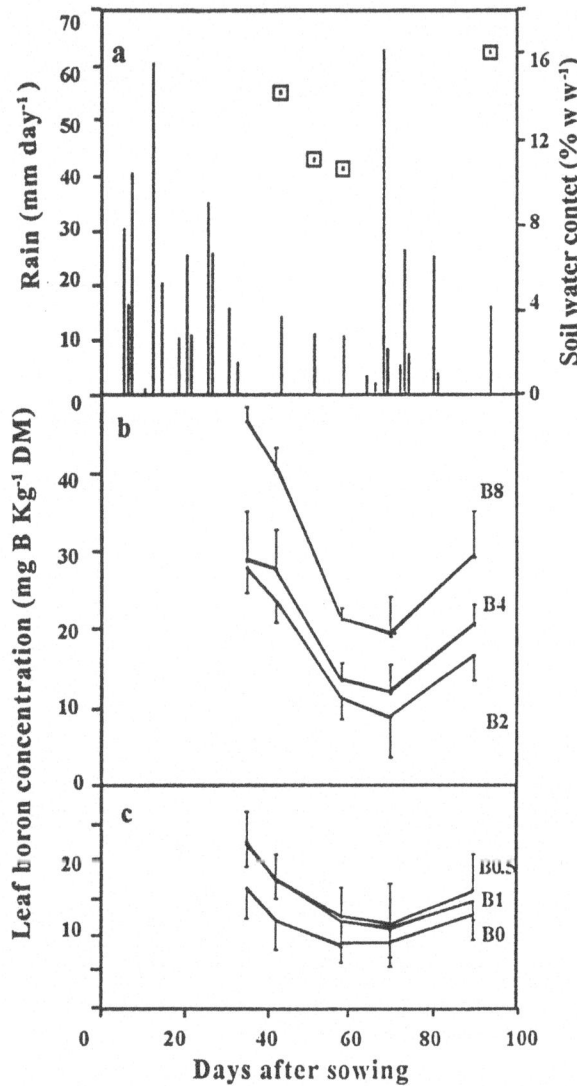

Figure 1. Changes with time in (a) rain fall, mean soil water content (□) and (b, c) mean B concentrations of the youngest fully expanded leaf of black gram. B0, B0.5, etc refer to the kg of borax applied per hectare at sowing on a silty loam in the field at Chiang Mai. Values are means of six replicates. Vertical lines represent standard errors where they exceed the height of symbols.

experimental plots in their soil water contents. Leaf B concentration increased with increasing soil water content in B0, B0.5, B1, B2, B4 (Figure 2). In addition, the changes in soil water contents with time show a pattern similar to that for B concentrations in the YFEL (Figure 1). A change of B concentration in leaflets with soil water content during plant development was also observed in Spanish peanut (Chrudimsky and Morrill, 1973). In alfalfa also, B uptake from soils has responded strongly to soil water content, being severe-

Table 1. Effect of fertilizer borax supply on number of pod bearing nodes and pods per plant and dry matter of seeds and pods (kg ha^{-1}) in *Vigna mungo* (cv. Regur) grown on a silty loam in the field at Chiang Mai. Values are means of six replicates. Means in a row followed by a common letter are not significantly different at $p<0.05$.

Days after Sowing	Parameters	Borax applied in kg ha^{-1}					
		0	0.5	1	2	4	8
70	Pod bearing node number	5a	7ab	9bc	11c	11c	10c
	Pod number	20a	31ab	32ab	41b	44b	40b
92	Pod weight	1170a	1544ab	1568ab	2188bc	2213bc	2823c
	Seed weight	18a	15a	944a	1269ab	1335ab	1823b

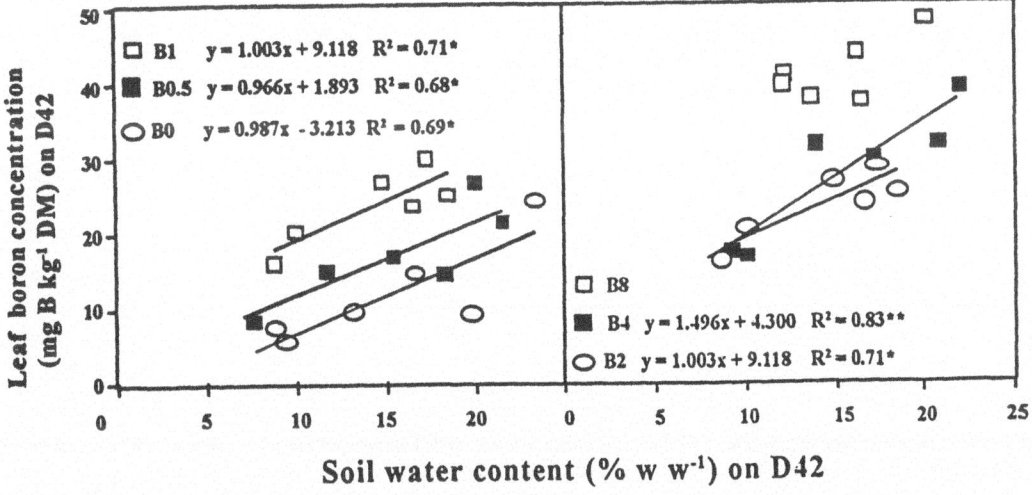

Figure 2. Relationships between soil water content and B concentrations in the youngest fully expanded leaf blade of black gram cv. Regur on D42, 42 days after sowing, for each of the six soil B treatments. B0, B0.5 etc refer to the kg of borax applied per hectare at sowing. Values are datum for a single plot. $*R^2$ significant at $p<0.05$. $**R^2$ significant at $P<0.01$.

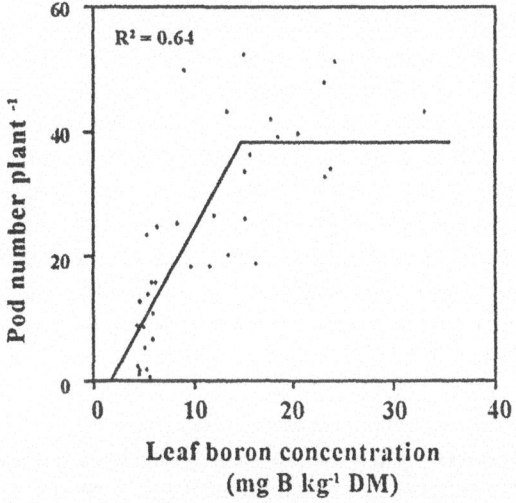

Figure 3. Relationship between pod number and B concentration in the youngest fully expanded leaf blade 70 days after sowing. The fitted line is obtained by the Two-phase linear model.

ly depressed by low soil water (Baker and Mortensen, 1966).

Diagnosis of boron deficiency

Since B deficiency reduced pod number by increasing the shedding of flowers and pods, diagnostic values were obtained from the correlation of B concentration in the YFEL with pod number. The critical B value in the YFEL for diagnosis of B deficiency for pod number on D70 estimated by the Two-phase linear regression (Smith and Dolby, 1977), was 15 ± 2 mg B kg^{-1} dry wt (Figure 3). This value is likely to be a reliable critical concentration since leaf B concentrations during the period from D58 to D70 were relatively constant (Figure 1). From a glasshouse experiment, leaf development is at least as sensitive to B deficiency as that of the flowers since the LBE of the young black gram (cv.

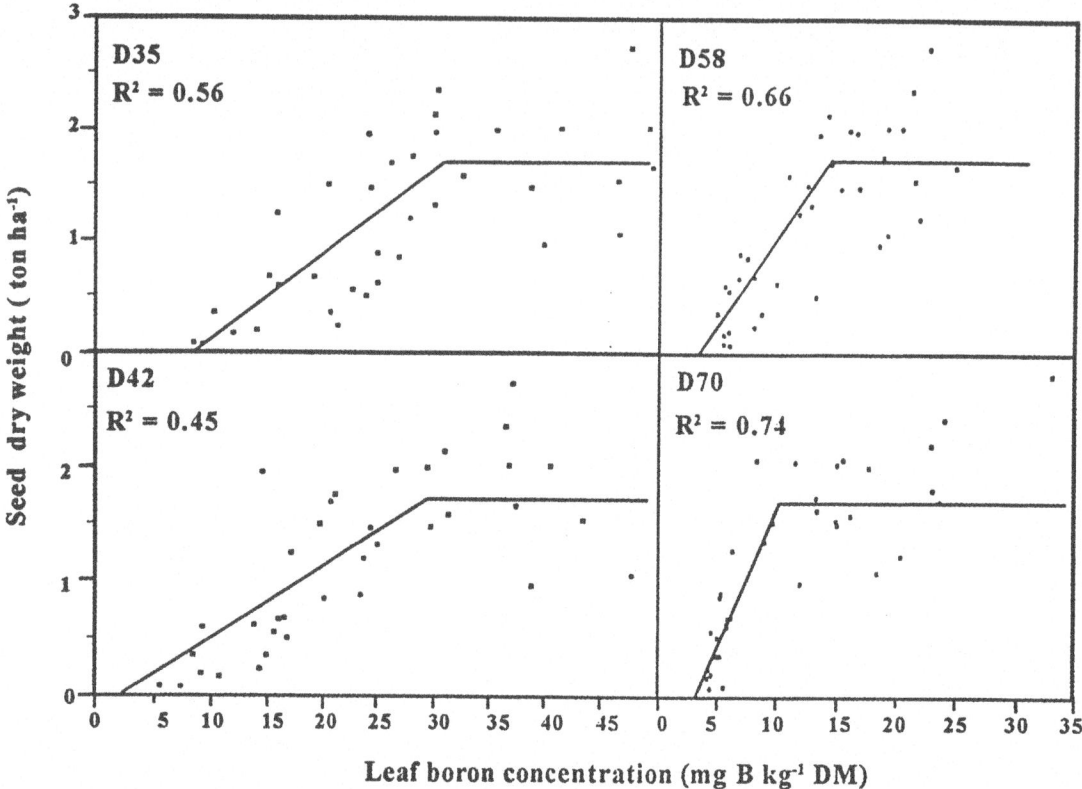

Figure 4. Relationship between seed dry weight at D92, 92 days after sowing, and B concentration in the youngest fully expanded leaf blade at D35, D42, D58 and D70. The fitted lines are obtained by the Two-phase linear model.

Regur) was depressed one day before the shedding of flowers (Noppakoonwong, 1991).

Thus, the above diagnostic value confirms that 12–18 mg B kg^{-1} dry wt in the YFEL, the functional B requirement for LBE, could be used as a critical range for diagnosis of B deficiency in black gram under field conditions. The results also support the hypothesis proposed by Kirk and Loneragan (1988) that the functional nutrient requirement of a specific plant part can be used as a critical value for diagnosis of nutrient deficiency if its responses is closely related to plant growth.

Indeed, this study appears to be the first to confirm the validity of the functional B requirement for LBE determined in glasshouse studies for diagnosis of B deficiency in the field. In addition to black gram, funtional B requirement for LBE are 10–12 and 14 mg kg^{-1} dry wt in soybean and green gram, respectively (Kirk and Loneragan, 1988; Bell et al., 1990). The present results of black gram give some confidence that functional B requirements for green gram and soybean should have application as a critical value for B deficiency diagnosis in those species in the field.

Prognosis of boron deficiency

Prognostic values obtained from the correlation of seed yield at maturity with B concentration in the YFEL on D35, D42, D58 and D70 by the Two-phase linear regression (Smith and Dolby, 1977) were 31 ± 3, 30 ± 4, 15 ± 1 and 9 ± 1 mg B kg^{-1} dry wt, respectively (Figure 4). All of the prognostic values except that on D70 were within or higher than the range of diagnostic values for LBE (12–18 mg B kg^{-1} dry wt). The unusually low prognostic value on D70 may be attributed to the fact that enhanced B uptake between D70 and D90 following rain was sufficient for moderately B deficient plants to recover fully and show no response to B in seed yield at maturity (Figure 1): on D70, plants with 9 mg B kg^{-1} dry wt in their YFEL had only 60% of maximum pod number (Figure 3), but still achieved close to maximum seed yield. Thus, a critical B concentration for prognosis of B deficiency in black gram can only be used as a guide to the possibility of B deficiency depressing final yield. Whether it does so or not will depend upon the post-sampling availability of B from the soil. Hannam et al. (1985) have suggested that prognostic tests for manganese (Mn) are likely to predict only the potential for the onset of Mn deficiency. Concentrations below the critical value or range can only give an indication that deficiency may occur because of the possibility of dramatic changes in availability from soil after sampling. Despite this limitation, the suggested critical concentration should prove useful as a guide to the likely development of B deficiency in black gram crops in the field.

Acknowledgment

The authors are grateful to the Australian Centre for International Agricultural Research (Projects 8329 and 8603) for financial support.

References

Baker A S and Mortensen W P 1966 Residual effect of single borate applications on Western Washington soils. Soil Sci. 102, 173–179.

Bell R W, McLay L, Plaskett D, Dell B and Loneragan J F 1989 Germination and vigour of black gram (*Vigna mungo* (L.) Hepper) seed from plants grown with and without boron. Aust. J. Agric. Res. 40, 273–279.

Bell R W, McLay L, Plaskett D, Dell B and Loneragan J F 1990 Internal boron requirements of green gram (*Vigna radiata*). *In* Plant Nutrition-Physiology and Application. Ed M L van Beusichem, p 275–280. Kluwer Academic Publisher, The Netherlands.

Chrudimsky W W and Morrill L G 1973 Uptake and distribution of boron in spanish peanuts. Agron. J. 65, 63–66.

Dible W T, Truog E T and Berger J C, 1952 Boron determination in soils and plants. Anal. Chem. 26, 418–421.

Hannam R J, Graham R D and Riggs J L 1985 Diagnosis and prognosis of manganese deficiency in *Lupinus angustifolius* L. Aust. J. Agric. Res. 36, 765–777.

Kirk G and Loneragan J F 1988 Functional boron requirement for leaf expansion and its use as a critical value for diagnosis of boron deficiency in soybean. Agron. J. 80, 758–762.

Noppakoonwong R N 1991 Diagnosis of Boron Deficiency in Black Gram. Ph D Thesis, Murdoch University, Western Australia.

Smith F W and Dolby G R 1977 Derivation of diagnostic indices for assessing the sulphur status of *Panicum maximm* var. *Trichoglume*. Commun. Soil Sci. Plant Anal. 8, 221–240.

R.W. Bell and B. Rerkasem (eds.), Boron in Soils and Plants, 95–99.
© 1997 *Kluwer Academic Publishers.*

Diagnosis and alleviation of boron deficiency causing flower and pod abortion in chickpea (*Cicer arietinum* L.) in Nepal

S. P. Srivastava[1], C. R. Yadav[1], T. J. Rego[2], C. Johansen[2] & N. P. Saxena[2]
[1] *Grain Legume Research Program (GLRP), Rampur, Chitwan, Nepal*
[2] *International Crops Research Institute for the Semi-Arid Tropics (ICRISAT), Patancheru Andhra Pradesh, 502 324, India*

Key words: B-deficiency, chickpea, diagnostic trial, flower abortion, molybdenum, pod set

Abstract

Symptoms of flower abortion and failure of pod set in chickpea in Chitwan, Nawalparasi and Makwanpur districts of Nepal suggested a nutrient disorder as a prime cause of the problem. Thus, a diagnostic nutrient trial with a susceptible chickpea variety, Kalika, was conducted in the 1994/95 season. Boron deficiency was established as the dominant nutritional problem causing flower and pod abortion. No pods or grains were formed in the absence of applied B. However, there was also a significant grain yield response to Mo. In the following season, a factorial combination of different rates of B and Mo was tested. Responses were less marked in this trial than previously, with no significant response to Mo application. Application of 0.5 kg B ha^{-1} was found to optimally correct the syndrome. It is suggested that B-deficiency may be a major factor limiting yields of grain legumes in Nepal; this is currently being assessed.

Introduction

Chickpea (*Cicer arietinum* L.) is the third most important legume of Nepal. It is mainly grown in the Terai and inner Terai regions of the country but varieties other than Dhanush do not perform well in the soils of Chitwan, Nawalparasi and Makwanpur districts. Flower abortion and consequent failure of pod set in large seeded varieties of chickpea are prevalent as a major constraint to production in these areas. Although these symptoms coincide with those of Botrytis gray mold (*Botrytis cinerea*) disease which is endemic in the region (Karki et al., 1993), the flower abortion and pod set problem persists even when Botrytis gray mold (BGM) cannot be detected. But affected plants show symptoms similar to those caused by boron deficiency. Abnormal plants are characterized by death of the terminal bud, increased lateral branching and progressive reduction in the size of the younger leaves, so called 'little leaves'. Similar symptoms have been described in B- deficient cereals (Gupta, 1979) and *Pisum sativum* (Piper, 1940).

Boron deficiency has also been reported from northern Bihar state of India, which is adjacent to the Terai region of Nepal, and susceptible varieties of chickpea and pigeonpea have been reported to respond to application of 1.5 and 2.5 kg B ha^{-1} on a B-deficient soil (Singh et al., 1991).

Therefore, in order to precisely diagnose the problem and then determine how best to alleviate it, two field experiments were conducted at the research station of the Grain Legume Research Program (GLRP), Rampur during 1994/95 and 1995/96 growing seasons.

Materials and methods

The soils of the two experimental sites are classified as 'Dystochrepts' according to the USDA Soil Taxonomy.

Diagnostic trial

Soil samples were collected from the experimental site before planting of the first experiment and were analyzed in the Soils and Agroclimatology Division (SACD) at ICRISAT Asia Center (IAC). The top 15 cm soil layer had pH (1:2 H_2O) 6.2; available P (Olsen) 15 mg kg^{-1} soil; ammonium acetate (NH_4OAc) – extractable K 26 mg kg^{-1} soil; NH_4OAc – extractable Ca 301 mg kg^{-1} soil; NH_4OAc – extractable Mg 34 mg kg^{-1} soil; $CaCl_2$ – extractable SO_4-S 21 mg kg^{-1} soil; cation exchange capacity (CEC) 6.7 cmol (+) kg^{-1}; organic carbon (OC) 9.5 g kg^{-1}; Diethylene triamine penta acetic acid (DTPA) – extractable Fe, Mn, Cu and Zn at 31, 2.2, 0.56 and 0.22 mg kg^{-1} soil respectively; and hot water soluble (hws) (Gupta, 1993) B 0.2 mg kg^{-1} soil. Sand, silt and clay contents were 660, 220, and 120 g kg^{-1} soil, respectively. The first experiment conducted for diagnostic purposes was a subtractive arrangement of treatments in a randomised complete block (RCB) design replicated four times. The nine treatments were zero control (no nutrients added), complete or adequate supply of all nutrients (kg ha^{-1}: 25 P, 50 K, 20 S, 0.5 B, 2.5 Zn, 0.15 Mo, 2.5 Cu, 2.5 Mn and 2.5 Fe), complete minus B, complete minus Zn, complete minus Mo, complete minus Fe, complete minus (Cu+Mn), complete minus (K+S), complete plus lime (1200 kg ha^{-1} as fine powdered $CaCO_3$). The various nutrients were supplied through different sources as follows: Ca (H_2PO_4).H_2O, KCl, Na_2SO_4, H_3BO_3, $ZnCl_2$, Na_2MoO_4.H_2O, $CuCl_2$.$2H_2O$, $MnCl_2$.$4H_2O$, $FeC_6H_5O_7$.$5H_2O$.

At the experimental site, the field was ploughed three times with a disc harrow and levelled with a wooden plank drawn by a tractor. The individual plot size was 1.8 × 2.1 m. Lime was applied 4 days before sowing and all other nutrient elements were mixed thoroughly in polythene bags and applied by broadcasting and then mixing into the top 10 cm of soil with a spade one day before sowing. Kalika variety was sown in opened furrows on 26 November 1994 at a rate of 2–3 seeds $hill^{-1}$ at 30 × 15 cm spacing. Plants were first thinned to 2 plants $hill^{-1}$ at 17 days after sowing (DAS) and then to 1 plant $hill^{-1}$ at 29 DAS. The crop was manually weeded at 26 DAS. At 131 DAS, 0.72 m^2 of net harvest area was sampled for data collection.

Rate of application trial

After detecting B deficiency as a dominant nutritional cause of the problem, followed by Mo deficiency, in the first experiment, another experiment with four levels of B (0, 0.5, 1.0 and 3.0 kg ha^{-1}) as boric acid and three levels of Mo (0, 0.15 and 0.3 kg ha^{-1}) as sodium

molybdate was conducted to establish optimum rates of application to alleviate the problem. The experiment was conducted adjacent (about 3 m away) to the site of the 1994/95 experiment. A factorial arrangement of treatments in RCB design, replicated three times, was used. The individual plot size was 3 × 3.6 m. Boric acid and sodium molybdate were mixed with dry soil and broadcast in respective plots followed by broadcasting of triple superphosphate (17.6 kg P ha^{-1}) and muriate of potash (16.6 kg K ha^{-1}). The fertilizers were then mixed into the top 10 cm of soil. Kalika variety was sown on 14 November, 1995 at 2 seeds $hill^{-1}$ at 30 × 15 cm spacing. Plants were thinned to 1 plant $hill^{-1}$ at 24 DAS. The crop was weeded manually at 21 and 57 DAS. Endosulphan was sprayed at 105 and 126 DAS for control of pod borer (*Helicoverpa armigera*). At 146 DAS, 6.48 m^2 of net harvest area was sampled for data collection.

In both the experiments, there was sufficient moisture in the soil as a result of residual soil water from the preceding rainy (monsoon) season and precipitation in the form of light winter rain and dew to ensure that crops did not suffer from drought stress during the crop growth period.

Results

In treatments omitting B, in both experiments, some plants showed symptoms of 'little leaf' from about 30 DAS, which persisted until crop maturity. Flowering occurred at around 80 DAS. There were many dead flowers (31 $plant^{-1}$) in the absence of applied B, but only 2 dead flowers $plant^{-1}$ resulted when 0.5 kg B ha^{-1} was applied (Table 1). When no nutrients (zero control) were applied only 15 dead flowers $plant^{-1}$ resulted. In the second experiment, application of 0.5 kg B ha^{-1} resulted in the lowest number of dead flowers (8 $plant^{-1}$) which is significantly lower than in the control (23 $plant^{-1}$) (Table 2). Application of the highest level of B, at 3.0 kg ha^{-1}, caused a significantly higher number of dead flowers compared to the application of 0.5 kg B ha^{-1} (Table 2). At the lower level of Mo (0.15 kg ha^{-1}), application of 0.5 kg B ha^{-1} resulted in 42% less flower drop than at the highest level of Mo (0.3 kg ha^{-1}). At the same level of Mo (0.15 kg ha^{-1}), application of 3.0 kg B ha^{-1} resulted in 72% more flower drop than 0.5 kg B ha^{-1}. At the highest level of Mo (0.3 kg ha^{-1}), 1.0 kg B ha^{-1} resulted in 30% less flower drop than 0.5 kg B ha^{-1}, but there was

Table 1. Effect of nutrient treatments on flower drop (at 111 DAS), total and filled pods plant^{-1} and pod and grain yield of Kalika variety of chickpea at harvest, Grain Legume Research Program (GLRP), Rampur, Nepal, 1994/95.

Treatments	Dead flowers plant^{-1a}	Total pods plant^{-1b}	Filled pods plant^{-1b}	Pod yield (kg ha^{-1})	Grain yield (kg ha^{-1})
Zero control (No nutrients applied)	15	1	0	29	7.1
Complete (P,K,S,B,Zn,Mo,Cu,Mn and Fe)	2	13	6	462	298
Complete -B	31	0	0	0.0	0.0
Complete -Zn	2	14	6	517	335
Complete -Mo	2	7	3	263	177
Complete -Fe	2	10	5	433	288
Complete - (Cu+Mn)	1	14	6	568	380
Complete - (K+S)	2	10	5	398	271
Complete + Lime	1	13	6	502	332
CV (%)	23.0	12.0	18.6	12.0	12.0
LSD (0.05)	2.6	1.5	1.1	23.9	15.7

[a] Mean of 5 plants and 4 replications
[b] Mean of 16 plants and 4 replications.

no significant difference in response between 0.5 and 3.0 kg B ha^{-1}.

No indication of BGM was observed in either experiment. When flowering began, in March of each year, minimum temperatures remained above 10° C. Thus, neither BGM nor low temperature appeared to be factors contributing to floral abortion.

In first experiment, because of the complete flower abortion no pods or grains formed in any plant when B was omitted from the complete treatment (Table 1). Omission of Mo from the complete nutrient treatment also caused significant reductions in total pods (46%), filled pods plant^{-1} (50%), and in pod (43%) and in grain yields (40%). Addition of lime did not significantly increase grain yield of Kalika.

In the second experiment, application of 0.5 kg B ha^{-1} resulted in the formation of 17 filled pods plant^{-1} which was significantly higher than in the control treatment (4 filled pods plant^{-1}) but similar to values at higher levels of B (Table 2). Maximum grain yield was obtained at the 0.5 kg B ha^{-1} treatment and this was significantly higher than in the control treatment (Table 2). The differences in grain yield response between 0.5 and 1.0 and 0.5 and 3.0 kg B ha^{-1} were not significant (Table 2). The response of any parameter to Mo was not significant in this experiment (Table 2).

Discussion

Soil properties indicate that the soil is acidic and deficient in secondary and micronutrients. The critical levels proposed are as follows: for NH_4OAc – extractable K, 50 mg kg^{-1} soil (Jackson, 1967), NH_4OAc – extractable Ca and Mg 1.5 and 1.0 cmol(+) kg^{-1} soil (Rao, 1993); hws B in acidic soil, 0.4 mg kg^{-1} soil and; DTPA – extractable Zn, Cu and Mn 0.6, 0.2 and 2.0 mg kg^{-1} soil, respectively (Gupta, 1993). Thus, possible deficiencies of K, Ca, Mg, Zn and B are indicated. A complex problem of primary, secondary and micronutrient deficiencies was also recognised in the soils of Chitwan valley by Khatri-Chhetri (1982). These deficiencies coupled with coarse texture, low pH and low organic matter content have contributed to this complex nutritional problem. The present study shows that deficiencies of B and to a lesser extent Mo, are the primary micro nutrient factors responsible for yield loss in chickpea. Phetchawee and Ratanarat (1989) also noted the severity of B deficiency mainly in coarse textured soils containing high sand fraction and low organic matter. Agarwala and Sharma (1979) also observed that B deficiency results in a marked decrease in the number of flowers and that the flowers of B-deficient chickpea plants lack pigmentation and fail to fruit, causing reductions in pod and grain yield. However, B applied at 3 kg ha^{-1} also increased flower drop in the present study. Sinha and Chatterjee (1994) have

Table 2. Effect of different levels of B and Mo on a) dead flowers per plant b) filled pods per plant and c) grain yield (kg ha^{-1}) of Kalika variety of chickpea at 124 DAS, GLRP Rampur, 1995/96.

Level of B (kg ha^{-1})	Level of Mo (kg ha^{-1})			
	0	0.15	0.3	Mean
Dead flowers plant^{-1}				
0	23.0	23.3	23.7	23.3
0.5	8.0	8.3	14.3	10.2
1.0	11.3	9.3	10.0	10.2
3.0	13.0	14.3	12.0	13.2
Mean	13.9	13.8	15.0	
Treatment	** LSD = 3.7 dead flower plant^{-1}			
B	** LSD = 1.5 dead flower plant^{-1}			
Mo	NS			
B × Mo	** LSD = 2.6 dead flower plant^{-1}			
Filled pods plant^{-1}				
0	4.0	4.0	5.3	4.4
0.5	17.0	20.7	18.7	18.6
1.0	17.3	23.0	19.0	19.7
3.0	20.0	20.7	15.0	18.5
Mean	14.5	17.0	14.5	
Treatment	** LSD = 6.1 dead flower plant^{-1}			
B	** LSD = 2.53 dead flower plant^{-1}			
Mo	NS			
B × Mo	NS			
Grain yield (kg ha^{-1})				
0	208	188	248	215
0.5	834	833	796	820
1.0	815	952	814	860
3.0	871	896	770	846
Mean	682	717	657	
Treatment	** LSD = 218 kg ha^{-1}			
B	** LSD = 89 kg ha^{-1}			
Mo	NS			
B × Mo	NS			

NS not significantly different $p<0.05$
** Significantly different at $p<0.05$

also reported the adverse effect of the excess supply of B on the inflorescence in pearl millet.

When a complete nutrient application minus B was applied to chickpea, as in the first experiment, the severity of B deficiency appeared more acute than when no other nutrients were applied at all (zero control). This may have been a result of dilution effects of other added nutrients on B uptake from an already low level of availability in the soil. Thus B requirement is likely to vary depending on supply and availability of other nutrient elements.

Boron responses also differed between years, being more severe in 1994/95. This may have been due to differences in soil availability of B between the sites of the two experiments, although they were adjacent to each other. Only composite soil samples from the general area were collected in 1994 and analyzed for B. Another reason may be differing weather patterns between years, affecting transpiration rates and therefore B uptake and transport to reproductive tissues. It was also noted that maximum grain yields in both seasons were relatively low considering yield potential of chickpea for the region. Thus factors other than nutrients may have been limiting yields and thereby interacting with B response. Nevertheless whether other yield limitations were operating, the results clearly indicate a severe B limitation in both years and a Mo deficiency in the first year.

It is well established that Mo deficiency increases in likelihood as soil pH declines (Evans et al., 1951). At the site of this study pH was only marginally acid and microsite differences in soil pH may have shifted the balance between Mo deficiency and sufficiency, between 1994/95 and 1995/96 respectively. However, the lack of response to lime application in 1994/95 suggested that the acid soil syndrome (including Mo deficiency as a factor) was not a major limitation. Another possible reason for the difference in Mo response between years may have been competitive effects on Mo availability or uptake of some of the additional nutrient ions added in 1994/95 but not 1995/96. Alternatively, seed Mo levels for the 1995/96 sowing may have been higher than the previous year, and sufficient to supply crop needs despite low soil Mo. Further diagnosis of incidence of Mo deficiency in chickpea in this region is warranted.

The present results suggest that basal soil application of 0.5 kg B ha^{-1} is appropriate to correct the severe B deficiency problem at this location. However, caution is required in avoiding an excessive dose as present results suggest that 3 kg B ha^{-1} may increase flower death in chickpea. Studies are in progress to determine how widespread this B deficiency problem in chickpea is in the region, particularly in Nawalparasi, Chitwan and Makwan districts. Simple treatments with and without B application are being imposed in small plots in farmers fields, along with soil analyses and symptom observations. Similarly, extent of B deficiency is being tested in other grain legumes grown in the region, such as pigeonpea, lentil and groundnut.

Acknowledgements

The authors are thankful to Mr. P Gautam, Executive Director and Mr. R P Sapkota, Director, Crops and Horticulture Research (NARC), Dr. C L L Gowda, Coordinator, and Dr. A Ramakrishna, Sr. Scientist, CLAN and Dr. S C Sethi, Sr. Scientist (ICRISAT) for keen interest, their visits to the plots and valuable advice. We express our sincere thanks to the analytical group of the Soil Fertility Laboratory, SACD, IAC for chemical analysis of soil samples. We thank Mrs. Sirisha (ICRISAT - SACD/Soil Chemistry) for word processing of this paper.

References

Agarwala S C and Sharma C P 1979 Recognising Micronutrient Disorders of Crop Plants on the Basis of Visible Symptoms and Plant Analysis. Prem Printing Press, Lucknow, India. 51 p.

Evans H J, Purvis E R and Bear F E 1951 Effect of soil reaction on availability of molybdenum. Soil Sci. 71, 117–124.

Gupta U C 1979 Boron nutrition of crops. Adv. Agron. 31, 273–307.

Gupta V K 1993 Soil analysis for available micronutrients. *In* Methods of Analysis of Soil, Plant, Water and Fertilizer. Ed H L S Tandon, pp 36–48. Fertilizer Development and Consultation Organisation, New Delhi, India.

Jackson M L 1967 Potassium determinations for soils. *In* Soil Chemical Analysis. pp 111–113. Prentice-Hall of India, New Delhi, India.

Karki P B, Joshi S, Chaudhary G and Chaudary R N 1993 Studies on Botrytis Gray Mold of chickpea in Nepal. *In* Recent Advances in Research on Botrytis Gray Mold of Chickpea: Summary Proceedings of the Second Work Group to discuss Collaborative research on Botrytis Gray Mold of Chickpea, Rampur, Nepal, 14–17 March 1993. Ed M P Haware, pp 11–13. ICRISAT, Patancheru, India.

Khatri-Chhetri T B 1982 Assessment of soil test procedures for available boron and zinc in the soil of Chitwan valley, Nepal. Ph.D. thesis, University of Wisconsin, Madison, U.S.A.

Phetchawee S and Ratanarat S 1989 Surveys on boron deficiency of groundnut in farmers fields in central and eastern Thailand in 1987. *In* Proceedings of the 7th Thailand National Groundnut Meeting, Pattaya, Chonburi, 16–18 Mar, 1988. Ed A Patanothai, pp 373–376. Khon Kaen University, Thailand.

Piper C S 1940 The symptoms and diagnosis of minor-element deficiencies in agricultural and horticultural crops. Empire J. of Exp. Agric. 8, 85–96.

Singh A P, Sakal R, Sinha R B and Bhogal N S 1991 Relative response of selected chickpea and pigeonpea cultivars to boron application. Ann. Agric. Res. 12, 20–25.

Sinha P and Chatterjee C 1994 Influence of boron on yield and grain quality of pearl millet. Indian J. Agric. Sci. 64, 836–840.

Subba Rao A 1993 Analysis of soils for available major nutrients. *In* Methods of Analysis of Soil, Plant, Water and Fertilizer. Ed H L S Tandon, pp 13–35. Fertilizer Development and Consultation Organisation, New Delhi, India.

R.W. Bell and B. Rerkasem (eds.), Boron in Soils and Plants, 101–104.
© 1997 *Kluwer Academic Publishers.*

Screening wheat for boron efficiency

P. Anantawiroon[1], K. D. Subedi[2] & B. Rerkasem[1]
[1] *Agronomy Department, Chiang Mai University, Chiang Mai, Thailand, 50200*
[2] *Lumle Agricultural Research Centre, Pokhara, Nepal*

Key words: boron deficiency, boron efficiency, wheat

Abstract

Boron deficiency causes grain set failure in wheat (*Triticum aestivum*), but a wide range of genotypic variation in boron (B) efficiency has been identified. This paper shows how wheat genotypes may be evaluated for B efficiency. Two sets of genotypes were screened in sand culture in Thailand and in the field in Nepal and Thailand. One set of genotypes, identified as Boron Screening Nursery 1992/93 (BSN 92/93), included 32 genotypes plus seven replicates each of the B inefficient cultivar SW41, and the more efficient Sonora 64, as checks. The other set, Thailand Observation Nursery 1995/96 (TON 95/96), included 56 genotypes and 8 replicates each of check cultivars Sonora 64, Insee 1 and the highly efficient Fang 60. Both the BSN 92/93 and TON 95/96 were evaluated in sand culture with two levels of B added to the nutrient solution, 0 and 1 μM (B0 and B+). BSN 92/93 was also grown on a low B soil (hot water soluble, HWS, B at 0.1 mg kg^{-1}) in the field in Nepal. Field evaluation of the TON 95/96 in Thailand was conducted on a sandy loam with two levels of B, 0.1 and 0.2 mg HWSB kg^{-1} (BL and BH). The response of genotypes to B was measured with the Grain Set Index (GSI). A wide range of B efficiency was expressed among the genotypes in both nurseries. An almost 1:1 relationship was found between GSI in B0 and GSI in B0 expressed as % of the GSI in B+, and similarly between GSI in BL and GSI in BL expressed as % of GSI in BH with field grown TON 95/96, R^2 0.95–0.98 (p <0.001). In sand culture with B0, the very inefficient genotypes set only a few grains per ear while grain set in efficient genotypes approached that in B sufficiency. The effect of B deficiency on grain set in the field was less severe, but the average GSI of genotypes in each group correlated closely (R^2 0.82–0.90) with their GSI in B0 sand culture. It is suggested that by including B efficient and B inefficient checks, and using GSI to measure plant response, wheat genotypes may be evaluated for their B efficiency without any need for a B sufficient control.

Introduction

Boron (B) deficiency which causes male sterility in field grown wheat (*Triticum aestivum*) is a major cause of yield loss in many wheat growing countries in Asia, e.g. China (Li et al., 1978), India (Tandon and Naqvi, 1992), Nepal (Subedi and Budhathoki, 1995), and Bangladesh (Rerkasem, 1995). A common feature among these reports is the wide range of genotypic variation in the response to low external B supply which has led to a suggestion that selection for B efficiency could be a solution to the problem of wheat production on low B soils (Rerkasem and Jamjod, 1997). For tolerance to B toxicity, a simple and quick method has been

developed for screening wheat which enables a large number of genotypes to be assessed at the seedling stage (Paull et al., 1992). However, the chance to develop a similar early screening method for B efficiency in wheat is limited by two related observations. Firstly, B efficiency in field grown wheat is expressed primarily through reproductive development and grain set. Secondly, there is a lack of correlation between reproductive efficiency and vegetative efficiency (Rerkasem et al., 1993). This paper shows how wheat genotypes may be evaluated for B efficiency, using the Grain Set Index (GSI, percentage of the 20 basal florets from 10 central spikelets with grain; Rerkasem and Loneragan, 1994), as the measure of plant response to B deficiency.

Materials and methods

Two sets of genotypes were screened in sand culture in Thailand and in the field in Nepal and Thailand. One set of genotypes, identified as Boron Screening Nursery 1992/93 (BSN 92/93), included 30 genotypes plus seven replicates each of the B inefficient cultivar SW41, and the more efficient Sonora 64, as checks. The other set, Thailand Observation Nursery 1995/96 (TON 95/96), included 56 genotypes and 8 replicates each of check cultivars Fang 60, Sonora 64 and Insee. Both the BSN 92/93 and TON 95/96 were evaluated in sand culture with two levels of B (0 and 1 μM, B0 and B+) added to the nutrient solution (Rerkasem and Loneragan, 1994). Ten plants were grown in each freely drained earthenware pot (diameter 30 cm, depth 30 cm) containing washed river quartz sand, with duplicate pots for each genotype and B level. The nutrient solution was applied twice daily.

BSN 92/93 was also grown in the field in Nepal, at a site where B deficiency had been previously observed and hot water soluble (HWS) B measured at 0.1 mg kg^{-1}. Field evaluation of the TON 95/96 in Thailand was conducted on a sandy loam with two levels of B, 0.1 (BL) and 0.2 (BH) mg HWSB kg^{-1}, with 1 kg B ha^{-1} applied in a previous year as borax. In the field, entries were sown in single rows 3 m long, 0.25 cm apart, with the check genotypes before the first row and then after every 7–10 rows of the genotypes being screened.

At maturity, the GSI was assessed on the first two ears from each plant in the sand culture and up to 20 randomly selected ears per plot in field grown plants.

Results and discussion

The level of external B appeared to be close to sufficiency for wheat in B+ and BH, as most genotypes set grain normally and GSI values were largely >80%. With low external B supply in B0 and BL, a wide range of B efficiency was expressed among genotypes in both nurseries. There was an almost 1:1 relationship ($p<0.001$) between the GSI in B0 and GSI in B0 expressed as percentage of the GSI in B+ (Figure 1), with the R^2 at 0.98 for the BSN 92/93 (Figure 1a) and 0.95 for the TON 95/96 (Figure 1b). A similarly close relationship (R^2 at 0.98; $p<0.001$) between GSI in BL and in BL relative to BH was found in the field with TON 95/96 (Figure 1c). These results suggest that wheat genotypes may be evaluated for B efficiency on

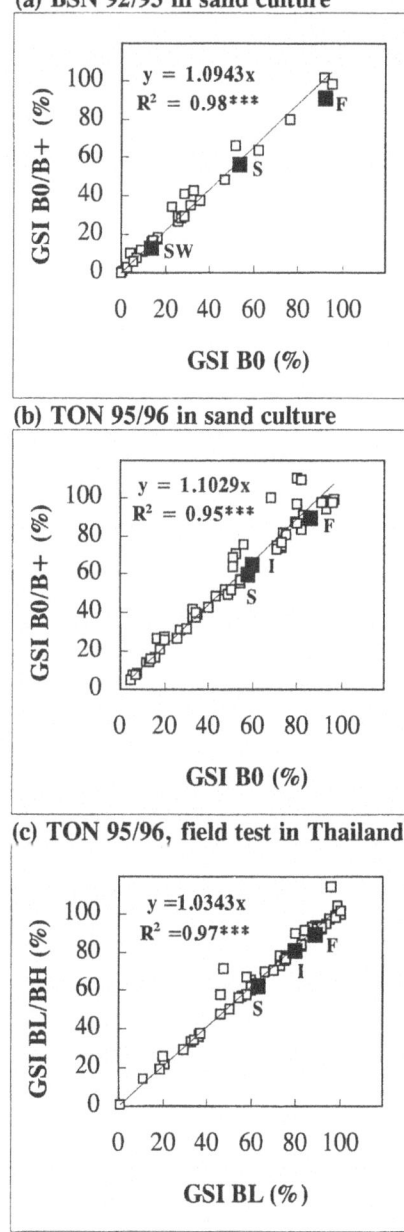

Figure 1. The relationship between grain set index (GSI) in boron (B) deficiency and the GSI expressed as the percentage of that in B sufficiency. (a) BSN 92/93 in sand culture, GSI in B0 and GSI in B0 relative to GSI in B+; (b) TON 95/96 in sand culture, GSI in B0 and GSI in B0 relative to GSI in B+; (c) TON 95/96 in soil, GSI in BL and GSI in BL relative to GSI in BH. Large solid squares denote check genotypes: F = Fang 60, S = Sonora 64, I = Insee 1 and SW = SW41. Each point is the mean of two replicates.

low B soils without the B sufficiency control. A measure of the intensity of B deficiency in the environment

(a) BSN 92/93, field test and sand culture screening

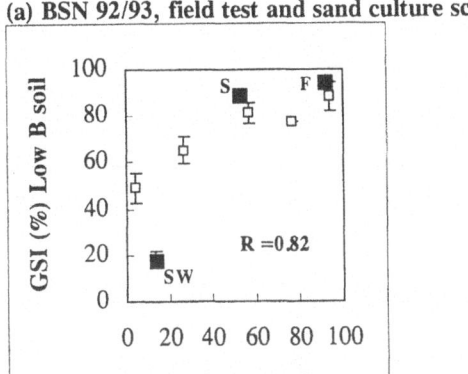

(b) TON 95/96, field test and sand culture screening

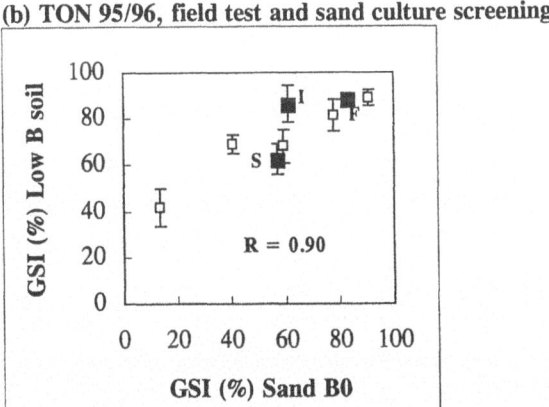

Figure 2. Correlation between grain set index (GSI) in B0 sand culture and GSI in low B soil for five classes of wheat genotypes with varying B efficiency, in two field tests, (a) Nepal, BSN 92/93, and (b) Thailand, TON 95/96. A small open square with error bars represents the average GSI for each class, large solid squares denote check genotypes: F = Fang 60, S = Sonora 64, I = Insee 1 and SW = SW41.

sand culture, with correlation coefficients of 0.82 for BSN 92/93 and 0.90 for TON 95/96 (Figure 2). Under the conditions of this study, the intensity of B deficiency in both field soils may have been much lower than that in the sand culture. Root geometry has been suggested as a mechanism for zinc efficiency in the wheat cultivar Excalibur and the reason for the difference in the expression of efficiency between solution culture and soil (Graham and Rengel, 1993). There is also a strong possibility that soil-root interactions, root growth and distribution as well as rhizosphere effects, (Marschner, 1995) could play a role in the expression of B efficiency of a genotype in the field. Direct evidence for this, however, is still unavailable.

In conclusion, this paper has shown how B efficiency of wheat genotypes may be assessed. Since B deficiency primarily affects reproduction in wheat, and there is no simple relationship between reproductive efficiency and vegetative efficiency (Rerkasem et al., 1993), it is at present essential that to screen for B efficiency plants must be grown until grain set. A better understanding of the efficiency mechanism(s), and their genetic control, especially at the molecular level, may lead to simpler and more economical methods for evaluation that can be carried out much earlier.

Acknowledgements

Part of this work was supported by a grant from AusAID to CIMMYT. The wheat genotypes in BSN 92/93 came from various nurseries introduced into Thailand by CIMMYT-Bangkok. The seed for TON 95/96 was supplied by Sutheera Moolsri of the Rice Research Institute of Thailand.

may be provided by the inclusion of check genotypes, preferably covering a range of B efficiency.

The wheat genotypes included in the sand culture experiments were classified into five B efficient groups according to their GSI in low external B supply, relative to the control genotypes. In B0, the most inefficient genotypes set only a few grains per ear (GSI < 20%), while grain set in the most efficient genotypes approached that in B sufficiency. Sand culture and field experiments were then compared on the basis of these B efficiency groups.

The effect of B deficiency on grain set in the field was less severe, but the average GSI of genotypes in each B efficiency class correlated with their GSI in B0

References

Graham R D and Rengel Z 1993 Genotypic variation in zinc uptake and utilization by plants. *In* Zinc in Soils and Plants. Ed A D Robson, pp 107–118. Kluwer Academic Publishers, Netherlands.

Li B H, Li W H, Kui M C, Chao W S, Jern H P, Li C R, Chu W J and Wang C L 1978 Studies on cause of sterility of wheat. J. Northeastern Agric. Coll. 3, 1–19. (In Chinese)

Marschner H 1995 Mineral Nutrition of Higher Plants 2nd Edition. Academic Press, London. 889 p.

Paull J G, Moody D B and Rathjen A J 1992 Genetics and breeding of wheat for boron toxicity. *In* Boron Deficiency in Wheat. Eds C E Mann and B Rerkasem, pp 90–97. CIMMYT Wheat Special Report No. 11. CIMMYT, Mexico.

Rerkasem B 1995 A general survey of the incidence of wheat sterility. *In* Causes of Sterility in Wheat. Eds R W Bell and B Rerkasem,

104

pp 1–8. Environmental Science Report 94/6, Murdoch University, Western Australia.

Rerkasem B and Jamjod S 1997 Boron deficiency induced male sterility in wheat (*Triticum aestivum* L.) and implications for plant breeding. Euphytica (in press ref. MS No EUPH 4247)

Rerkasem B and Loneragan J F 1994 Boron deficiency in two wheat genotypes in a warm, subtropical region. Agron. J. 86, 887–890.

Rerkasem B, Netsangtip R, Lordkaew S and Cheng C 1993 Grain set failure in boron deficient wheat. Plant and Soil 155/156, 309–312.

Subedi K D and Budhathoki C B 1995 Wheat sterility problem in Nepal: research at Lumle Agricultural Research Centre. *In* Causes of Sterility in Wheat. Eds R W Bell and B Rerkasem, pp 44–56. Environmental Science Report 94/6, Murdoch University, Western Australia.

Tandon J P and Naqvi S M A 1992 Wheat varietal screening for boron deficiency in India. *In* Boron Deficiency in Wheat. Eds C E Mann and B Rerkasem, pp 76–78. CIMMYT Wheat Special Report No. 11. CIMMYT, Mexico.

R.W. Bell and B. Rerkasem (eds.), Boron in Soils and Plants, 105–109.
© 1997 *Kluwer Academic Publishers.*

Chromosomal variations for boron tolerance in wheat (*Triticum aestivum*)

Y. Chantachume[1], K. W. Shepherd[2], J. G. Paull[2] & A. J. Rathjen[2]
[1] *Nakhon Sawan Field Crops Research Center, Takfa, Nakhon Sawan, Thailand 60190; E-mail: yodsapct mozart.inet.co.th*
[2] *Department of Plant Science, Waite Agricultural Research Institute, Glen Osmond, South Australia 5064*

Key words: boron tolerance, chromosomal variation, wheat

Abstract

Monosomic plants of a moderately boron (B) tolerant wheat variety (Condor of homoeologous groups four and seven) were identified cytologically and then crossed as the female parent to six B tolerant wheat varieties; G 61450, Halberd, India 126, Benventuto Inca, Lin Calel and AUS 4041. Monosomic F_1 hybrid plants of each cross were identified and used as both male and female parents in backcrosses to the Condor monosomic parental lines. Each pair of the 68 reciprocal families was then compared for the differences in response to B between the homologous chromosomes of the tester Condor monosomic lines and the tested varieties. The backcross reciprocal monosomic analysis indicated that there was chromosomal variation between the different varieties. The significant increase in root length resulting from the presence of chromosome 7B of Halberd and 4A of Benventuto Inca, G 61450 and India 126 were responsible for tolerance to B. The significant difference between the mean root length of the critical family for chromosome 4A from India 126 and that of the family for chromosome 4A from Condor was less than expected, indicating that there was probably another chromosome also responsible for tolerance to B in India 126.

Introduction

The establishment of the chromosomal location of genes controlling agronomic characters is essential for the success of intraspecific chromosome manipulation techniques in wheat improvement (Law et al., 1981). Although the F_2 monosomic method is the most commonly used method because it is easily applied and is particularly efficient (Macer, 1966), it cannot be used with characters having continuous variation where discrete phenotypes are not discernible in the F_2 because of the confounding effects of allelic variations and chromosome dosage (Snape et al., 1983).

Reciprocal monosomic crossing, which allows the comparison between the two F_1 monosomics from the reciprocal crosses between the homologous monosomics of two varieties (McEwan and Kalsikes, 1970), overcomes this problem. However, the reciprocal monosomic method is limited to varieties for which a monosomic series has been developed.

A more flexible method of chromosome assay which overcomes the deficiencies of both of the above methods is the backcross reciprocal monosomic method described by Snape and Law (1980). The backcross reciprocal method makes it possible to compare a particular monosomic chromosome from one variety with same monosomic chromosome from another variety (Law et al., 1983).

The results of F_2 monosomic analysis indicated that chromosomes 7B and 4A are the locations of genes controlling boron (B) tolerance in Halberd and G 61450, respectively (Chantachume et al., 1994). As it was not feasible to undertake backcross reciprocal monosomic analysis for all chromosomes for several varieties because of the time required for the cytological examination of the monosomic plants, only the effects of chromosomes of groups four and seven were tested here. The objective of this experiment was to identify whether chromosomes other than those in groups four and seven controlled B tolerance in four other tolerant exotic varieties. Halberd and G61450

106

Table 1. Varieties used in experiments, Australian winter Cereals Collection (AUS) accession numbers, their pedigrees, origins and responses to boron toxicity.

Variety	AUS	Pedigree	Origin	B response[a]
India 126	4743	Unknown	India	T[b]
G 61450	6141	Mentana/Kenya/Quaderna	Greece	T
Benventuto Inca	1929	Mentana/Lin Calel M.A.	Argentina	T
AUS 4041	4041	Abyssinia 10	Ethiopia	T
Lin Calel	2881	Unknown	Argentina	T
Halberd	11612	(Scimitar × Kenya C6042) × (Bobin) × (Insignia 49)	Australia	MT
Condor	16036	Penjamo 62/4* Gabo 56/2/ TZPP/Nainari 60/4/2* Lerma Rojo/2/Norin 10/Brevor (Seln. 14)/3/3* Andes	Australia	MS

[a] Data of boron response were derived from: Moody et al. (1988) and Paull et al. (1991).
[b] T = tolerant, MT = moderately tolerant, MS = moderately sensitive, S = sensitive, VS = very sensitive.

were also included to check the results described by Chantachume et al. (1994).

Materials and methods

The seeds of monosomics for homoeologous groups 4 and 7 of CSP 44, a selection of Condor, were sown in standard potting mix in a glasshouse and the monosomic plants identified cytologically by determining the chromosome complements of pollen mother cells at metaphase I as described in Chantachume (1995). These were crossed as the female parents to six B tolerant wheat varieties; G 61450, Halberd, India 126, Benventuto Inca, Lin Calel, and AUS 4041 (Moody et al., 1988). Pedigree, B response and the origin of these varieties is presented in Table 1.

Two monosomic F_1 plants of each cross were identified but, where possible, only one plant was used as both male and female parent in backcrosses to the Condor monosomic parental line. The second plant was crossed only if there was insufficient pollen produced by the first plant. Each pair of the 68 BCR families (BC_1F_1) (4 of the families lost during the experiment) was then compared. The seeds of the backcross reciprocal families of Lin Calel and AUS 4041 for chromosome 4A were not available because their F_1 seeds were damaged by insects and not viable.

The 34 pairs of BCR families were tested for B tolerance in a randomized complete block design with three replicates using the filter paper technique (Chantachume et al., 1995) with a B treatment of 100 mg

L^{-1}. The experimental procedures, including pretreatment conditions, were described by Chantachume et al. (1995). Five seeds of each family of each reciprocal pair plus two seeds of each parent were included in a single filter paper. Each replicate contained 34 filter papers. The lengths of the longest root of each seedling were measured after 12 days.

The statistical analysis was calculated by MSTAT microcomputer program version 4.0 written at Michigan State University.

Results and discussion

Families with chromosome 4A from three of the tolerant varieties, Benventuto Inca, G61450 and India 126 and for chromosome 7B of Halberd, exhibited increased root lengths relative to those families in which most of their progenies carry the Condor homologues. The difference between the families with predominately chromosome 4A from Benventuto Inca and G61450 and that of Condor was 4.14 and 3.80 cm, respectively (Table 2), which was highly significant. There was also a significant difference of 2.15 cm between the chromosome 4A family of India 126 compared with that of Condor (Table 2). The difference between the chromosome 7B family of Halberd compared with that of Condor was a significant 2.13 cm increase in length (Table 3). Differences between the other BCR families were all smaller and non-significant.

Table 2. Mean length of roots (cm) and differences between backcross reciprocal families for chromosomes of group 4 derived from the Condor monosomics in crosses with six tolerant genotypes.

| Genotype | Root length (cm) | | | | | | | | |
| | 4A | | | 4B | | | 4D | | |
	Disomic[a]	BCR[b]	Diff[c]	Disomic	BCR	Diff	Disomic	BCR	Diff
G61450	9.9	8.38[d]	3.8**	9.5	6.00	0.12	8.8	5.99	1.32
Condor	4.4	4.58[e]		4.1	5.88		3.9	4.67	
Halberd	7.0	5.04	0.55	7.1	4.83	−0.01	6.5	5.18	−0.20
Condor	3.6	4.49		3.9	4.84		4.0	5.38	
India 126	10.2	8.78	2.15*	9.2	6.51	0.57	11.3	7.67	1.48
Condor	3.5	6.63		3.5	5.94		3.7	6.19	
Benventuto, Inca	9.2	8.39	4.14**	9.4	5.81	−1.46	10.6	5.91	0.38
Condor	3.5	4.25		4.4	7.27		3.9	5.53	
Lin Calel	NA[f]	NA	NA	10.6	4.95	0.75	10.7	7.32	−0.28
Condor				3.8	4.20		4.4	7.60	
AUS 4041	NA	NA	NA	10.1	6.35	1.06	12.5	6.32	−0.27
Condor				4.0	5.29		4.2	6.59	

[a] Disomic parent
[b] Backcross reciprocal families
[c] Differences between backcross reciprocal families
*,** Significance of differences: *$p<0.05$; ** $p<0.01$, tested by Duncan's New Multiple Range Test
[d] Mean root length of a family derived from crossing between F_1 monosomic of (Condor monosomic 4A × G61450) as male and Condor monosomic 4A as female
[e] Mean root length of a family derived from crossing between F_1 monosomic of (Condor monosomic 4A × G61450) as female and Condor monosomic 4A as male
[f] NA = not available.

Table 3. Mean length of roots (cm) and differences between backcross reciprocal families for chromosomes of group 7 derived from the Condor monosomic in crosses with six tolerant genotypes.

| Genotype | Root length (cm) | | | | | | | | |
| | 7A | | | 7B | | | 7D | | |
	Disomic[a]	BCR[b]	Diff[c]	Disomic	BCR	Diff	Disomic	BCR	Diff
G61450	8.7	5.75[d]	0.09	9.1	6.12	−0.80	10.9	5.51	0.71
Condor	3.4	5.66[e]		3.4	6.92		4.5	4.80	
Halberd	7.1	4.97	0.81	6.4	6.42	2.13*	8.1	4.93	0.48
Condor	3.8	4.16		3.4	4.29		3.7	4.45	
India 126	10.2	6.62	1.18	12.8	8.13	1.28	12.7	5.52	−0.65
Condor	3.6	5.44		4.0	6.85		3.5	6.17	
Benventuto, Inca	9.8	6.88	0.44	8.3	6.36	1.60	9.9	5.82	0.54
Condor	4.2	6.44		3.5	4.76		3.8	5.28	
Lin Calel	10.4	5.51	−1.17	9.1	5.55	1.12	9.6	5.32	−0.04
Condor	4.0	6.68		4.0	4.43		4.2	5.36	
AUS 4041	12.8	5.93	0.59	12.1	4.88	1.08	12.4	6.74	0.55
Condor	4.0	5.34		3.6	3.80		3.8	6.19	

[a] Disomic parent
[b] Backcross reciprocal families
[c] Differences between backcross reciprocal families
*,** Significance of differences: *$p<0.05$; ** $p<0.01$, tested by Duncan's New Multiple Range Test
[d] Mean root length of a family derived from crossing between F_1 monosomic of (Condor monosomic 7A × G61450) as male and Condor monosomic 7A as female
[e] Mean root length of a family derived from crossing between F_1 monosomic of (Condor monosomic 7A × G61450) as female and Condor monosomic 7A as male.

(a)

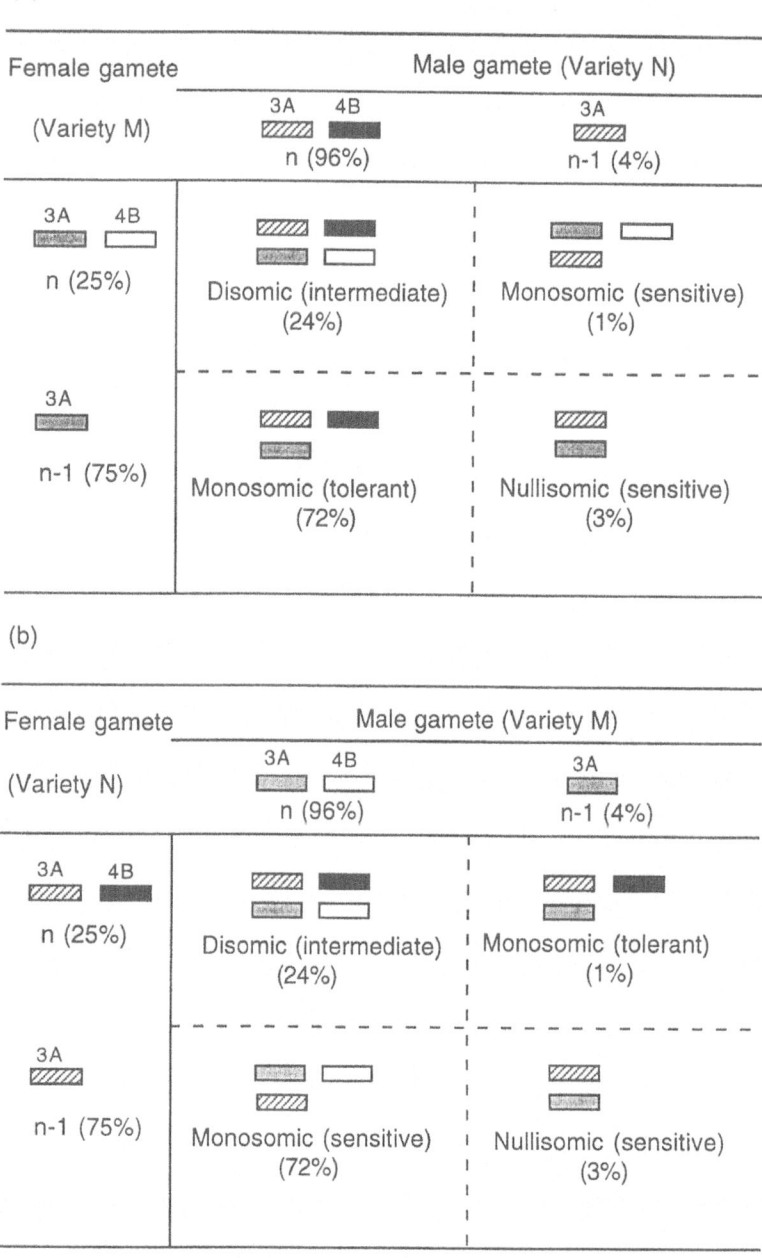

(b)

Figure 1. Diagram of reciprocal families derived from crossing between monosomics carrying critical chromosome 4B of varieties M (sensitive) and N (tolerant). The scheme is simplified to show only two pairs of homoeologous chromosomes 3A and 4B and the frequency of disomic, monosomic and nullisomic progenies derived from the reciprocal crosses. (a) a family derived from crossing between monosomic 4B of variety M (as female) and monosomic 4B from variety N (as male), (b) a family derived from crossing between monosomic 4B of variety N (as female) and monosomic 4B of variety M (as male) n = haploid number of chromosome = 21; figures in brackets are percentage values.

For the critical families of chromosomes of the tolerant varieties and those of Condor, the expected segregation ratio would be 24 tolerant-intermediate: 1 sensitive (Figure 1a) and 1 tolerant-intermediate: 3 sensitive (Figure 1b), respectively.

The root length of the individual plants within each family was compared with that of the disomic parents

Table 4. Chi-square analysis of the observed and expected segregation ratio of F_1 backcross reciprocal families carrying critical chromosome of G61450, Halberd, India 126, Benventuto Inca and Condor, tested in filter paper at 100 mg B L^{-1}.

Family		Frequency Model	Tol-int[a]	Sens[b]	c_1^2
G61450 4A[c]	Obs[e]		13	2	
	Exp[f]	24:1	14.40	0.60	3.40
Condor 4A[d]	Obs		4	11	
	Exp	1:3	3.75	11.25	0.03
Halberd 7B	Obs		13	1	
	Exp	24:1	13.44	0.56	0.36
Condor 7B	Obs		0	13	
	Exp	1:3	3.25	9.75	4.33*
India 4A	Obs		15	0	
	Exp	24:1	14.4	0.60	0.63
Condor 4A	Obs		4	9	
	Exp	1:3	3.25	9.75	0.23
Benventuto	Obs		14	1	
Inca 4A	Exp	24:1	14.4	0.60	0.28
Condor 4A	Obs		3	12	
	Exp	1:3	3.75	11.25	0.20

[a] Tolerant-intermediate,

[b] Sensitive,

[c] Family derived from crossing between the F_1 monosomic of (Condor monosomic 4A × G61450) as male and Condor monosomic 4A as female,

[d] Family derived from crossing between the F_1 monosomic of (Condor monosomic 4A × G61450) as female and Condor monosomic 4A as male,

[e] Observed value,

[f] Expected value,

* Significant difference at $0.01 < p < 0.05$.

and classified into categories. A plant with a root length within the range of the sensitive parent was classified as sensitive. Whereas a plant with a root length longer than the range of the sensitive parent was classified as tolerant-intermediate.

Chi-square analysis indicated that the segregation of root length of the plants within the critical families for chromosomes of G 61450, Halberd, India 126 and Benventuto Inca was consistent with the expected ratio of 24 tolerant-intermediate: 1 sensitive (Table 4). Whereas the segregation of the critical families for chromosomes of Condor was consistent with the ratio of 1 tolerant-intermediate: 3 sensitive with the exception of the critical family for chromosome 7B (Table 4). The deviation from the expected ratio was principally due to a deficiency of the tolerant-intermediate plants. These results confirmed that chromosome 4A of G 61450, India 126 and Benventuto Inca and 7B of

Halberd were responsible for tolerance to B relative to Condor.

Acknowledgement

The financial support of the Australian government is gratefully acknowledged.

References

Chantachume Y 1995 Genetic studies on the tolerance of wheat to high concentrations of boron. PhD Thesis. University of Adelaide, Adelaide.

Chantachume Y, Shepherd K W, Paull J G and Rathjen A J 1994 Chromosomal location of genes in wheat controlling tolerance to high concentrations of B. *In* Proc. Seventh Assembly Wheat Breeding Society of Australia. Eds J G Paull, I S Dundas, K W Shepherd and G J Hollamby, pp. 27–30. University of Adelaide, South Australia.

Chantachume Y, Smith D, Hollamby G J, Paull J G and Rathjen A J 1995 Screening for B tolerance in wheat (*T. aestivum*) by solution culture in filter paper. Plant and Soil 177, 249–254.

Law C N, Snape J W and Worland A J 1981 Intraspecific chromosome manipulation. Phil. Trans. R. Soc. Lond. B 292, 509–518.

Law C N, Snape J W and Worland A J 1983 Quantitative genetic studies in wheat. *In* Proc. Sixth International Wheat Genetic Symposium. Ed S Sakamoto, pp. 539–547. Kyoto University, Japan.

Macer R C F 1966 The formal and monosomic genetic analysis of stripe rust (*Puccinia striiformis*) resistance in wheat. Hereditas Supplement Vol. 2, 127–142.

McEwan J M and Kaltsikes P J 1970 Early generation testing as a mean of predicting the value of specific chromosome substitutions into common wheat. Can. J. Genet. Cytol. 12, 711–723.

Moody D B, Rathjen A J, Cartwright B, Paull J G and Lewis J 1988 Genetic diversity and geographical distribution of tolerance to high levels of soil B. *In* Proc. Seventh International Wheat Genetics Symposium. Eds T E Miller and R M D Koebner, pp. 859–865. Institute Plant Science Research, Cambridge.

Paull J G, Rathjen A J and Cartwright B 1991 Major genes control of tolerance of bread wheat to high concentrations of soil boron. Euphytica 55, 217–228.

Sears E R 1953 Nullisomic analysis in common wheat. Am. Nat. 87, 245–252.

Snape J W and Law C N 1980 The detection of homologous chromosome in wheat using backcross reciprocal monosomic lines. Heredity 45, 187–200.

Snape J W, Parker B B and Gale M D 1983 Use of the backcross reciprocal monosomic method for evaluating chromosomal variation for quantitative characters. *In* Proc. Sixth International Wheat Genetics Symposium. Ed S Sakamoto, pp. 367–373. Kyoto University, Japan.

R.W. Bell and B. Rerkasem (eds.), Boron in Soils and Plants, 111–115.
© 1997 *Kluwer Academic Publishers.*

Genetic variation in the tolerance of durum wheat (*Triticum turgidum* L. var *durum*) to high concentrations of boron

S. Jamjod[1,2], J. G. Paull[1], B. J. Brooks[1] & A. J. Rathjen[1]
[1] *Department of Plant Science, Waite Agricultural Research Institute, University of Adelaide, Glen Osmond, SA 5064, Australia*
[2] *Present Address: Agronomy Department, Faculty of Agriculture, Chiang Mai University, Chiang Mai 50200, Thailand*

Key words: Boron, durum wheat, tolerance, toxicity, variation

Abstract

The genetic range of tolerance to a high concentration of boron (B) was examined for 300 genotypes of durum wheat (*Triticum turgidum* L. var *durum*). Genotypes were screened in filter paper soaked with boric acid solution at the concentration 100 mg B L^{-1}, assessed for root growth of seedlings and compared to four bread wheat (*Triticum aestivum*) checks of contrasting levels of tolerance to B. All varieties/lines and accessions originating from Australia, Italy, CIMMYT and African countries were rated as either sensitive or moderately sensitive to B. Seven accessions or 2% of the population, originating from India, China and Iraq, were classified as either moderately tolerant or tolerant to B. The results indicated substantial genetic variation existed for tolerance to high B in durum wheat.

Introduction

High levels of boron (B) are found in the subsoils of many regions of southern Australia (Cartwright et al., 1984; 1986). The B restricts root growth of wheat (Holloway and Alston, 1992) and results in reduced grain yields of cereals (Cartwright et al., 1984; Rathjen et al., 1987). As removal of excess B from soil is not feasible in southern Australia, the use of tolerant varieties is the best option to overcome this problem (Rathjen et al., 1987). A large range of genetic variation for response to high concentrations of B has been reported in bread wheat, barley and peas and genetic variation for B tolerance was found among the commercial varieties of these crops cultivated in South Australia (Paull et al., 1992a; Jenkin, 1993; Bagheri et al., 1992). Significant increases in grain yield of bread wheat have been achieved by incorporating B tolerance genes into sensitive varieties (Moody et al., 1993).

Durum wheat is an important crop in the Mediterranean basin and West Asia, the plains of North Dakota (USA), Saskatchewan (Canada), USSR and central India (Bozzini, 1988). In contrast, durum wheat in Australia is a minor crop, compared to bread wheat. Brooks (1991) examined the adaptation of durum wheat to South Australia and found that when durum was grown at a high B site, Two Wells, all durums showed severe symptoms of B toxicity and concluded that the effect of B toxicity was one of major problems for adaptation of durum to South Australia. Yau et al. (1995) also reported that most durum varieties were identified as sensitive to very sensitive to B. Therefore, there was a need to identify B tolerant durum germplasm and incorporate the tolerance into varieties for cultivation in problem areas.

The major aim of this experiment was to identify sources of B tolerance from germplasm collected from areas where B tolerance predominated in hexaploid wheat (Moody et al., 1988), barley (Boyd, pers comm.), and pea (Bagheri et al., 1992). Screening a large number of accessions should enable the identification of the geographical origin of tolerant accessions and allow a comparison of the range of variation within durum and bread wheat.

Materials and methods

Genotypes

A total of 249 accessions collected from North Africa and Asia, and 51 varieties or advanced lines from Australia, Italy and CIMMYT were tested for response to B. Seeds of all accessions were obtained from the Australian Winter Cereals Collection, Tamworth, New South Wales. Seeds of varieties and advanced lines were obtained from a collection maintained at the Waite Institute. Four bread wheat genotypes with known levels of response to B: G 61450 (tolerant), Halberd (moderately tolerant), Schomburgk (moderately sensitive) and Kenya Farmer (sensitive) (Paull et al., 1992a) were included as checks. All seed was imbibed on petri-dishes at 4°C for two days and at 18°C for one day before sowing to ensure uniform germination.

Screening procedure

A filter paper bioassay described by Chantachume et al. (1995) was used to study root growth in response to B. Solutions containing 500 μM Ca(NO$_3$).4H$_2$O and 2.5 μM ZnSO$_4$.7H$_2$O and boric acid at the rates of 0 and 100 mg B L^{-1} (designated as B0 and B100, respectively) were prepared. Sheets of filter paper (Ekwip$^®$ grade R6 size 36 × 42 cm) were soaked in these solutions for 2 min and allowed to drain for 2 min. Germinated seeds were placed in a single row along the centre of each paper with a spacing of 2 cm between each seed. The paper was then rolled up, covered with aluminium foil and stored vertically at 18°C. After 12 days, the length of the longest root of each seedling was measured and the average seedling root length of each genotype was calculated and compared with the checks.

Initial screening

The 300 genotypes were screened at the B100 treatment. Seven to ten germinated seeds of each genotype were sown on each paper. The screening consisted of two replicates. Three bread wheat checks, G 61450, Halberd and Schomburgk, were included every fifty entries.

Second screening

Six durum accessions identified as tolerant, moderately tolerant, moderately sensitive and sensitive were selected and the seedlings were transplanted to normal soil for seed multiplication. To obtain confirmation of response and compare the range of response to B with the bread wheat genotypes, seeds harvested from accessions were tested for response to B in filter paper at B0 and B100 treatments and compared to four bread wheat checks. The experiment was arranged as a split plot design with three replications. Fifteen seeds were grown in each B treatment and genotype combination.

Results

The root lengths of the three bread wheat checks Schomburgk, Halberd and G 61450 ranged between 3.2–4.6, 5.8–7.2 and 8.1–8.8 cm, respectively (data not shown) at B100. According to the response of the bread wheat checks, durum accessions were classified into four categories, namely, sensitive (≤ 3 cm), moderately sensitive (>3–5 cm), moderately tolerant (>5–8 cm) and tolerant (>8 cm).

A narrow range in response to B was measured among the adapted genotypes and most of the accessions of durum wheat. All varieties/lines and accessions originating from Australia, Italy, CIMMYT and African countries were rated as either sensitive or moderately sensitive to B (Table 1). Only seven lines, or 2% of the total population, originally collected from China (2), India (3) and Iraq (2), were classified as moderately tolerant or tolerant. Two tolerant, two moderately tolerant, one moderately sensitive and one sensitive accessions were selected for further study (Table 2).

When selected accessions were progeny tested and compared with the bread wheat checks, significant differences in root length and a significant B treatment × genotype interaction were found in both durum and bread wheats (Figure 1). At B0, the root length of durum and bread wheat ranged between 11 and 14 cm. At B100, genotypes were classified into four distinct classes, namely, tolerant: G 61450, AUS 10110 and AUS 10105, moderately tolerant: Halberd, AUS 10344 and AUS 14010, moderately sensitive: Schomburgk, AUS 10348 and Yallaroi and sensitive: Kenya Farmer and AUS 13244.

Discussion

A large range in root lengths of seedlings grown in a high concentration of B occurred among the durum accessions originating from diverse geographical locations. The B treatment used for screening was able to

Table 1. Distribution of response to B in durum genotypes, according to countries of origin, based on root lengths in filter papers containing 100 mg B L^{-1}.

	Response to B					
Country	S	MS	MT	T	No. of lines	% MT-T
Africa						
Algeria	12	17	0	0	29	0
Ethiopia	10	13	0	0	23	0
Morocco	8	10	0	0	18	0
Tunisia	6	7	0	0	13	0
Total	36	47	0	0	73	0
Asia						
Afghanistan	6	8	0	0	14	0
China	2	8	2	0	12	17
India	5	5	1	2	13	23
Iran	3	10	0	0	13	0
Iraq	6	6	2	0	14	14
Syria	24	30	0	0	54	0
Turkey	15	31	0	0	46	0
Total	61	98	5	2	166	4.2
Others						
Australia	1	23	0	0	24	0
Italy	3	4	0	0	7	0
Mexico	12	8	0	0	20	0
Total	16	35	0	0	51	0
Total	113	180	5	2	300	2.3
Frequency (%)	37.7	60.0	1.6	0.7		2.3

Table 2. Australian Wheat Collection accession number, name, country of origin and mean root length in B100 for selected durum accessions.

Accession No.	Name	Country of origin	Root length (cm)^a
AUS 10110	AUS 10110	India	8.7
AUS 10105	AUS 10105	India	8.2
AUS 10344	Niloticum	Iraq	5.6
AUS 14010	Lingzhi Baimong Baidamai	China	5.3
AUS 10348	Erythrospermum	Iraq	3.7
AUS 13244	Gandum	Afghanistan	1.5

^{a} Root length grown at 100 mg B L^{-1} in initial screening (Table 1)

discriminate between tolerant and sensitive genotypes. For example, the root length of accessions classified as tolerant (AUS 10110) and sensitive (AUS 13244) at B100 were 7.0 and 2.0 cm, respectively, and a similar level of discrimination was achieved for the bread wheat checks (Figure 1).

Genetic variation in response to B was demonstrated among the exotic durum germplasm with seven accessions (2% of the population), originating from India, Iraq and China, being classified as moderately tolerant or tolerant. Boron tolerant material from other crops has also been identified from these regions. For example, Moody et al. (1988) screened 1579 accessions of bread wheat and classified a high proportion of lines from Iraq (37%) and India (58%) as moderately tolerant or tolerant. A number of reports in the liter-

114

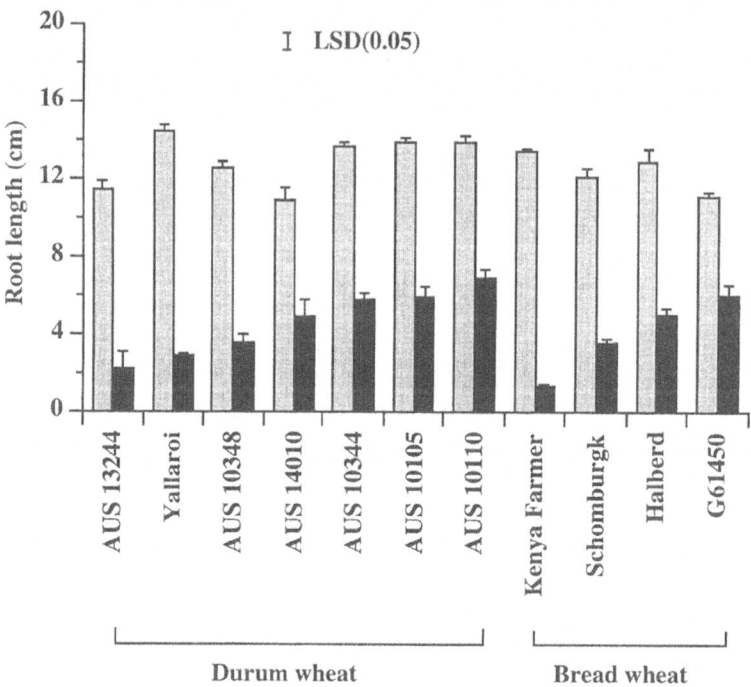

Figure 1. Root length of durum and bread wheat genotypes tested in filter papers at B0 (▨) and B100 (■) treatments in the second screening. Vertical bars represent the standard error of three replicates.

ature have described areas of India having B toxicity problems (e.g. Chauhan and Asthana, 1981; Chauhan and Powar, 1978). The results of the durum screening were also consistent with screening of other germplasm collections including medics (Paull et al., 1992b), peas (Bagheri et al., 1994) and barley (Jenkin, 1993; Boyd, pers. comm.) with a majority of B tolerant accessions originating in Eastern Mediterranean and Asian countries. This suggests that high concentrations of B may be present in soil and have a major influence upon the adaptation of plants in these regions. It can be proposed that germplasm of other crops originating from these regions is likely to provide a high level of tolerance to B.

Durum wheat had a smaller range than bread wheat for variation in response to B. The most tolerant durums were rated the same as G 61450. Moody et al. (1988) identified some overseas genotypes as more tolerant than G 61450. Furthermore, the most sensitive durum, AUS 13244, had significantly shorter roots than Schomburgk, but slightly longer than Kenya Farmer (Figure 1). Paull et al. (1992a) concluded that the variation in response to B found in bread wheat was controlled by at least four genes. The smaller range of response to B in durum wheats may be due to response being controlled by fewer genes than in bread wheat,

possibly a consequence of being a tetraploid species whereas bread wheat is hexaploid.

Limited genotypic variation was found among the varieties of durum wheat agronomically adapted to Australia or from other modern breeding programs. All varieties or advanced lines originating from NSW, Australia or CIMMYT were rated as moderately sensitive or sensitive. This was in contrast to the findings in bread wheat where higher levels of tolerance were identified within historically important Australian varieties such as those derived from Federation and Currawa, which were selected and grown in high B areas (Paull et al., 1992a). It is probable that the limited genetic variation in response to B among adapted durum wheat varieties, could be due to:

  – breeding and selection in areas where high concentrations of B do not occur,
  – minimal genetic variation for B tolerance in the parents used in the breeding programs,
  – a shorter and less intensive effort in breeding durum wheat, compared to bread wheat, in

Australia, reducing the opportunity to develop varieties well adapted to the range of Australian environments.

There was also little variation in response to B among the Mediterranean Basin durum varieties. This is likely to have resulted from most of the varieties having been grown and selected on deep soils in high rainfall conditions or with supplementary irrigation; intense selection for other characters, e.g. quality, having reduced the genetic variation in breeding populations or landraces, and; lack of variation for tolerance to B in the wild progenitor of durum wheat (*T. dicoccoides*) collected from the Israel/Jordan region (A.J. Rathjen, unpublished). Collections of *T. dicoccoides* from further north and east in the fertile crescent have not been tested.

This experiment showed that genetic variation for response to high concentrations of B does exist in durum wheat. The solution for overcoming B toxicity would be breeding and selection of B tolerant varieties. However, more information is needed on the mechanism of tolerance in this species and how it can be incorporated into intolerant but otherwise adapted varieties to facilitate the breeding process.

References

Bagheri A, Paull J G, Rathjen A J, Ali S M and Moody D B 1992 Genetic variation in the response of pea (*Pisum sativum* L.) to high soil concentrations of boron. Plant and Soil 146, 261–269.

Bozzini A 1988 Origin, distribution, and production of durum wheat in the world. *In* Durum Wheat: Chemistry and Technology. Eds G Fabrini and C Lintas, pp 1–16. American Association Cereal Chemists, Minnesota, U.S.A.

Brooks B J 1991 The adaptation of *Triticum turgidum* L. var. durum (durum wheat) to South Australia. Honours Thesis, University of Adelaide, South Australia.

Cartwright B, Zarcinas B A and Mayfield A H 1984 Toxic concentrations of boron in red-brown earth at Gladstone, South Australia. Aust. J. Soil Res. 22, 261–272.

Cartwright B, Zarcinas B A and Spouncer L R 1986 Boron toxicity in South Australian barley crops. Aust. J. Agric. Res. 37, 351–359.

Chantachume Y, Smith D, Hollamby G J, Paull J G and Rathjen A J 1995 Screening for boron tolerance in wheat (*T. aestivum*) by solution culture in filter paper. Plant and Soil 177, 249–254.

Chauhan R P S and Asthana A K 1981 Tolerance of lentil, barley and oats to boron in irrigation water. J. Agric. Sci. 97, 75–78.

Chauhan R P S and Powar S L 1978 Tolerance of wheat and pea to boron in irrigation water. Plant and Soil 50, 145–149.

Holloway R E and Alston A M 1992 The effect of salt and boron on growth of wheat. Aust. J. Agric. Res. 43, 987–1001.

Jenkin M J 1993 The genetics of boron tolerance in barley. PhD Thesis, University of Adelaide, South Australia.

Moody D B, Rathjen A J and Cartwright B 1993 Yield evaluation of a gene for boron tolerance using backcross-derived lines. *In* Genetic Aspects of Plant Mineral Nutrition. Eds P J Randall, E Delhaize, R A Richards and R Munns, pp 363–366. Kluwer Academic Publishers, The Netherlands.

Moody D B, Rathjen A J, Cartwright B, Paull J G and Lewis J 1988 Genetic diversity and geographical distribution of tolerance to high levels of soil boron. *In* Proceeding 7th International Wheat Genetics Symposium. Eds T E Miller and R M D Koebner. Cambridge Laboratory, IPSR, Cambridge.

Paull J G, Nable R O and Rathjen A J 1992a Physiological and genetic control of the tolerance of wheat to high concentrations of boron and implications for plant breeding. Plant and Soil 146, 251–260.

Paull J G, Nable R O, Lake A W H, Materne M A and Rathjen A J 1992b Response of annual medics (*Medicago* spp.) and field peas (*Pisum sativum*) to high concentrations of boron: genetic variation and the mechanism of tolerance. Aust. J. Agric. Res. 43, 203–213.

Rathjen A J, Cartwright B, Paull J G, Moody D B and Lewis J 1987 Breeding for tolerance of mineral toxicities in Australian cereals with special reference to boron. *In* Priority in Soil/Plant Relations Research for Plant Production. Eds P G E Searle and B G Davey, pp. 110–130. School of Crop Sciences, The University of Sydney, Australia.

Yau S K, Nachit M M, Ryan J and Hamblin J 1995 Phenotypic variation in boron-toxicity tolerance at seedling stage in durum wheat (*Triticum durum*). Euphytica 83, 185–191.

R.W. Bell and B. Rerkasem (eds.), Boron in Soils and Plants, 117–120.
© 1997 *Kluwer Academic Publishers.*

Variation in boron-toxicity tolerance in a durum wheat core collection

Sui-kwong Yau, Miloudi M. Nachit & John Ryan
International Center for Agricultural Research in Dry Areas (ICARDA), Aleppo, Syria

Key words: grain yield, seedling growth, shoot boron concentration, toxicity symptoms

Abstract

High boron (B) soils, which occur in arid and semi-arid areas, can cause B toxicity and reduce yield. In this study we assessed the variation of B-toxicity tolerance in an ICARDA core collection of 125 durum wheat (*Triticum durum*) entries. Seedlings were screened in a plastic house in a B-treated soil (100 mg B kg^{-1} soil). Seven entries with least B-toxicity symptoms and best growth were then grown to maturity in pots with 50 mg B kg^{-1} soil or without adding B. There were significant differences in B-toxicity symptoms and seedling growth between the 125 entries. With added B, grain yield of the seven entries was not significantly different from Halberd, the tolerant bread wheat check, although they had higher symptom scores and shoot B concentrations. These B-toxicity tolerant durum entries, though they absorbed more B from the soil than Halberd, seemed able to localize the excess B in leaf tips without serious effects on grain yield. Results suggest that symptom score and shoot B concentration should not be used to directly compare durum entries with bread wheat checks in future screening for B-toxicity tolerance at the seedling stage.

Introduction

West Asia and North Africa (WANA) is a dry region with a Mediterranean climate of winter rainfall, cool to cold winters, and hot, dry summers. Only about 30% of the land in WANA is arable, half of which receives <350 mm of annual rainfall. Soils of the region are generally calcareous.

Boron (B) toxicity occurs mainly in arid and semi-arid areas. It was reported in the cereal-growing drylands of South Australia, and in India, Iraq, Pakistan, Peru and USA (Nable, 1992). Globally the Mediterranean region has high topsoil B concentrations (Sillanpaa, 1982), and B toxicity is increasingly being recognized as a problem (Yau et al., 1994).

Soil treatment against excess B, for example by leaching, is usually not economically feasible, and is technically difficult in dry areas having limited water resources. Selecting or breeding varieties having tolerance to B toxicity is the most promising approach. Genetic tolerance to high soil B levels exists in bread wheat (Paull et al., 1988), and is currently being incorporated into otherwise adapted cultivars in South Australia (Moody et al., 1993).

Durum wheat is an important crop of WANA and Mediterranean Europe, and is grown mainly under dryland conditions in WANA. Screening of advance durum lines at ICARDA revealed only low/moderate tolerance to B toxicity, suggesting the need for a more extensive search for tolerant materials (Yau et al., 1994; 1995). In this study we investigated the variation of B-toxicity tolerance in a durum wheat collection.

Materials and Methods

The Core Collection of the Durum Wheat Project at ICARDA was screened in winter in a plastic house at 4–20° C under natural sunlight (Experiment I). The collection contained 125 entries, which included landraces, breeding lines and cultivars from West Asia, North Africa, southern Europe, USA, and Canada, and elite lines developed by the project. Three bread wheat lines: G 61450, which is tolerant to B toxicity (Paull et al., 1991), Halberd (moderately tolerant; Paull et al.,

1991), and Schomburgk (moderately sensitive; Ralph, 1992), were included as checks.

Five seeds of each entry were sown as hill plots (7 cm between plots) in a large box ($2 \times 1 \times 0.25$ m) of soil-mix to which boric acid was uniformly incorporated at the rate of 100 mg B kg^{-1} soil [hot water soluble B (HWSB) 23 mg kg^{-1} soil]. There were two replications. Plants were watered regularly as needed. Forty days after sowing (Schomburgk at 4 leaf stage), plants were rated visually for tolerance to excess B by scoring growth (scale 1 to 5, 1 = good, 5 = poor) and B-toxicity symptom expression on the first leaf (scale 0 to 5, where 0 = no symptom and 5 = symptoms covered whole leaf). The two scores were added up to form the combined symptom + growth score.

Seven entries with the best symptom + growth scores were then tested (Experiment II). Halberd was used as the B-toxicity tolerant bread wheat check. Based on results of our earlier studies, Oued Zenati, a durum landrace from Algeria, was used as the moderately tolerant check. Cham-1, a Syrian durum variety, was used as the sensitive check. Plants were grown to maturity in pots in a plastic house. The experiment was in a randomized complete block design with two replications and two B treatments: control (0.7 mg HWS B kg^{-1} soil) and 50 mg B kg^{-1} soil (12 mg HWS B kg^{-1}). For each pot (3 kg dried soil-mix), eight seeds of each entry were sown; seedlings were thinned to six after 20 days. The pots were watered once or twice per week, and nitrogen and phosphorus were added as needed. The plastic house temperature was set to simulate field conditions, initially at 5–15° C, and later increased to 10–20 and 20–30° C.

Four seedlings were harvested from each pot 8 weeks after sowing, and dried and weighed. Shoot B concentration was measured colorimetrically by the Azomethine-H method after dry-ashing. Foliar B-toxicity symptoms based on the percentage of lower leaves being affected (0 = no symptoms; 9 = >75% symptom coverage on 3rd leaf) were scored one week later. Days-to-heading, i.e., two of the main shoots in a pot had their heads half extruded, were recorded. At maturity, the whole shoot was harvested, dried, weighed, and threshed for grain yield.

Results

There were significant differences in B-toxicity symptoms ($p < 0.05$), seedling growth ($p < 0.01$) and the symptom + growth score ($p < 0.001$) between entries

in Experiment I; phenotypic CVs were 20.8, 9.7, and 11.2%, respectively. The result for symptom scores corroborated an earlier study (Yau et al., 1995). The tolerant bread wheat check, G 61450, had the lowest symptom+growth score. Two entries, Jord and Cand, had a lower score than Halberd, but 27 entries had poorer scores than Schomburgk, the moderately-sensitive bread wheat check. For the seven selected entries (Table 1), symptom and growth scores were better than Schomburgk, and comparable with those of Halberd.

At tillering stage, symptom scores and shoot B concentrations of the seven entries under the +B treatment in Experiment II were comparable with the moderately tolerant check, Oued Zenati, but were significantly higher than Halberd's values (Table 2). The sensitive check, Cham-1, had the highest symptom score and B concentration. Five entries, and Oued Zenati, headed out one month later than Halberd and Cham-1. With added B, the seven entries had grain yields which were not significantly different from Halberd and Oued Zenati. Cand and Sena, had the highest yields. The sensitive check had the lowest grain yield, and was the only entry with a significant ($p=0.10$) yield reduction from the control. Cand and Sena also had high straw yield. Compared with the control, Jord and Halberd had significantly lower straw yield with added B. Harvest indices of Jord and Chah were significantly higher when B was added, but that of Cham-1 was significantly lower.

Discussion

Based on yield performance, this study showed that we have found B-toxicity tolerant durum lines. Old cultivars such as Candeal de Grao Escuro no. 7746 and Senatore Cappelli had high grain and straw yield similar to Halberd in the +B treatment. Although Senatore Cappelli is still being grown in WANA's marginal areas, these two late cultivars with low yield potential would only be suitable as parental materials. Unlike them, Chahba-88, is an early line bred recently by the durum wheat project at ICARDA, and gives high yield in many dry areas of WANA.

In the +B treatment, some durum lines yielded as well as Halberd with no yield reduction from the control, although they had more symptom expression and higher shoot B concentrations than Halberd at tillering. A similar result was obtained by Yau et al. (1996) using a limited number of advanced durum and bread

Table 1. The names, abbreviations, and B-toxicity symptom and growth scores of the seven selected entries from the Durum Core Collection in comparison with bread wheat checks.

Names	Abbreviation	Symptom score[a]	Growth score[b]
Durum wheat entries:			
Kishk	Kish	2.8	1.5
North Dakota 86 line no. 10	Dako	2.8	1.5
Candeal de Grao Escuro no. 7746	Cand	2.8	1.3
Jordan Collection 86 no. 44	Jord	3.0	1.0
Senatore Cappelli	Sena	3.0	1.3
Chahba-88	Chah	2.5	1.8
Siliana	Sili	2.5	1.8
Bread Wheat Checks:			
G61450		2.0	1.8
Halberd		2.8	1.5
Schomburgk		3.3	2.0

[a] Rated on a 0–5 scale: 0 = no symptoms, 5 = symptoms covered whole leaf
[b] Rated on a 1–5 scale: 1 = good, 5 = poor

Table 2. Performance of the seven selected entries and checks in a high-B soil and in the control.

Entry[a]	Symptom at +B[b]	Shoot B at +B ($mg\ kg^{-1}$)	Days to heading	Grain yield +B ($g\ plant^{-1}$)	diff[c] ($g\ plant^{-1}$)	Straw yield +B ($g\ plant^{-1}$)	Diff ($g\ plant^{-1}$)
Kish	2.25	992	149	3.58	−0.63	12.2	0.4
Dako	2.75	1007	149	3.11	−0.54	13.0	1.4
Cand	2.75	1046	147	3.99	−0.60	12.3	−1.7
Jord	2.0	1010	159	2.60	−0.47	8.9	5.1*
Sena	2.0	1073	159	3.99	−0.64	14.4	−1.1
Chah	2.25	1139	105	3.69	0.19	5.7	2.4
Sili	2.5	1109	109	3.22	−0.58	8.9	−1.9
Checks:							
Halberd (BW)	1.0	573	112	3.89	0.69	7.2	3.8*
Oued Zenati	2.25	945	116	3.40	−0.55	10.9	2.0
Cham-1	3.25	1375	122	1.49	1.28+	7.55	0.9
LSD[d] (p<0.05)	0.67	279	10.4	1.75		7.78	
CV%	13	12	5	23	17	15	

[a] For full names, see Table 1
[b] Rated on a 0–9 scale: 0 = symptoms, 9 = >75% symptom coverage on 3rd leaf
[c] Control minus +B
[d] For within column comparisons
+ Significantly different between control and +B at $p = 0.10$ based on LSD
* Significantly different between control and +B at $P = 0.05$ based on LSD.

wheat lines. Thus, when screening for B-toxicity tolerance at seedling stage, we should no longer directly compare durum lines with bread wheat checks on symptom severity and shoot B concentration. This will avoid drawing the wrong conclusions that there are no B-toxicity tolerant durum lines.

Results of the study suggest that unlike the B-toxicity tolerant bread wheat, the tolerant durum entries used a different strategy to adapt to high-B soil. The higher shoot B concentrations and B content (data not presented) in the 7 durum entries indicated that they took up more B than Halberd. But they prob-

120

ably were able to localize B in the leaf tips, resulting in a higher B-toxicity symptom expression, but a similar grain yield as Halberd. This strategy appeared to work as well as restricting B-uptake as employed by Halberd.

References

Moody D B, Rathjen A J and Cartwright B 1993 Yield evaluation of a gene for boron tolerance using backcross-derived lines. *In* Genetic Aspects of Plant Mineral Nutrition. Eds P J Randall, E Delhaize, R A Richards and R Munns, pp 363–366. Kluwer Academic Publishers, The Netherlands.

Nable R O 1992 Mechanism of tolerance to boron toxicity in plants. *In* Boron Deficiency in Wheat. Ed C E Mann and B Rerkasem, pp 98–109. Wheat Special Report No. 11, CIMMYT, Mexico D F, Mexico.

Paull J G, Cartwright B and Rathjen A J 1988 Response of wheat and barley genotypes to toxic concentrations of soil boron. Euphytica 39, 137–144.

Paull J G, Rathjen A J and Cartwright B 1991 Major gene control of tolerance of bread wheat (*Triticum aestivaum* L.) to high concentrations of soil boron. Euphytica 55, 217–228.

Ralph W 1992 Boron toxicity in southern cereals. Rural Res. 130, 25–27.

Sillanpaa M 1982 Micronutrients and the Nutrient Status of Soils: A Global Study. FAO, Rome.

Yau S K, Hamblin J and Ryan J 1994 Phenotypic variation in boron toxicity tolerance in barley, durum and bread wheat. Rachis 13, 20–25.

Yau S K, Nachit M, Ryan J and Hamblin J 1995 Phenotypic variation in boron toxicity tolerance at seedling stage in durum wheat (*Triticum durum*). Euphytica 83, 185–191.

Yau S K, Saxena M C, Ryan J, Ortiz Ferrara G and Nachit M 1996 Comparing durum with bread wheat for boron-toxicity tolerance. *In* Book of Abstracts, 5th International Wheat Conference, June 10–14, 1996, pp 218–219. Ankara, Turkey.

R.W. Bell and B. Rerkasem (eds.), Boron in Soils and Plants, 121–123.
© 1997 *Kluwer Academic Publishers.*

Effects of foliar applications of boron on citrus fruit and on foliage and soil boron concentration

Antonio E. Boaretto[1], Carlos S. Tiritan[2] & Takashi Muraoka[1]
[1] *Soil Fertility and Plant Nutrition Laboratory, Centro de Energia Nuclear na Agricultura, Universidade de São Paulo, Piracicaba, 13400-970, Brazil*
[2] *Universidade Oeste Paulista, Presidente Prudente, Brazil*

Key words: amino acid, boron, foliar application, citrus

Abstract

An experiment was conducted in a 5 year old citrus orchard, for four years, to study the effect of boron (B) containing spray mixtures (nitrogen, phosphorus, potassium + amino acids) on fruit yield and quality and on the foliage B concentration. The foliar fertilization increased the leaf B concentraction, initially low, to within the adequate range. However, there was no corresponding effect on the fruit yield and juice chemical characteristics. The addition of amino acids, urea and potassium chloride to the foliar fertilizer did not affect the absorption of B by the orange (*Citrus sinensis*) leaves. A second experiment was carried out in orange orchards of different ages, ranging from 3 to 20 years old, in order to determine the effects of repeated foliar applications of B containing spray on the soil B concentration. The soil B concentration increased with increasing orchard age, indicating that part of the foliar sprayed solution fell on the soil and it was higher in soil samples taken under the citrus canopies than between the rows.

Introduction

Among the micronutrients, boron (B) is frequently deficient in Brazilian citrus orchard soils, together with Zn. To prevent it, application of foliar fertilizers containing micronutrients has been a routine practice in the orange (*Citrus sinensis*) orchards.

Little information exists on the foliar absorption of B, especially when applied with organic substances as chelating agents. The addition of nitrogen (N) and potassium (K) to foliar sprays is known to increase the absorption of micronutrients by the leaves. Amino acids (a.a.) and vitamins, usually food industry residues, have also been added for this purpose, but results have been controversial (Mello et al., 1983; Ashmead, 1986; El-Fouly et al., 1988; Miller and Warnick, 1986; Castro, 1991).

It is common to spray citrus orchard with 2000 L ha^{-1} of solution containing micronutrients; however, a considerable proportion drops from foliage to the soil surface.

The objectives of this study were: a) to verify the effect of foliar applications of a product containing N, K and some a.a. on the uptake of B, on fruit yeld and juice chemical characteristics; b) to determine the effect of foliar application of B containing fertilizers on soil B concentration.

Materials and methods

Two experiments were conducted in Mogi Mirim, SP-Brazil (Latitude 22°42′ south and longitude 42°25′ west; altitude 680 m; humid tropical climate with dry winter (June–August) and rainy summer (December–February); annual average relative humidity, 74%; summer and winter average temperature, 24° C and 17.2° C) in a dark red podzolic, clay texture soil.

The first experiment, in a randomized block design with 6 replicates, was carried out in a five year old orange orchard, grafted on rangpur lemon (*Citrus limon*), with the following treatments: 1- control; 2-

liquid fertilizer (LF) (4 L ha^{-1}); 3- LF (8 L ha^{-1}); 4-LF (12 L ha^1); 5- LF (16 L ha^{-1}); 6- LF (20 L ha^{-1}); 7- LF (24 L ha^{-1}); and 8-conventional treatment.

The LF is an organo mineral fluid foliar fertilizer (Ajifol-L^{TM})[1], density 1.5 g mL^{-1}, total N=10% (w/w), water soluble K$_2$O=2%; Zn=5%; Mn=2%; B=1%; total a.a.= 8%; total organic matter=45%; Na=2%; S=23%; Cl=2%; Ca= 630 mg kg^{-1}; Mg=600 mg kg^{-1}; Fe= 500 mg kg^{-1}; Cu=2 mg kg^{-1}; and Ni=2 mg kg^{-1}. Boron was included, as boric acid. Part of N is in the urea form and the rest is in the a.a. form, derived from the sodium glutamate manufacture residues. The analysis of this product revealed the following a.a.: glutamic acid, asparagine, alanine, valine, leucine, isoleucine, glycine, threonine, arginine, lysine, serine and phenylalanine.

The conventional treatment is a micronutrient formulation (kg ha^{-1}) recommended for citrus plantation (GPCAC (1994): 1.32 (Zn), 1.0 (Mo): 0.272 (B) and 4.5 (N). It is applied twice a year, first at flowering and the other at vegetative flushing. The fertilizer treatments were diluted in water and sprayed at the rate of 2000 L ha^{-1}, four times: 30.10.1993; 21.02.1994; 14.11.1994 and 28.2.1995. Leaf samples (approximately 6 months old third or fourth leaves from fruiting shoots, 4 per plant, at about 1.5 m above ground level at points facing north, east, south and west on the tree's periphery, according to GPACC (1994), were taken twice: two months after the second and one month after the last application, for the nutrient analysis.

The fruits were harvested in September 1994 and in August 1995, for: yield, size (diameter and length), determination of bagasse % and juice content, and chemical analysis (brix, acidity and brix/acidity ratio).

In the second experiment, 3, 4, 6, 7, 8, 11, 13, 14 and 20 years old citrus orchards were chosen. These plants had received foliar spraying of products containing micronutrients twice a year until the 12th year and once a year thereafter until 18th , when foliar fertilizer was no longer applied. Soil samples were taken, at 0 to 10 cm depth, from: a) 10 cm from the plant trunk; b) below the canopy edge and c) between the rows (8 m). The soil B concentration was determined after extraction with boiling 0.1% CaCl$_2$.2H$_2$O (2:1), colorimetrically (Cruz and Ferreira, 1984). Soil samples were also taken from the nearby area without orange plants.

---

[1] AJIFOL-L^{TM} = Ajinomoto do Brasil S.A.

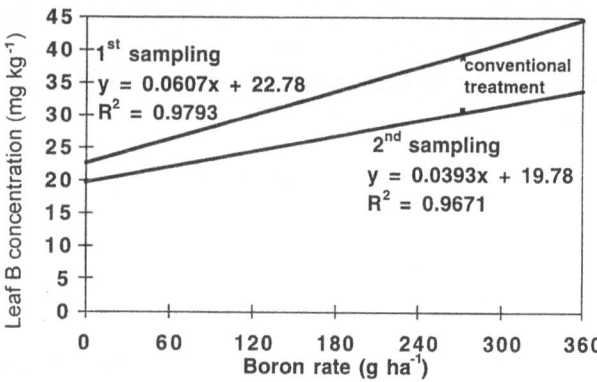

Figure 1. Leaf boron concentration (1st and 2nd sampling) as affected by increasing rate of foliar sprays and by conventional treatment spray.

Results and discussion

Analysis of leaf samples taken, in both periods, indicated no significant effect of foliar spraying on N, K, calcium, magnesium and sulfur concentrations, as were expected, considering that the amounts applied were relatively small. The macronutrient concentrations were within the level considered adequate (GPCAC, 1994). With regard to leaf B concentration, there was a linear increase with increasing rate of applied B (Figure 1). But, an adequate leaf B level (36–100 mg kg^{-1}, according to GPACC, 1994) was obtained only in the 2nd year of the experiment, after four applications of foliar fertilizers. The B concentrations in the leaves from the conventional treatment were approximately equal to those from the commercial fertilizer applied at the same B rate. This suggests that the urea, KCl and a.a. added to the fluid fertilizer probably did not increase the foliar B uptake in citrus plant.

The average fruit yields were 49.9 t ha^{-1} (1994 harvest) and 36.5 t ha^{-1} (1995 harvest), considered satisfactory for the age (5 and 6 year) of the orchard.

There were no difference due to the treatments on yield, the fruit size (diameter and length), juice content, and on quality parameters (acidity, brix and ratio).

There were no correlations between the leaf B concentration and respective fruit yields or any other analyzed parameter. Therefore, for prognosis of B status in citrus in Brazil, current standards for leaf analysis need to be reevaluated. This observation confirms the previous report for Zn and Mn (Embleton et al., 1990). In fact it is not uncommon for orange orchards with leaves exhibiting B visual deficiency symptoms to contain leaf B concentrations within the range con-

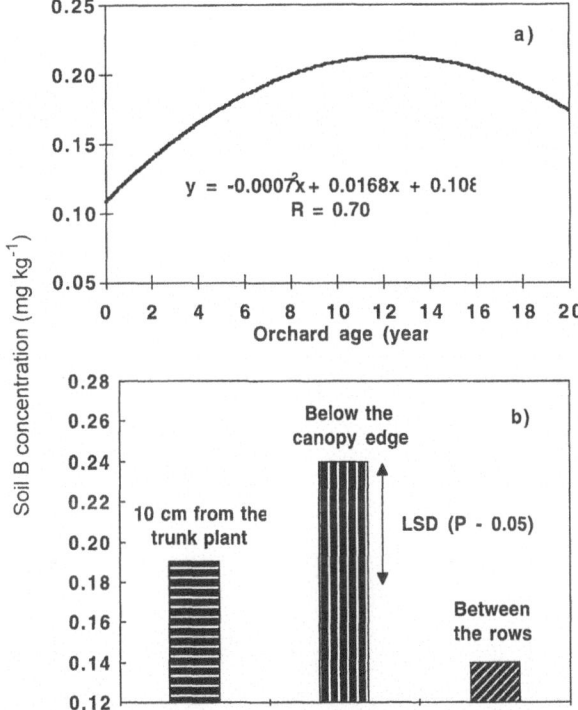

Figure 2. Soil B concentration as affected by orchard age (a) and soil sampling position (b). The data on b) were averaged for all orchards.

taken below the canopy edge, followed by that taken near the tree, presented, in most of orchards, higher B concentrations, compared to that taken from between the rows. The results of B concentrations obtained in soil samples taken at three distances from the orange trunks are presented in Figure 2b.

Conclusions

The foliar fertilization increased the leaf B concentration, initially low, to within the adequate level. However, there was no corresponding effect on the fruit yield and juice quality. The addition of a.a., urea and potassium chloride to the foliar B fertilizer did not affect the absorption of B by the orange leaves. The soil B concentration increased under plant canopies with increasing orchard age, indicating that part of the foliar sprayed solution fell on the soil. It may also have been caused by the dropping of sprayed leaves on the ground and the return of foliar B to the soil after litter decomposition.

References

Ashamed H D 1986 The absorption of amino acid chelates by plant cells. *In* Foliar Feedings of Plants with Amino Acid Chelates. Ed H D Ashamed, H H Ashamed and G W Miller, pp 219–235. Noyes Publications, Park Ridge.

Castro A M C 1991 Adubação foliar e tratamento de sementes de feijoeiro (*Phaseolus vulgaris* L.) com nutrientes, vitamina B_1 e metionina. (M.Sc. dissertation). Faculdade de Ciências Agronômicas. Botucatu, UNESP. 97 p.

Cruz M C P and Ferreira M F 1984 Seleção de métodos para avaliação do boro disponível em solos. Pesq. Agropec. Bras., Brasília 19, 1457–1464.

El-Fouly M M, Firgany A H and Fawzi A F A 1988 The effect of amino acid chelates on yield of different crops in Egypt. *In* International Citrus 6, Tel Aviv, Citriculture. pp 6–10. Balabean Publ. Rehovot, Israel.

Embleton T W, Matsumura M and Khan I A 1990 Citrus zinc and manganese nutrition revised. Horticultural Abstracts 60, 653.

GPACC – Grupo Paulista de Adubação e Calagem para Citros 1994 Recomendações de adubação e calagem para citros no Estado de São Paulo. Laranja, edição especial, Cordeirópolis, 3. ed. Rev. Atul., 27 p.

Malavolta E 1980 Elementos de Nutrição Mineral de Plantas. Ed Ceres, S. Paulo, Brazil. 251 p.

Mello W J, Fornasieri Filho D and Vitti G C 1983 Efeito de um ativador biológico à base de cisteina sobre a cultura do milho (*Zea mays* L). Revista de Agricultura. 58, 159–167.

Miller G W W and Warnick K 1986 Foliar applications of micronutrients in correcting deficiencies in oats. *In* Foliar Feeding of Plants with Amino Acid Chelates, Ed. H D Ashmead, H H Ashmead and G W Miller, pp 300–320. Noyes Publications, Park Ridge.

sidered adequate. One possible explanation would be the difficulty to remove the applied B which remained in the leaf cuticle or bounded in the pectin layer of cell walls, without fulfilling a metabolic function, but causing an overestimation of the leaf B status.

The suggestion that there is an accumulation of B in the soil with increasing orchard age was substantiated (Figure 2a). The soil B concentration was very low and varied from 0.1 to 0.3 mg kg^{-1} and it is noted that levels increased with increasing orchard age until 12 years and then decreased as spraying in older orchards were less intensive than in the younger ones. The amount of B applied per spraying (360 g ha^{-1} application in the highest doses) was too low to give substantial increase in soil B concentration. Besides, part of applied B was hopefully taken up by the trees and exported as harvested fruit and part could be leached beyond 10 cm depth where soil samples were collected. According to Malavolta (1980), this micronutrient is easily leached through the profile in the near neutral soil (pH 5.5–7.0). The average soil pH of the orchards was 5.9 and annual rainfall mean of the region is 1500 mm.

The position of soil sampling in relation to the orange trees show that foliar application probably affected the soil B concentration. The soil samples

R.W. Bell and B. Rerkasem (eds.), Boron in Soils and Plants, 125–129.
© 1997 *Kluwer Academic Publishers.*

Boron and fruit quality of apple

R. H. Dong[1,2], R. N. Noppakoonwong[1], X. M. Song[3] & B. Rerkasem[1]
[1] *Agricultural Systems Programme, Faculty of Agriculture, Chiang Mai University, 50200, Thailand*
[2] *Yunnan Agricultural University, Kunming 650201, P.R. China*
[3] *Dabaiqiao Horticultural Farm, Kunming 650211, P.R. China*

Key words: apple, fruit abscission, fruit boron concentration, fruit quality, seed number

Abstract

A field experiment was conducted in Yunnan, P. R. China to examine effects of boron (B) fertilizer on fruit yield and quality of apple (*Malus domestica* cv. Golden Delicious). Four rates of borax (0, 40, 80, 160 g tree^{-1}) were applied to the soil at full bloom, with 4 replicated trees in each treatment. A survey of commercial apple orchards was also conducted to establish the relationship between fruit B concentration and quality. In the field experiment, increasing rates of B application, increased fruit B concentration but decreased the rate of fruit drop by the end of the second wave of fruit drop in June. Fruit began to drop when fruit B concentrations were about 12 and 8 mg B kg^{-1} dry wt at the pea-size (1.5 cm fruit diameter) and June drop (3.5 cm fruit diameter), respectively. In both the field experiment and the survey, increasing fruit B concentration was correlated with increased fruit size and number of seeds, decreased firmness and titratable acidity, but unrelated to vitamin C concentration. It can be concluded that low B in the fruit is one of the factors involved in fruit drop. Optimal fruit quality was obtained when fruit B contained 14.7 mg B kg^{-1} dry wt corresponding with an application of 80 g borax tree^{-1}. It is suggested that B may affect fruit quality through its influence on seed production.

Introduction

Boron (B) is an essential micronutrient required for optimal yield and quality of apple fruit (Shear, 1980; Shorrocks and Nicholson, 1980). Symptoms of B deficiency in apples include internal and external cork formation in the fruit and the development of small, deformed fruits (Shorrocks and Nicholson, 1980). In Yunnan Province, China , most of these symptoms are commonly seen in orchards and on fruit for sale in the market.

Low B soils cover large areas in China especially in the east, northeast and southeast (Liu et al., 1981). Yunnan Soil and Fertilizer Station (1991) reported that low B levels were found in 90% of soils in the agricultural land of Yunnan province. Furthermore, our preliminary data suggested that B concentration in the youngest fully expanded leaves of Golden Delicious apple was 15 mg kg^{-1}, being lower than a marginal concentration of B in apple leaf (20 mg kg^{-1}) (Robin-

son, 1986). Thus, low B status in apple trees might limit apple fruit quality in Yunnan. However, there has been no systematic study of the effects of B deficiency on apple quality in Yunnan. This study was conducted in order to examine effects of B fertilizer on fruit quality and to establish a relationship between fruit B concentration and quality.

Materials and methods

A field experiment was carried out at Dabaiqiao Horticultural Farm, Kunming P.R. China. The soil was a red earth with the following properties: pH, 5.5; organic matter, 20.2 g kg^{-1}; total N, 600 mg kg^{-1}; available P, 8.5 mg kg^{-1}; exchangeable K, 85 mg kg^{-1}; exchangeable Ca, 211 mg kg^{-1}; and hot water soluble, B 0.28 mg kg^{-1} soil. The spacing of 15 year old Golden Delicious apple trees was 7 m × 7 m. Boron was applied as borax at the rates of 0, 40, 80 and 160 g tree^{-1} (equivalent to 0, 8.2, 16.3, 32.6 kg borax ha^{-1}). The B treatments were imposed on four replicate trees in a randomized complete block design. Each experimental unit was one tree, with a guard tree on all four sides. The borax was spread uniformly on the soil under the tree canopy at full bloom.

There were three waves of fruit drop: pea-size drop (1.5 cm fruit diameter), June drop (3.5 cm fruit diameter) and pre-harvest drop (Frank, 1986). The dropped fruit were collected and their number counted in the first two waves. Fruits were thinned to 25% of total fruit set at the end of second wave. In addition, fruit that set and dropped were sampled (30 fruit sample^{-1}) and analyzed for B. At maturity, ten fruit from each tree were sampled to determine B concentration, size (diameter, length, weight), physical firmness, fruit chemical composition (soluble solids, total sugar, titratable acidity, vitamin C) and number of seeds. Flesh firmness and soluble solids were measured by using a Magness-Taylor penetrometer and a hand refractometer, respectively. Total sugar in 50 mL extracted fruit juice was titrated with Feilin's reagent; titratable acidity (% malic acid equiv.) in 20 mL extracted fruit juice was titrated to pH 7 with 0.1N NaOH; and Vitamin C in 10 ml extracted fruit juice was titrated with 0.001N KIO$_3$. For B concentration, the fresh fruits were washed with distilled water, cross-cut into thin slices (0.2 cm) excluding the core and oven dried at 65° C. Dried fruit samples were dry ashed at 500° C for 8 h and analysed for B using azomethine-H method (Gong and Zhou, 1982).

A survey was conducted in three major apple production areas in the temperate highlands of Yunnan, P. R. China: Zhaotong (elevation: 1950 m); Kunming (elevation: 1900 m); and Lijiang (elevation: 2393 m). Six mature Golden Delicious fruit were collected from five trees from each of twenty orchards in each surveyed area. The fruit properties (weight, diameter, length, firmness, soluble solids), seed number and B concentration were determined.

Results and discussion

Effect of boron application on fruit abscission

Without B application, 81% of the apple fruit dropped from the experimental trees by the end of the second wave. With borax at 40 g tree^{-1}, the fruit drop decreased slightly to 78%, and fruit drop decreased further to 68% with increasing rate of B application to

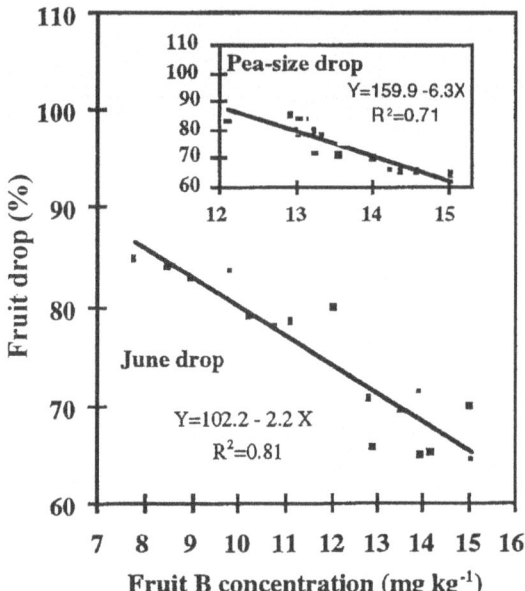

Figure 1. The relationship between boron (B) concentration in set apple fruit and percentage of fruit dropped at pea-size (F^{a} = 47.9∗∗) or June drop. (F^{a} = 74.3∗∗) [^{a} F-test F$_{0.05}$(1,15) = 4.54, F$_{0.01}$(1,15) = 8.68]

160 g tree^{-1} (Table 1). By the end of the second wave of fruit drop, the setting rate should ideally be about 25% for optimum fruit quality (Zhou et al., 1981). With higher setting rates, thinning is considered essential. In practical apple production, the thinning generally occurs after pea-size drop. However, in this study, fruit set with no borax added was only 19%. With increasing B rates, the set rate increased to 22% for 40 g, 30% for 80 g and 32% for 160 g borax tree^{-1}. Thinning was only imposed on the 80 and 160 g borax tree^{-1} treatments at the end of the second wave to maintain 25% fruit set. The final fruit numbers per tree were 315, 351, 392 and 395 for 0 to 160 g borax tree^{-1}, respectively (Table 1). If thinning was imposed at pea-size stage, numbers of fruit retained per tree may have been greater because of reduced competition among fruit after thinning.

Effect of boron application on fruit boron concentration

Fruit B concentration increased with B application, at growth stages corresponding to the pea-size drop, June drop and maturity (Tables 1 and 2). The dropped fruits had lower B concentration than set fruits at all levels of B application ($p < 0.05$) (Table 1). Fruit began to drop when fruit B concentrations were about 12 and 8 mg

Table 1. Effect of boron (B) application on B concentration in set and dropped fruit at pea-size and June drops and on the percentage of fruit drop and final fruit numbers in 'Golden Delicious' apples[a].

Borax Rate (g tree^{-1})	Sampling time				Fruit drop (%)	Final fruit number tree^{-1}
	Pea-size drop		June drop			
	Fruit B (mg kg^{-1} dry wt)					
	Set	Dropped	Set	Dropped		
0	12.8a	10.2a	8.8a	7.0a	81a	315a
40	13.1ab	10.3a	11.1b	8.1b	78b	351b
80	13.6b	11.7b	13.8c	8.8c	70c	392c
160	14.5c	12.4c	14.0c	9.0c	68c	395c
Means[b]	13.5e	12.1c	11.9e	8.2c		

[a] Means within a column with the same letter do not differ significantly at $p<0.05$ (LSD)
[b] For each sampling time, means within row with the same letter do not differ significantly at $p<0.05$ (LSD)

Table 2. Effect of boron (B) application on B concentration in fruit (dry weight basis), fruit size and fruit firmness at fruit maturity of 'Golden Delicious' apples[a].

Borax rate (g tree^{-1})	Fruit B (mg kg^{-1})	Fruit size			Firmness (kg cm^{-2})
		Diameter (cm)	Length (cm)	Weight (g)	
0	9.5a	7.39a	7.15a	159.4a	12.7c
40	13.4b	7.44a	7.19a	164.7a	11.9b
80	14.7c	7.31b	7.31b	176.0b	11.7b
160	15.8d	7.86b	7.59c	181.0b	10.3a

[a] Means within a column with the same letter do not differ significantly at $p<0.05$ (LSD)

kg^{-1} dry wt at the pea-size and June drop, respectively. A higher fruit B concentration for fruit development at pea-size drop compared with June drop might be attributed to a higher accumulated starch in fruit at the later stage.

Relationship between fruit boron concentration and its abscission or quality

The fruit drop rate and the B concentration in the set fruit were closely correlated at pea-size ($R^2 = 0.71$) and June drop ($R^2 = 0.81$) ($p <0.05$) (Figure 1). Although the waves of apple fruit abscission are a natural physiological phenomenon, low B in the fruit might be one of the factors associated with fruit drop.

Fruit diameter, length and weight generally increased with increasing fruit B concentration, but fruit firmness decreased (Table 2). Similar relationships were also found in the survey (Table 3). When fruit B concentration increased to 14.7 mg B kg^{-1} dry

wt (80 g borax tree^{-1}), fruit weight and fruit diameter were maximized, but fruit length was maximized when fruit B concentration was 15.8 mg B kg^{-1} dry wt (Table 2). By contrast, when fruit B concentration increased to 15.8 mg B kg^{-1} dry wt, the titratable acidity and soluble solids in the fruit decreased (Table 4). There was no effect of increased fruit B concentration on vitamin C in the apple fruit.

The non-linear model, $Y=e^{a+bx}$ (r negative) of the form described by Huguet and Borioli (1990) was fitted to the relationship between fruit B concentration (X) and soluble solids (Y). Data from the survey (Table 3) confirmed that this model described the relationship between fruit B concentration and soluble solids, $Y=e^{2.875-0.020x}$ ($R^2=0.52$). Thus, high B fertilizer may change internal chemical composition of apple fruit.

Wang (1992) established the criteria for ascertaining high fruit quality in apples in China. The highest rank of Golden Delicious apple quality was given to

Table 3. The relationship between fruit boron (B) concentration (fruit dry weight basis) and fruit diameter, length, weight, firmness, soluble solids or number of seeds (Survey data).

Quality parameter (Y)	Regression equation between fruit B $(mg\ kg^{-1})$ (X) and quality parameter (Y)	F[a]
Fruit diameter (cm)	$Y = 6.101 + 0.105X$ ($R^2 = 0.55$)	72.2**
Fruit length (cm)	$Y = 5.518 + 0.109X$ ($R^2 = 0.58$)	79.9**
Weight (g fruit^{-1})	$Y = 96.464 + 5.306X$ ($R^2 = 0.53$)	65.2**
Fruit firmness (kg cm^{-2})	$Y = 15.964 + 0.276X$ ($R^2 = 0.59$)	83.1**
Fruit soluble solids (%)	$Y = e^{2.875 - 0.020X}$ ($R^2 = 0.52$)	64.2**
Number of seed fruit^{-1}	$Y = -0.817 + 0.740X$ ($R^2 = 0.76$)	184.9**

[a] F-test $F_{0.05}$ (1,59) = 3.84, $F_{0.01}$ (1,59) = 6.64.

Table 4. The relationship between fruit boron (B) concentration (fruit dry weight basis) and number of seeds or fruit chemical composition at maturity of 'Golden Delicious' apples[a].

Fruit B $(mg\ kg^{-1})$	Number of seed fruit^{-1}	Chemical composition[b]				
		SS	TS	TA	TS:TA ratio	V.C.
9.5a	9.0a	13.7c	9.7a	0.32d	31	2.29a
13.4b	9.5a	13.3b	9.8a	0.30c	33	2.23a
14.7c	11.5b	12.9b	10.2b	0.27b	37	2.33a
15.8d	12.3c	12.6a	10.2b	0.24a	43	2.22a

[a] Means within a column with the same letter do not differ significantly at $p < 0.05$ (LSD)
[b] SS – Soluble solids (%), TS = Total sugar (%), TA = Titratable acidity (% malic acid aquiv.), V.C. = Vitamin C (mg per 100 g fresh fruit).

fruit with diameter >75 mm, firmness >11 kg cm^{-2}, soluble solids >12%, and a sugar:acidity ratio ranging from 29 to 40. In this study, fruit with 14.7 and 15.8 mg B kg^{-1} dry wt had diameters >75 mm, fruit with 9.5, 13.4, and 14.7 mg B kg^{-1} dry wt had a firmness > 11 kg cm^{-2}, and a sugar:acidity ratio within the range 29 to 40, and all treatments (fruit with 9.5–15.8 mg B kg^{-1}) produced fruit with >12% soluble solids. Therefore, in the present study fruit quality was optimised in fruit from the 80 g borax tree^{-1} treatment with a fruit B concentration of 14.7 mg kg^{-1} dry wt and the thinning of fruit to 25% of total fruit set.

Huguet and Borioli (1990) defined internal quality according to an apple quality index (QI = Total sugar + 10 × Malic acidity) and found that optimal quality corresponded with an index of 13.8 and 2.7 mg B kg^{-1} fruit fresh wt (FW). In this study, a fruit B concentration of 14.7 mg B kg^{-1} dry wt (equivalent to 2.8 mg B kg^{-1} FW) corresponded to QI = 12.9. Thus, when fruit B concentration = 2.7–2.8 mg B kg^{-1} FW, QI = 12.9–13.8 was obtained.

Fruit boron concentration and seed number

Increasing fruit B concentration, increased the number of seeds per fruit. Maximum seed number was reached when fruit B was 15.8 mg kg^{-1} dry wt (160 g borax tree^{-1}) ($p < 0.05$) (Table 3). Similarly, in the survey it was found that the fruit B concentration was positively correlated to the number of seeds ($R^2 = 0.76$; $p < 0.05$) (Table 4). Boron is important for seed production in many crops. Under B deficiency, the number of seeds may be reduced drastically or entirely suppressed (Gupta, 1993).

Apple belongs to the pome fruit, which forms fruit from the receptacle. The receptacle needs endogenous hormone induction as well as endogenous hormone stimulation (Yang, 1993). Hoad (1978) pointed out that the developing seed was the site of endogenous hormone production. Fruit size and shape are often related to seed number and distribution within the fruit. Cytokinin, auxin and gibberellin produced by the developing seed may be responsible for these effects (Chen, 1969; Frank, 1986; Yang, 1993).

The present results suggested that the effect of B on fruit quality might occur indirectly through effects on seed production. However, the mechanism(s) by which B plays such a role in fruit quality of apple is still unclear. More research is needed for a better understanding of the mode of action of B on fruit production and quality.

Acknowledgments

The first author is grateful for financial support from the Ford Foundation to conduct the present study as part of a M.Sc. degree at Chiang Mai University and the Australian International Development for financial support under its International Seminar Support Scheme. We are grateful for valuable comments and suggestions from Dr. R.W. Bell.

References

Chen J C 1969 The role of hormones in fruit set and development. Hortsci. 4, 108–111.

Frank G 1986 Apple. *In* Handbook of Fruit Set Development. Ed S P Monselise, pp 1–40. CRC Press, Boca Raton, Florida.

Gong Y A and Zhou H J 1982 Diagnosis Method of Fruit Nutrition. pp 102–106. Agricultural Press, Beijing (In Chinese).

Gupta U C 1993 Responses to boron on field and horticultural crop yields. *In* Boron and Its Role in Crop Production. Ed U C Gupta, pp 177–183. CRC Press, Boca Raton, Florida.

Hoad G V 1978 The role of seed-derived hormones in the control of flowering in apple. Acta Hort. 80, 93–103.

Huguet C and Borioli P 1990 Leaf and fruit boron contents in French apple orchards: Their relationship to apple tree nutrition and eating quality. *In* Behavior, Function and Significance of Boron in Agriculture. Report on an International Workshop St. John's College, England 23–25 July 1990. Ed V M Shorrocks, pp 31–35. Micronutrient Bureau, UK.

Liu Z, Zhu Q and Tang L 1981 Boron-deficient soils and their distribution in China. Soil Research Report No. 5. Institute of Soil Science, Academia Sinica, Nanjing China.

Robinson J B 1986. Fruit. *In* Plant Analysis. An Interpretation Manual . Eds D J Reuter and J B Robinson, pp 123–124, Inkata Press, Melbourne.

Shear C B 1980 Interaction of nutrition and environment on mineral composition of fruits. *In* Mineral Nutrition of Fruit Trees. Eds D Atkinson, J E Jackson, R O Sharples and W M Waller, pp 41–50. Butterworths, London.

Shorrocks V M and Nicholson D D 1980 The influence of boron deficiency on fruit quality. *In* Mineral Nutrition of Fruit Trees. Eds D Atkinson, J E Jackson, R O Sharples and W M Waller, pp 103–108. Butterworths, London.

Wang Y 1992 Market demand for fresh apple. *In* The Technique of Fruit Management. Ed Agricultural Bureau of Chinese Agricultural Ministry, pp 54–55. Food Stuff Press, China (In Chinese).

Yang H J 1993 Effect of seeds on fruit size and fruit quality in Starling apple. Shaixi Fruit 1, 7–4 (In Chinese).

Yunnan Soil and Fertilizer Station 1991 Soil available boron. *In* The Nutrient Distribution in Yunnan Soil. pp 26–28. Yunnan Press, Kunming (In Chinese).

Zhou H J Gong Y A and Yu D J 1981 Diagnosis of boron nutrition in apple at upland orchard. *In* Scientific Study Report, pp 77–90. Agricultural Press, Beijing (In Chinese).

R.W. Bell and B. Rerkasem (eds.), Boron in Soils and Plants, 131–133.
© 1997 *Kluwer Academic Publishers.*

Boron deficiency of avocado. 1. Effects on pollen viability and fruit set

Timothy E. Smith[1,2], Russell A. Stephenson[2], Colin J. Asher[1] & Suzan E. Hetherington[1]
[1] *Department of Agriculture, The University of Queensland, Brisbane, Q 4072, Australia*
[2] *Maroochy Research Station, Department of Primary Industries, Nambour, Q 4560, Australia*

Key words: boron, fruit retention, fruit set, Hass avocado, pollen viability

Abstract

The effect of boron (B) supply on pollen viability was studied using Hass avocado trees (*Persea americana*) grown in a glasshouse pot experiment with B-deficient soil. Seven rates of B were applied to the soil 12 months prior to anthesis. The number of pollen grains germinating in an artificial minus-B pollen growth medium increased approximately 16-fold when B equivalent to 0.8 or 1.6 g m^{-2} was applied to the soil.

In a field experiment with 5-year old Hass avocado trees of marginal B status (15–29 mg B kg^{-1} dry wt of summer flush leaves), spraying Solubor® onto fully developed panicles at the beginning of anthesis caused a 42% increase in fruit set, but this was not accompanied by a significant increase ($p=0.05$) in the number of fruit retained following the first fruit drop. Spraying panicles at earlier stages of development had no detectable effect on either fruit set or fruit retention.

Introduction

The importance of boron (B) in plant reproduction is well recognised (see review by Rerkasem, 1995). Only limited research has been conducted on the effects of B deficiency on the reproductive biology of avocado. Coetzer and Robbertse (1987) concluded that the B status of the avocado stigma was more important than the B status of the pollen since application of pollen from low-B trees to B-sufficient pistils resulted in higher pollen germination than when B-sufficient pollen was applied to low-B pistils. The aim of this experiment was to assess the effect of B supply to the tree on the ability of pollen to germinate without the assistance of B from stigmatic secretions, in a minus-B pollen tube growth medium under controlled laboratory conditions.

In Queensland, Australia, avocado flowering often coincides with a period of low rainfall and occasional cool weather, both of which have a detrimental effect on B supply to flowers and fruit set. Direct application of B through foliar sprays provides a possible solution to low B supply to flowers. Several studies have investigated the effect of foliar B sprays on pollen tube growth and fruit set in avocados. In South Africa, Robbertse et al. (1992) applied foliar B sprays to tree canopies at panicle emergence, anthesis, and at both panicle emergence and anthesis. Fruit set tended to increase when B sprays were applied at anthesis and both panicle emergence and anthesis, but the effect was not significant. In California, Jayanath and Lovatt (1995) reported that when B sprays were applied to the canopy when 50% of panicles were at the early expansion stage, an increased number of pollen tubes reached the ovule ($p=0.05$) and increased ovule viability. In the present paper, individual panicles were assessed and treated with B sprays at a specific panicle development stage in order to determine the optimum timing of B sprays to panicles for maximum fruit set and retention.

Materials and methods

Pollen germination experiment

Hass avocado trees (*Persea americana*) were grown in pots containing 19 kg of air dry sandy loam soil of low B status (<0.05 mg hot water extractable B kg^{-1}),

132

to which B was added as borax, at rates equivalent to 0, 0.025, 0.05, 0.1, 0.2, 0.4, 0.8, or 1.6 g B m^{-2} on a pot surface area basis. There were 4 replications per treatment. The soil was limed to pH 5.7 and basal nutrients were applied to all pots. The trees flowered after 12 months of growth in the pots.

Pollen sacs (dehisced anthers) were collected from terminal panicles at anthesis. The pollen grains were dispersed into a droplet of minus-B pollen tube growth medium (approx. 2 mg pollen mL^{-1}) on a glass microscope slide which was then inverted in accordance with the hanging drop method (Vasil, 1960) and placed in a growth cabinet at 25° C. After 24 h, the number of pollen grains with an emerging germ tube was assessed under a light microscope. The pollen tube growth medium (pH 5.5) consisted of: 200 mg MgSO$_4$.7H$_2$O L^{-1}; 15%w/v Polyethylene Glycol 4000; 100 mg KNO$_3$ L^{-1}; 700 mg Ca(NO$_3$)$_2$.4H$_2$O L^{-1}; 25 mM MES and 20% sucrose.

Fruit set and retention experiment

A row of 8 commercially grown 5 year old Hass trees (1993) at Montville, Queensland, Australia, displaying typical B deficiency symptoms were selected for the experiment. The trees had summer flush leaf B concentrations (15–29 mg kg^{-1} dry wt) within the marginal range (10–50 mg kg^{-1} dry wt) as defined by Reuter and Robinson (1986). The experimental design treated each tree as a separate randomised block. Each tree had 4 spray treatments and 8 replicate panicles/treatment. Treatments were: distilled water control; B applied to terminal immature panicles (approx. 4 cm in length); B applied to fully developed terminal panicles (>10 cm in length) prior to anthesis; and B applied to fully developed terminal panicles at the start of anthesis. Boron was applied as 1 g Solubor® L^{-1} (0.2 g B L^{-1}). A non-ionic surfactant, Shirwet 600,® was added to each spray at 0.1 mL L^{-1}. The sprays were administered using a fine mist hand sprayer and applied only to the target panicle. This was achieved by enclosing the panicle in a plastic bag during application to prevent drift and to collect excess spray. Fruit set per panicle was assessed 1 month after the end of flowering and fruit retention per panicle was determined the following month, after fruit drop.

Table 1. Effect of soil boron treatments on germination of pollen from Hass avocado trees in an artificial pollen tube growth medium.

B Treatment (g B m^{-2})	Germinated pollen (grains 100 counts^{-1})
0	3.2
0.025	5.0
0.05	9.5
0.1	52.6
0.2	24.3
0.4	44.1
0.8	54.3
1.6	47.3
LSD$_{p=0.05}$	18.7

Table 2. Effect of boron sprays and timing on fruit set and fruit retention of commercially grown avocado trees of marginal boron status.

Treatment	Timing of B spray	Fruit set (no. panicle^{-1})	Fruit retained (no. panicle^{-1})
Control		16.6	0.53
+B	Partially developed panicles (<4 cm)	15.6	0.53
+B	Fully developed panicles (>10 cm) before anthesis	20.2	0.51
+B	Fully developed panicles (>10 cm) at start of anthesis	23.5	0.77
LSD$_{p=0.05}$		4.58	NS

Results

Pollen germination

Pollen from trees receiving optimum B applications, when germinated in minus-B pollen tube growth medium, showed a 16 fold increase in germination compared to pollen from control trees (Table 1). The pollen exine was visibly thinner in the minus-B pollen grains compared to pollen from optimum B treatments. In addition, sloughing off of the pollen exine was observed in some pollen grains from control trees. This resulted in a naked intine which would be susceptible to either desiccation or to excessive imbibition of fluids from the stigmatic surface.

Fruit set and retention

Fruit set increased by 42% when fully developed panicles were sprayed with a solution of 1 g Solubor® L^{-1} at the start of anthesis, compared to the control (Table 2). The treatment differences were significant at $p=0.01$. Boron sprays to partially and fully developed panicles before the start of anthesis did not significantly affect numbers of fruit set compared to the controls. The data suggested a 45% increase in fruit retention for panicles treated with B at the start of anthesis (Table 2), but this difference was not significant ($p=0.05$) due to high variability.

Discussion

Boron plays an important role in the structure and germination of pollen grains. In the absence of B, pollen grains and tubes may burst, due to either excessive water imbibition or to mechanical weakness of the cell wall (Vasil, 1963). In our experiment there was less extensive development of the pollen exine in control trees and compared to control pollen, germination was reduced by 94%. Therefore pollen from low B anthers is less likely to achieve fertilisation of ovaries, due to low germination, compared with pollen formed under adequate B supply.

When B was applied to panicles of low-B trees in the present study, a significant increase in fruit set occurred only when B was applied at the start of anthesis. At this stage, the pollen would have been fully formed, so the treatment was unlikely to have affected pollen viability. The results obtained are consistent with the earlier conclusion of Coetzer and Robbertse (1987) that B status of the stigma was more important than the B status of the pollen for avocado fruit set.

Avocado trees produce a vastly greater number of flowers than fruit (Addicott, 1978). Developing fruit undergo direct competition for resources with the spring leaf flush resulting in the abscission of large numbers of fruit. Additional adjustment of fruit numbers can also occur during the development of the summer leaf flush. In our experiment, 3% of the initial fruit set was retained after the first fruit drop. There was a trend towards increased fruit retention on panicles treated with B at anthesis compared to controls. However, there was considerable variation between panicles in the number of fruit they were able to sustain, rendering the treatment effect non-significant. The role of B in fruit set and retention may be more important under adverse weather conditions, such as unseasonably low temperatures at anthesis which act to reduce the pollen tube growth rate (Sedgely, 1977) and in trees of lower B status than those used in the present study. Further research is needed to determine whether or not fruit retention on low B avocado trees can be increased by application of B to the panicles.

Acknowledgements

The authors thank the Department of Primary Industries for providing funding, facilities and assistance to conduct the research. The authors would also like to convey their gratitude to J. & M. Saranah for the use of their orchard and assistance.

References

Addicott F T 1978 Abscission strategies in the behaviour of tropical trees. *In* Tropical Trees as Living Systems. Eds. P B Tomlinson and M H Zimmermann, pp 381–398. Cambridge University Press, London.

Coetzer L A and Robbertse P J 1987 Pollination biology of *Persea americana* Fuerte. Ybk. Sth. Afric. Avocado Grow. Assoc. 10, 43–45.

Jayanath I and Lovatt C J 1995 Efficacy studies on prebloom canopy applications of boron and/or urea to Hass avocados in California. *In* Proc. of Third World Avocado Congress Tel Aviv, Israel. In Press.

Rerkasem B 1995 Boron and plant reproductive development. *In* Sterility in Wheat in Subtropical Asia: Extent, Causes and Solutions. Ed. H M Rawson and K D Subedi, pp. 32–35. ACIAR Proc. 72.

Reuter D J and Robinson J B 1986 Plant Analysis: an Interpretation Manual. Inkata Press, Melbourne, Australia.

Robbertse P J, Coetzer L A and Bessinger F 1992 Uptake by avocado leaves and influence on fruit production. *In* Proc. of Second World Avocado Congress. pp 173–178.

Sedgley M 1977 The effect of temperature on floral behaviour, pollen tube growth and fruit set in the avocado. J. Hortic. Sci. 52, 135–141.

Vasil I K 1960 Studies on pollen germination of certain Cucurbitaceae. Am. J. Bot. 47, 239–247.

Vasil I K 1963 Effect of boron on pollen tube germination and pollen tube growth. *In* Pollen Physiology and Fertilisation. Ed HF Linskens, pp 107–119. North Holland, Amsterdam.

R.W. Bell and B. Rerkasem (eds.), Boron in Soils and Plants, 135–137.
© 1997 Kluwer Academic Publishers.

Boron deficiency of avocado. 2. Effects on fruit size and ripening

Timothy E. Smith[1,2], Colin J. Asher[1], Russell A. Stephenson[2] & Suzan E. Hetherington[1]
[1] Department of Agriculture, The University of Queensland, Brisbane, Q 4072, Australia
[2] Maroochy Research Station, Department of Primary Industries, Nambour, Q 4560, Australia

Key words: boron, deficiency, fruit size, ethylene, ripening, Hass avocado

Abstract

Soil boron (B) applications were made to avocados (Persea americana cv. Hass) growing on a low B Krasnozem soil with a clay loam texture. The B treatments had positive effects on fruit size and postharvest characteristics. Addition of B produced a 13–16% increase in average fruit size. Fruit harvested from control trees ripened earlier and had higher ethylene production at 20° C compared to +B trees. Days to eating softness and peak ethylene production were reduced by approx. 4.4 days in fruit from control trees.

Introduction

Boron (B) deficiency of avocado is a problem in Australia (Smith et al., 1995) and worldwide (Whiley et al., 1996). Boron deficiency symptoms in avocados are expressed as splitting of leaf veins, reduced shoot length, swelling of shoot nodal regions, loss of apical dominance, reduced root growth (Haas, 1943). In addition to these symptoms, we have observed interveinal crinkling of leaves, prostrate growth of shoots (loss of geotropism), and deformity of fruit. Prior to the current research, there was little knowledge of the effects of B treatments on fruit yield and post-harvest fruit characteristics for avocados grown under Australian conditions.

Materials and methods

The trial was conducted from 1993–1996 on a commercial orchard at Maleny, Queensland, Australia. The Hass avocado (Persea americana cv. Hass) trees were growing on a Kraznozem soil of low B status and a clay loam texture. The trees displayed typical B deficiency symptoms of shothole in leaves, loss of apical dominance in shoots, prostrate growth habit and fruit deformities as described in Whiley et al. (1996). Their 1992 spring flush leaf B concentrations were 18–25 mg

kg^{-1} dry wt, indicating a marginal B status according to Reuter and Robinson (1986). There were 28 trees used in the experiment with single tree plots. The trees were 5 years old at the start of the trial (1993). The treatments consisted of 7 rates of B as borax, equivalent to 0, 0.03, 0.09, 0.23, 0.73, 2.1, and 5.9 g B m^{-2} applied annually to the soil surface directly under the tree canopy. There were 4 replications of each treatment. All fruit was harvested, weighed and counted to determine total fruit yield, total fruit number and average fruit weight. In this paper we compare the means of the control (-B) with the highest treatment of 5.9 g B $m^{-2} y^{-1}$ (+B) for the 1994, 1995 and 1996 harvests.

Postharvest assessment

Forty-five fruit per tree were harvested from selected B treatments for postharvest assessments in August, 1995. The fruit were dipped in 0.55 mL Prochloraz L^{-1} to reduce the spread of anthracnose and stem-end rots. In this experiment we selected 3 fruit at random from each set of 45 and placed these into individual fruit containers for ethylene determinations.

The individual fruit were held at 20° C until eating softness ($\sim$7N or less resistance to compressive force), which was determined by hand assessment. The ethylene production from each fruit was monitored every 6 hours using gas chromatography. The automatic sam-

Table 1. Effects of soil boron treatments on fruit yield, fruit number, and average fruit weight of Hass avocado trees for the 1994, 1995 and 1996 seasons.

Year	B Treatment	Fruit yield (kg tree^{-1})	Fruit number	Average fruit weight (g fruit^{-1})
1994	-B	110	506	219
	+B	97	396	246
	LSD$_{p=0.05}$	NS	NS	26.7
1995	-B	26	100	270
	+B	47	162	297
	LSD$_{p=0.05}$	NS	NS	NS
1996	-B	142	721	198
	+B	117	528	224
	LSD$_{p=0.05}$	NS	NS	NS
Mean	-B	68	303	226
94+95	+B	72	279	258
	LSD$_{p=0.05}$	NS	NS	28.2
Mean	-B	84	410	206
95+96	+B	82	345	239
	LSD$_{p=0.05}$	NS	NS	29.4

pling system used for monitoring fruit ethylene production was detailed in Smith et al. (in press). The gas chromatograph used an activated alumina column at 110° C, with injector and detector temperatures at 120 ° C. The flow rate of gas through the system was 300–400 mL min^{-1}. This paper reports peak ethylene production, the time from harvest to reach peak ethylene production, and the time taken to reach eating softness for the control (-B) and +B treatments.

Results

Fruit yield and average fruit size

Avocados tend to be biennial bearing trees with a year of high fruit yield usually followed by a year of low fruit yield. The trial trees expressed strong biennial bearing with up to 5 fold differences in fruit yield between years (Table 1). The results have been expressed both as annual averages and means between consecutive years to logically address the complication of biennial bearing. There were no significant effects of B application on fruit yield or fruit number in any individual year or combination of consecutive years. However, soil B applications significantly increased fruit size in 1994 and in means of consecutive years 1994+1995 and 1995+1996 (Table 1) by 13, 14, and 16%, respectively. There were trends of increasing fruit size in

Table 2. Effects of soil boron treatments on ethylene production and days to eating softness of Hass avocado fruit held at 20°C in individual containers.

	-B	+B	LSD$_{p=0.05}$
Peak ethylene production (μL kg^{-1} h^{-1})	179	120	49
Days to ethylene peak (d)	6.7	11.1	2.2
Days to eating soft (d)	9.4	13.7	2.0

1995 and 1996. However, tree-to-tree variation and non-synchronised biennial bearing rendered the effects non-significant.

Ethylene production and days to eating softness

Fruit harvested from control trees produced greater quantities of ethylene and had significantly fewer days to ethylene peak production and to eating softness (approx. 4.4 d) compared to fruit from the +B treatment (Table 2). Fruit in both B treatments reached eating softness approximately 2.6 days after the ethylene peak.

Discussion

In Australia, there is a market preference for large avocado fruit, and small fruit size is recognised as a common problem in Hass avocados (Moore and Wolstenholme, 1996). Hence the tendency for the application of B fertiliser to increase average fruit weight of Hass avocados (Table 1) is of commercial significance. Blumenfeld and Gazit (1970) showed that fruit size in avocado was related to longevity of the seed coat, cell division in the mesocarp slowing considerably on death of the seed coat. Avocado seed coats shrivel and turn brown as part of the normal fruit maturation process. However Harkness (1959) showed that early browning and death of the seed coat was associated with B deficiency. Thus in trees with leaf B concentrations <15 mg kg^{-1}, he observed 32–100% brown seeds, whereas in trees with 25–50 mg kg^{-1}, 0–31% of seeds were brown. In the present study, seed coats were not assessed prior to fruit maturity, but in a concurrent trial at Childers, Queensland, we observed a higher incidence of brown seeds prior to maturity in the lower B treatments (data not presented). The premature browning was possibly due to increased levels of phenolic compounds in the seed coat (Sitrit et al., 1987).

The respiration rate of avocado decreases after harvest until a climacteric rise in both ethylene production and respiration, which are associated with ripening in mature fruit (Adato and Gazit, 1977). In our experiment, the control fruit had fewer days to eating softness, had higher respiration rates (data not shown) and reached peak ethylene production 4.4 days before +B fruit. Ethylene production was increased by 49%. Similar responses in ethylene production of avocados were found by Eaks (1985) in response to Ca treatment of the fruit: increased Ca supply increasing days to ripening, and reduced respiration and ethylene production. The physiological link between B deficiency and increased ethylene production in avocados is yet to be established. However, interactions between B and Ca in avocado fruit may hold the key. This possibility requires further investigation.

Acknowledgments

The authors thank the Department of Primary Industries (DPI), Queensland, the Australian Avocado Growers Federation, and the Horticultural Research and Development Corporation for providing funding, and DPI Maroochy Research Station and DPI Horticulture Postharvest Group for providing facilities and assistance to conduct the research. The authors would also like to convey their gratitude to R. & C. Maywald for the use of their orchard and assistance during the field trials, and S. Dear, J. Mercer and G. Walters for technical assistance in various aspects of the research.

References

Adato I and Gazit S 1977 Role of ethylene in avocado fruit development and ripening. J. Exper. Bot. 28, 644–649.

Blumenfeld A and Gazit S 1970 The role of the seed coat in avocado fruit growth and maturity. Calif. Avocado Soc. Yrbk. 54, 100–104.

Eaks I L 1985 Effect of calcium on ripening, respiratory rate, ethylene production and quality of avocado fruit. J. Amer. Soc. Hort. Sc. 110, 145–148.

Haas A R C 1943 Boron content of avocado trees and soils. Calif. Avocado Soc. Yrbk. 28, 41–52.

Harkness R W 1959 Boron deficiency and alternate bearing in avocados. Proc. Florida State Hort. Soc. 72, 311–317.

Moore G C and Wolstenholme B N 1996 Effect of composted pinebark mulching on Persea americana Mill. cv. Hass fruit growth and yield in a cool subtropical environment. J. Southern African Soc. for Hort. Sc. 6, 23–26.

Reuter D J and Robinson J B 1986 Plant Analysis: an Interpretation Manual. Inkata Press, Melbourne, Australia.

Sitrit Y, Blumenfeld A and Riov J 1987 Ethylene biosynthesis in tissues of young and mature avocado fruits. Physiol. Plant. 69, 504–510.

Smith L G, Hofman P J, Jordan R A, and Lee C 1997 An inexpensive, low maintenance, multiple controlled atmosphere system for research on perishable products. Post-Harvest Biology and Technology. In Press.

Smith T E, Stephenson R A, Asher C J and Hetherington S E 1995 Boron nutrition of avocado – effects on fruit size and diagnosis of boron status. In Proc. Aust. Avocado Grow. Fed. Conf. 1–3, April 1995 pp 159–165. Department of Agriculture, Ed A McCarthy, South Perth.

Whiley A W, Smith T E, Wolstenholme B N and Saranah J B 1996 Boron nutrition of avocados. Sth African Avocado Grow. Yrbk. 1996. In Press.

R.W. Bell and B. Rerkasem (eds.), Boron in Soils and Plants, 139–143.
© 1997 *Kluwer Academic Publishers.*

Influence of boron spray on boron concentration, fruit set and calcium related disorders in apple (*Malus domestica*) cv. 'Elstar'/M26

Manuela Zude[1], Alvin Alexander[2] & Peter Lüdders[1]

[1] *Humboldt University Berlin, Institute of Horticulture, Section Fruit Science, Albrecht-Thaer Weg 3, 14195 Berlin, Germany*
[2] *Aglukon GmbH, Heerdter Landstr. 199, 40549 Düsseldorf, Germany*

Key words: apple, boron, calcium, fruit set, nitrogen, skin cracking

Abstract

To investigate effects of boron (B) sprays applied either during flowering or 4 weeks after petal fall, the present study was carried out in 2 sites of 5-year-old apple (*Malus domestica*) cv. 'Elstar' on M26 rootstock. In the first trial trees were treated with B or with a combination of B and nitrogen (N) at the balloon stage of the central bud. In the second trial two different B-fertilisers were sprayed 4 weeks after petal fall.

At balloon stage of the central bud, B spray increased B concentration in the apple inflorescences. B-treated trees exhibited less fruit drop compared to control trees, whereas fruit set was not improved.

Boron application 4 weeks after petal fall enhanced B concentration of the fruit. The bitter pit risk, estimated from K/Ca ratio, was not affected by B application. After a 5-month period in cold-storage, bitter pit on fruits was slightly reduced by B. Increasing B concentration in the fruit led to a significant decrease of fruit skin cracking on B-treated trees.

Introduction

Previous studies reported on possibilities to improve flower boron (B) fertilisation (Lewis, 1980; Marschner, 1995; Thompson and Batjier, 1950). The exact time of B application during the flowering period should be a compromise between the synchrony of flowering, leaf area and the time of pollination. Application at mouse ear stage of flower buds did not result in increased B concentration of flower tissue in apple and pear (v.d. Maas, pers. comm.). In the present study B was applied when the central bud in inflorescences reached balloon stage and its influence on fruit set and fruit drop was measured. Beneficial effects of combined foliar application of B and nitrogen (N) on B uptake, fruit set and fruit drop should be investigated. It is assumed that nitrogen increases B uptake via foliar application whereas in case of combined fertilisation via soil (Johnson and Samuelson, 1990) N causes a decrease in B uptake.

The influences of B application to improve fruit storability in connection to calcium (Ca) related disorders have been studied (Dixon et al., 1973; Granelli et al., 1989; Peryea and Drake, 1991). Additionally an inhibitory influence of B on Ca precipitation by oxalic acid was pointed out (Anversa, 1980). Therefore foliar applied B resulting in enhanced B concentrations on fruit surfaces might lead to better Ca availability in plant tissue. The effects of B fertilisation on Ca concentrations and related disorders as well as magnesium (Mg) and potassium (K) concentrations in fruit were investigated in the present study.

Material and methods

The study was carried out on 'Elstar' at the Humboldt University Berlin, Germany. The site was a five-year-old orchard with 250 trees on M26 rootstocks. Soil pH was 4.8. Air temperature, relative humidity, soil temperature and soil water potential via tensiometer at 20 and 50 cm depths were measured continuously.

Boron was applied when the central bud in inflorescences reached balloon stage. Control trees were sprayed with water. Boron treatments were carried out with two applications of 150 mg B L^{-1} via either Wuxal-Folibor (carrier=boron-ethanol-amine, liquid, AgrEvo GmbH, Germany) or Wuxal-Folibor combined with Azolon (methylene urea, liquid, AgrEvo GmbH, Germany). In another set of trees B was applied 4 weeks after petal fall, i.e. fruit approximately 2 cm large. Trees with B application were treated two times with 300 mg B L^{-1} by spraying either Solubor (carrier=$Na_2B_8O_{13}.4H_2O$, Borax Consolidated Ltd., United Kingdom) or Wuxal-Folibor. Each variant was repeated six times.

For chemical analyses flower (2 weeks after spraying) and middle section fruit samples (Kohl, 1967) were freeze dried. After micro wave digestion in H_2O_2, B analyses were made by inductively coupled plasma atomic emission spectrometry at 249.6 nm. Total Ca and Mg concentrations in fruits were measured by atomic absorption spectroscopy and K by emission spectroscopy. Analyses were carried out 6 times for each treatment with 2 replicates. Fruit set (after petal fall) and fruit drop (after fruit fall in June) were counted in 4 branches from each tree. All fruits were assessed for skin cracking and subsamples of 25 fruits were cold stored for further assessment of bitter pits. After 3 and 5 month, fruits with bitter pit symptoms were counted.

Results

Application during flowering

Boron fertilisation when the central bud in inflorescences reached balloon stage led to an increase in B concentration in flowers (p=0.05) (Figure 1). In trees with combined B and N treatment an even higher increase (p=0.01) in B concentration was evident. Fruit set in trees with B application and in control trees indicated no significant differences (Figure 2). By constrast, fruit drop was reduced in trees which were fertilised with B. Combined B and N spray increased fruit set and fruit drop so that the number of harvested fruit was unchanged compared to control trees.

Application 4 weeks after petal fall

Boron application 4 weeks after petal fall led to increased B concentrations in fruits compared to controls. Calcium, Mg and K concentrations in fruit did not change significantly due to enhanced B supply (Figure 3). Therefore the bitter pit risk, estimated from K/Ca ratio, did not change. After 3 month cold-storage no bitter pit symptoms were found while after a 5-month period about 8% of fruit exhibited bitter pits in control trees (Figure 4). Foliar applied B induced a marked decrease in fruit skin cracking (Figure 5). The different B fertilisers had no significant effect on the appearance of bitter pit or skin cracking in the present investigation where the same B concentrations in fertiliser solution were applied.

Discussion

Application during flowering

Particular weather conditions during flowering may promote translocation of B from leaves into flowers (Hanson and Breen, 1985). This is supported in the present study by the fact that even flowers of control trees showed high B concentrations. However, B concentration increased significantly in apple inflorescences after spraying at the balloon stage of central bud. Therefore this developmental state is suitable for B application in apple cv. 'Elstar' in contrast to earlier stages of flower development (Maas, pers. comm.).

Results of the present study indicate low fruit set in trees after B spraying as well as in control trees. This was possibly due to unfavourable weather conditions (heavy wind) during flowering which reduced insect activities necessary for pollination. However, increased B decreased fruit drop. This may be attributed to the role of B in auxin metabolism and transport. It has been reported (Bohnsack and Albert, 1977; Marschner, 1995) that B enhances auxin decomposition. Auxin concentration has an important influence on fruit drop in apple. Decreased auxin concentrations after B application might have led to reduced fruit drop in our study.

Combined application of B and N during balloon stage of central bud increased B concentration in flowers as well as fruit set and fruit drop. Similar results are reported by Yogaratnam and Greenham (1982). We assume, that fruit set during previous year is an important factor, since alternate bearing is a well known

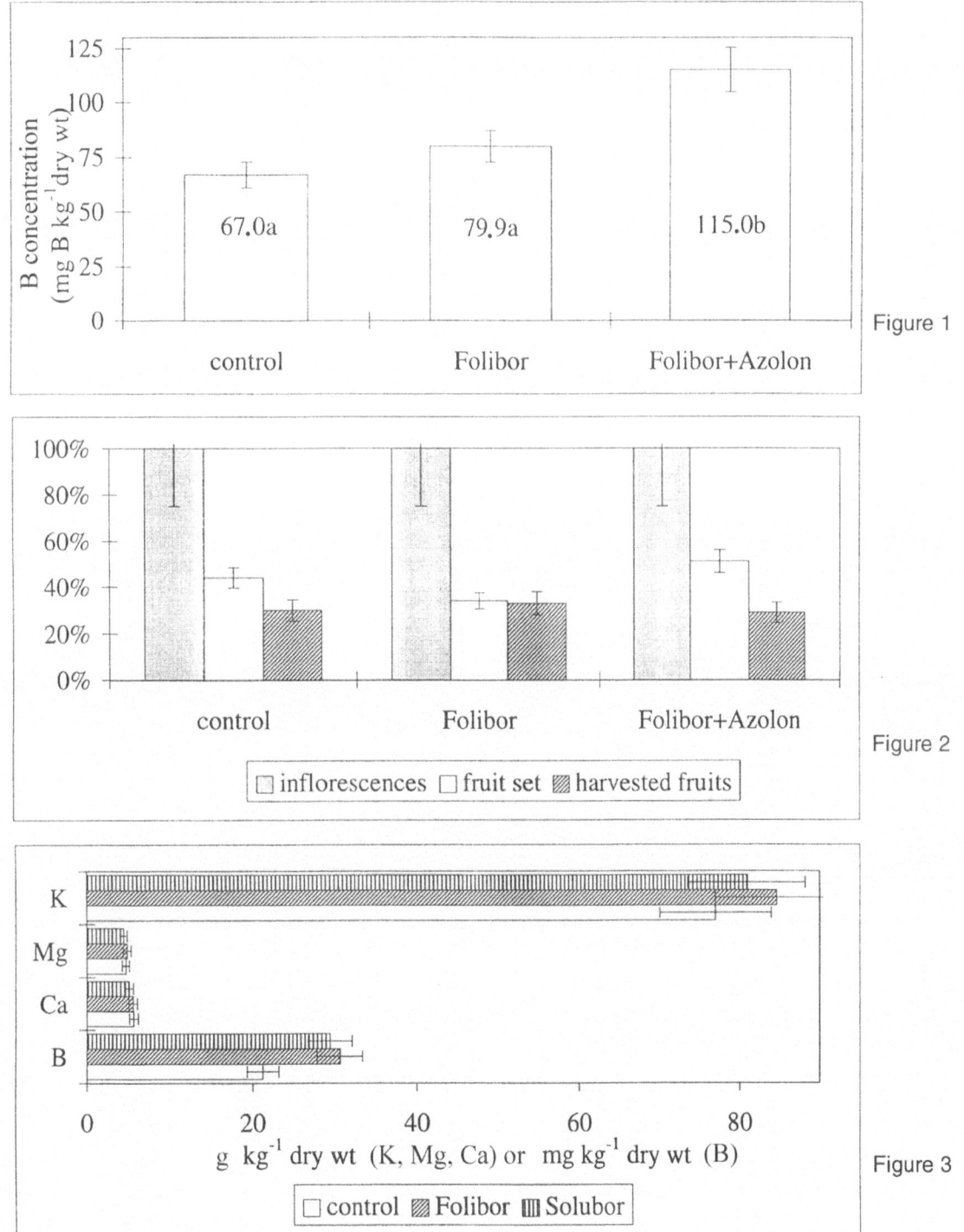

Figure 1

Figure 2

Figure 3

Figures 1-3. (1.) Influence of foliar boron (Folibor) and combined boron and nitrogen application at balloon stage of central bud in inflorescences on boron concentration in inflorescences of apple trees cv. 'Elstar'. Different letters means significant difference (*p*=0.01).
(2.) Influence of foliar boron and combined boron and nitrogen application during flowering on fruit set and amounts of harvested fruits expressed as percentage of inflorescences (inflorescences are made up of 4–5 single flowers) of apple trees cv. 'Elstar'.
(3.) Influence of boron sprays via different fertilisers 4 weeks after petal fall on mineral concentrations in fruit dry weight of apple trees cv. 'Elstar'.

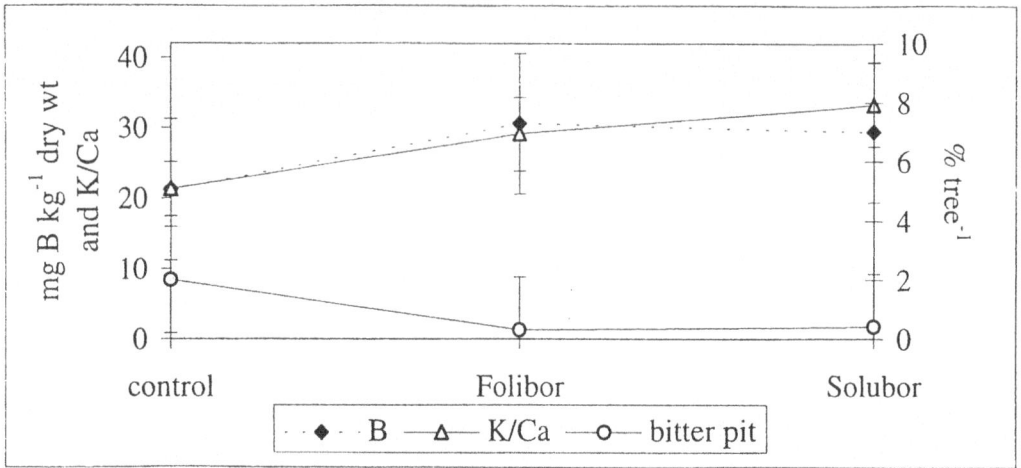

Figure 4. Influence of boron sprays via different fertilisers 4 weeks after petal fall on boron concentration in fruit dry weight, bitter pit risk (K/Ca<30 indicate reduced risk) as well as measured bitter pit amounts after 5 month cold-storage of apple cv. 'Elstar'.

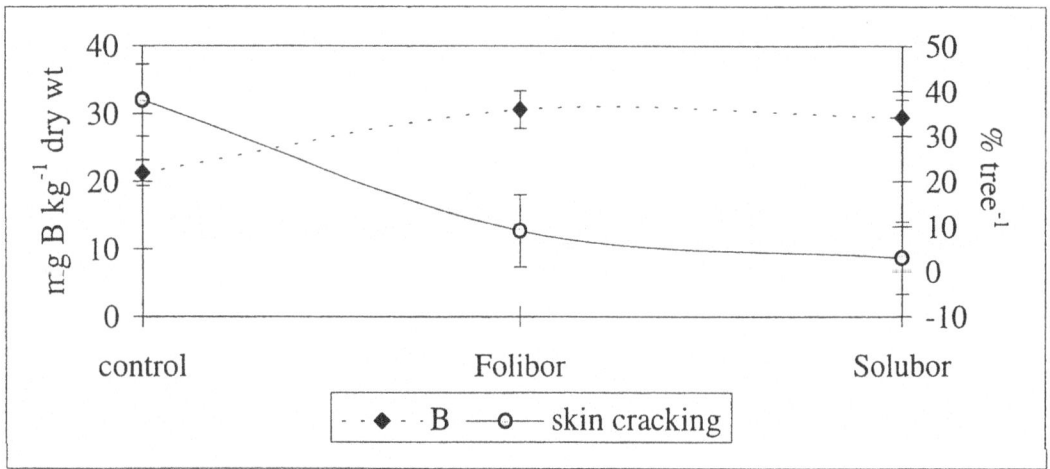

Figure 5. Influence of boron sprays via different fertilisers 4 weeks after petal fall on boron concentration in fruit dry weight and percentage of fruits exhibiting skin cracking expressed as percentage of harvested fruits per tree of apple trees cv. 'Elstar'.

problem in cultivar 'Elstar' and present at the experimental site. In the previous year, trees were unable to store adequate amounts of minerals due to very high fruit load. Under conditions of low mineral storage the additional N application might have had beneficial effects on metabolism resulting in enhanced B uptake. Nitrogen application is supposed to enhance vegetative growth and increase the production of phytohormones. Therefore, increased fruit drop might be attributed to enhanced auxin concentrations in the present investigation.

Application 4 weeks after petal fall

In 1996 the fruit skin cracking was a wide-spread disease in the Berlin-Brandenburg fruit growing area due to changing water potential of fruits during the season. In our study, B application reduced skin cracking significantly in agreement with the results from earlier investigations (Dixon et al., 1973; Yogaratnam and Johnson, 1982).

Expected alterations in Ca mobility resulting in enhanced Ca concentration in fruit due to B supply were not detected. However, improved fruit epidermis quality and slightly reduced bitter pit disorder is attributed to foliar applied B. It seems reasonable that B spraying during early fruit development maintains texture quality even in situations of physical stress caused by fluctuations in fruit water potential.

Acknowledgements

This work was supported by Aglukon GmbH, 40549 Düsseldorf, Germany. The authors thank Dr. G Ebert for critical review of this manuscript.

References

Anversa M 1980 Störungen der Entwicklung höhere Pflanzen durch unharmonische Bor und Calzium – Angebote und der Wert komplexer Symptome für die Diagnose. Dissertation FU-Berlin, Section Biology.

Bohnsack C W and Albert L S 1977 Early effects of boron deficiency on indole acetic oxidase levels of squash root tips. Plant Physiol. 59, 1047–1050.

Dixon B, Sagar G R and Shorrocks V M 1973 Effect of calcium and boron on the incidence of tree and storage pit in apple of the cultivar 'Egremont Russet'. J. Hort. Sci. 48, 403–411.

Granelli G, Ughini V and Minguzzi V 1989 Harvest and postharvest apple quality influenced by boron application. Acta Hort. 258, 405–412.

Hanson E J and Breen P J 1985 Effects of fall boron sprays and environmental factors on fruit set and boron accumulation in 'Italian' prune flower. J. Amer. Soc. Hort. Sci. 110, 389–392.

Johnson D S and Samuelson T J 1990 Short-term effects of changes in soil management and nitrogen fertilizer application on 'Bramley's Seedling' apple trees. I. Effects on tree growth, yield and leaf nutrient composition. II. Effects on mineral composition and storage quality of fruit.. J. Hort. Sci. 65, 489–502.

Lewis D H 1980 Are there inter-relations between the metabolic role of boron, synthesis phenolic phytoalexins and the germination of pollen?. New Phytol. 84, 261–270.

Marschner H 1995 Mineral Nutrition of Higher Plants. Academic Press, London. 889 p.

Peryea F J and Drake S R 1991 Influence of mid-summer boron sprays on boron content and quality indices of 'Delicious' apples. J. Plant Nutr. 14, 825–840.

Thompson A H and Batjer L P 1950 The effect of boron in the germination medium on pollen germination and pollen tube growth for several deciduous tree fruits. Proc. Amer. Soc. Hort. Sci. 55, 227–229.

Yogaratnam N and Greenham D W P 1982 The application of foliar sprays containing nitrogen, magnesium, zinc and boron to apple trees. I. Effects on fruit set and cropping. J. Hort. Sci. 57, 151–158.

Yogaratnam N and Johnson D S 1982 The application of foliar sprays containing nitrogen, magnesium, zinc and boron to apple trees. II. Effects of the mineral composition and quality of the fruit. J. Hort. Sci. 57, 159–164.

R.W. Bell and B. Rerkasem (eds.), Boron in Soils and Plants, 145–149.
© 1997 *Kluwer Academic Publishers.*

Boron toxicity in sunflower

F P C Blamey, C J Asher & D G Edwards
Department of Agriculture, The University of Queensland, Brisbane, Queensland 4072, Australia

Key words: boron, deficiency, *Helianthus annuus* L., sunflower, toxicity

Abstract

It has been hypothesised that there is a narrow range of supply between boron (B) deficiency and toxicity in plants. Two solution culture experiments were conducted to study the response of sunflower (*Helianthus annuus* L.) cv. Hysun 31 to potentially-deficient and potentially-toxic supplies of B. The concentration of B in the youngest mature leaf blade (YMB) of 46-day-old plants increased linearly up to a maximum of 1870 mg kg^{-1} dry wt in response to increasing B supply. There was a five-fold increase in plant dry mass as B in the YMB increased from 22 to 290 mg kg^{-1} dry wt. The critical concentration for B toxicity (90% of maximum yield) was determined as 1130 mg kg^{-1} in the YMB. Dry matter yield declined by only 25% with 1870 mg B kg^{-1} in the YMB. Results of this study indicated that there is not a narrow range of supply between B deficiency and toxicity in sunflower, and that this species is tolerant of high B concentration in leaf tissue.

Introduction

It is generally accepted that there is a narrow range of supply between boron (B) deficiency and toxicity in plants (Marschner, 1995). This probably stems from the early sand culture studies of Eaton (1944) who found that B toxicity symptoms developed in plants supplied with high B.

In some species, e.g. soybean (*Glycine max* (L.) Merr.), there may be a narrow range of supply between B deficiency and toxicity. Chapman et al. (1997), however, have shown that many crop species are tolerant of high B supply, and no generalisation should be made in this regard.

Sunflower (*Helianthus annuus* L.) is one of the most sensitive field crops to low B supply, and B deficiency in this crop has been reported from around the world (Blamey et al., 1997). Deficiency symptoms first become evident on the younger leaves which have a bronze colour, and become hardened, malformed and necrotic (Schofield et al., 1940; Blamey, 1976; Blamey et al., 1987). The upper nodes are shortened, and the upper stem becomes corky, which may result in stem break (Fernandez et al., 1985). The capitulum is often malformed with poor seed set (Blamey, 1976). The critical concentration for B deficiency (i.e. associated with 90% maximum seed yield) in the youngest mature leaf blade (YMB) at flowering (i.e. Growth Stage R5.1) (Schneiter and Miller, 1981) has been reported as 34 mg kg^{-1} dry wt (Blamey et al., 1979), with 31–140 mg kg^{-1} (Fernandez et al., 1985; Bergmann, 1992) considered adequate. Large increases in seed yield have been reported in response to the application of B fertiliser on soils low in B (Blamey et al., 1997).

In contrast with research on B deficiency, little work has been reported on sunflower grown with high B supply. Symptoms of B toxicity in sunflower plants grown in solution culture initially include a slight chlorosis along the margins of the lower leaves, followed by a dark-brown, mottled necrosis along the margin and between the major veins (Blamey et al., 1987). These symptoms progress from the lower to the upper leaves as the disorder increases in severity. The necrotic areas are sites of B accumulation, with up to 2177 mg B kg^{-1} in the worst-affected regions of the leaves (Scott, 1960). Concentrations associated with B toxicity have been reported as 100–700 (Aitken and McCallum, 1988), >500 (Cerda et al., 1981), 925 (Bergmann, 1992) or 1150 mg B kg^{-1} in whole leaves (Blamey et al., 1987). Aitken and McCallum (1988) reported

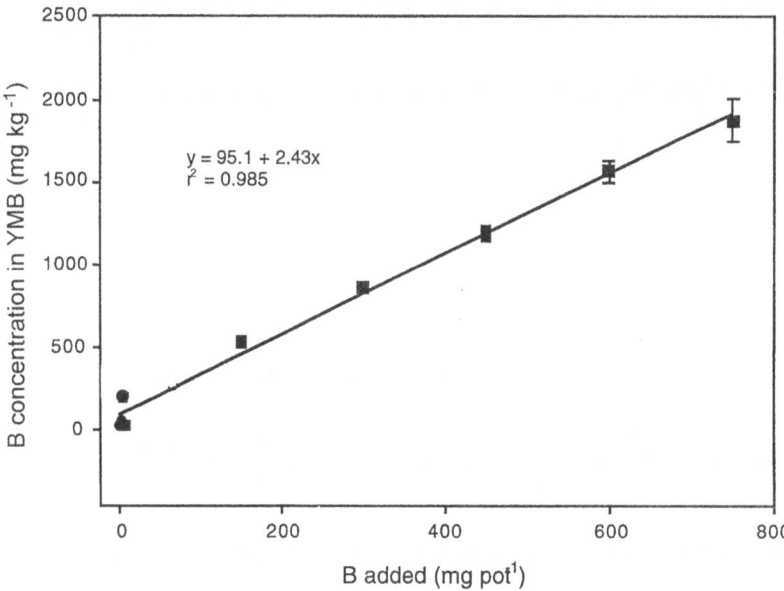

Figure 1. Effects of B supply on B concentration in the youngest mature leaf blade (YMB) of sunflower cv. Hysun 31 grown in Experiment 1 (●) and Experiment 2 (■). Vertical bars, where larger than the size of the symbol, indicate the standard error of the mean.

that sunflower growth decreased above a threshold of 176 μM B in the soil solution, plant dry mass decreasing by 40% with 2930 μM B in the soil solution (i.e. >16-fold increase in B supply).

Given the scarcity of information available on the growth response of sunflower to high B, a solution culture study was conducted to investigate the contention that there is a narrow range of supply between B deficiency and toxicity in this crop.

Materials and methods

Two concurrent solution culture experiments were conducted using the Programmed Nutrient Addition technique (Asher and Blamey, 1987). Experiment 1 evaluated the response of sunflower to low B supply, while Experiment 2 focused on the potentially-toxic range of B supply.

Sunflower seeds (cv. Hysun 31) were germinated in trays, and four young seedlings transplanted into 22 L pots containing nutrient solution when the radicle was ca. 20 mm long. The initial composition of the nutrient solution was as follows (μM): 562 Ca, 310 N, 83 K, 29 Mg, 29 S, 16 Fe, 11 P, 0.2 Mn, 0.1 Zn, 0.03 Cu and 0.01 Mo. Seedlings were thinned to two per pot 4 d after transplanting. In the control treatment of each experiment, sunflower seedlings at transplanting were estimated to have a dry mass of

0.07 g, and exponential growth for the first 20 d at a relative growth rate of 0.155 g g^{-1} d^{-1}. Thereafter, linear growth was expected until harvest, 46 d after transplanting when plants were at Growth Stage R2 (Schneiter and Miller, 1981). Nutrients were supplied twice weekly on the basis of the expected growth and nutrient concentrations in the whole plant of (%): 5.0 N, 0.3 P, 4.0 K, 2.0 Ca, 1.0 Mg and 0.2 S, and (mg kg^{-1}) 100 Fe, 50 Mn, 10 Zn, 5 Cu and 0.2 Mo. The optimum B concentration was assumed to be 40 mg kg^{-1} for the purposes of calculating the twice-weekly additions of B. Solutions were maintained at ca. pH 6.0.

The six treatments in each experiment were imposed with B supplied as proportions of the control treatment, viz. 0.05, 0.10, 0.25, 0.50, 1.00 and 2.00 in Experiment 1, and 1, 5, 10, 15, 20 and 25 in Experiment 2. A total of 0.23, 0.45, 1.13, 2.25, 4.50 and 9.00 mg B pot^{-1} was added in Experiment 1. Higher quantities of B than originally planned were added in Experiment 2 starting 17 d after transplanting because of the absence of B toxicity symptoms during early growth. A total of 4.5, 150, 300, 450, 600 and 750 mg B pot^{-1} was added for the six treatments in Experiment 2. There were four replications in each experiment, and treatments were arranged as a randomised, complete block design. The two experiments were analysed separately, and linear and non-linear regression techniques subse-

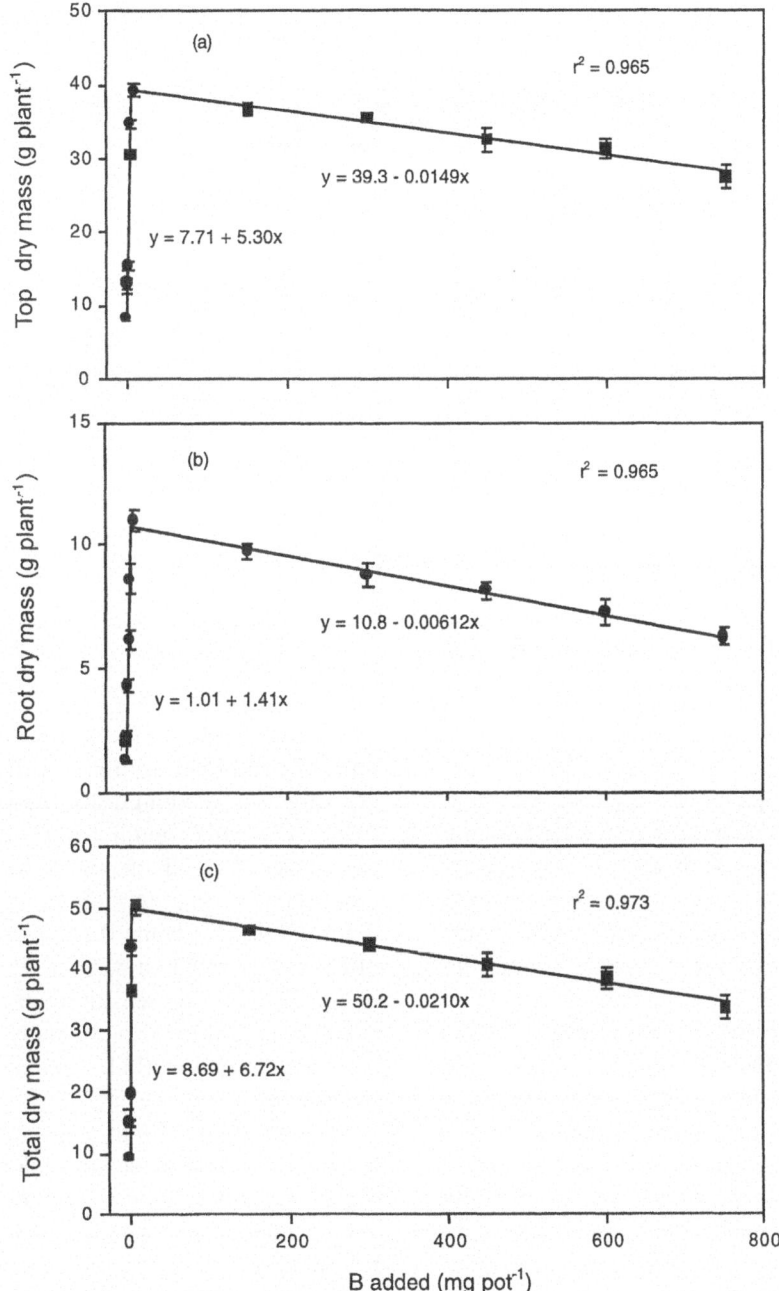

Figure 2. Effects of B supply on (a) top, (b) root and (c) total plant dry mass of sunflower cv. Hysun 31 grown in Experiment 1 (●) and Experiment 2 (■). Vertical bars, where larger than the size of the symbol, indicate the standard error of the mean. The join point of the two linear curves in each plot were determined using the method of Hudson (1966), and coefficients of determination (r^2) calculated from the correlation between the observed values and those estimated from the two regression equations.

quently used to interpret the combined data from the two experiments.

At harvest, the plants were divided into the YMB, tops and roots. These were placed in a dehydrator at 70° C for 7 d, and weighed. Thereafter, the YMB was

ground, dry ashed at 500° C, and the B concentration determined by inductively coupled plasma atomic emission spectroscopy (ICPAES).

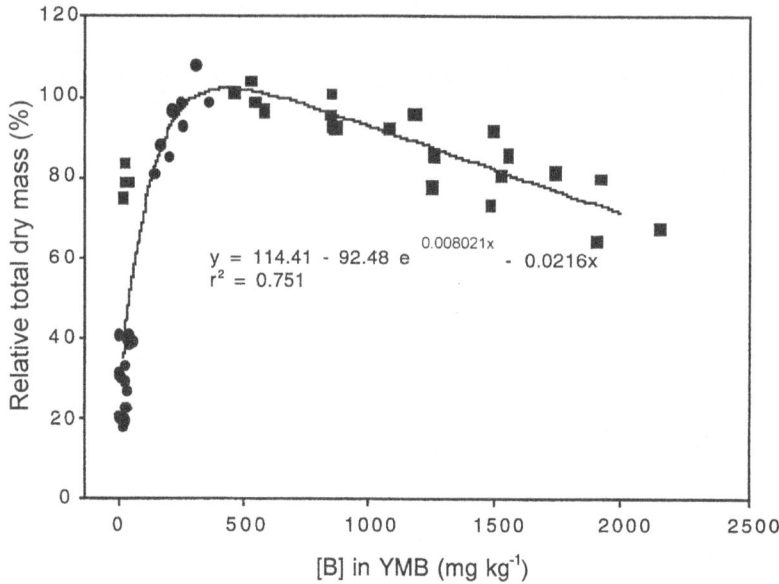

Figure 3. Relationship between relative total plant dry mass and B concentration in the youngest mature leaf blade (YMB) of sunflower cv. Hysun 31 grown in Experiment 1 (●) and Experiment 2 (■).

Results and discussion

Boron deficiency symptoms as seen in the field (Blamey, 1976) were first evident 14 d after thinning in plants grown in solutions with low B supply (Experiment 1). The youngest leaves were distorted, hard and leathery, and had a downward cupping. The plants had an overall darker appearance than those receiving adequate B.

In Experiment 2, the addition of high rates of B initially had little visible effect. Additional B supply from 17 d after transplanting resulted in B toxicity symptoms that were first evident 37 d after transplanting at the two highest rates of B. There was initially a mild, mottled chlorosis of tissue ca. 2 mm in from the margin of the lower leaf blades, followed by a necrosis of this region and the area between the major veins. However, there was only a slight decrease in plant growth.

The B concentration in the YMB increased significantly with increase in B supply in both experiments. Over the ranges studied, B in the YMB increased from 22 to 290 mg kg^{-1} in Experiment 1 and from 30 to 1870 mg kg^{-1} in Experiment 2. By combining the data from the two experiments, it was evident that there was a linear increase in B concentration in the YMB with increase in B supply (Figure 1), suggesting little internal control of B uptake by sunflower in response to high B concentration in the root environment.

In Experiment 1, there was a marked increase in top, root and total dry mass with increase in B supply (Fig. 2). Maximum dry mass was attained with twice the assumed optimum rate of B supply (i.e. 9 mg pot^{-1} instead of 4.5 mg pot^{-1}). Root growth appeared more sensitive than top growth to low B in solution, the root:top ratio being 0.15 at the three lowest rates of B supply compared with ca. 0.25 in the other treatments. Overall, the sensitivity of sunflower to low B supply was confirmed in this study, with total plant dry mass increasing five-fold with increase in B in the root environment.

The addition of high rates of B in Experiment 2 had relatively little effect on plant dry mass (Figure 2). There was an initial increase in dry mass as B supply was increased from 4.5 to 150 mg pot^{-1}. This was followed by a linear decline in dry matter production as B supply increased to 750 mg pot^{-1}. However, total dry matter yield decreased by only 25% with this 83-fold increase in B supply above that which produced maximum yield in Experiment 1 (i.e. 9 mg pot^{-1}). This was in spite of a B concentration of 1870 mg kg^{-1} in the YMB at the highest level of B supply. Roots and tops appeared to respond in the same manner to high B supply, there being no difference among treatments in root:top ratio which averaged 0.25.

The data from the two experiments were combined to determine the relationship between B concentration in the YMB and total plant dry mass (Figure 3).

There was a marked increase in total dry mass with an increase in B in the YMB above 22 mg kg^{-1}, followed by a slight, linear decline with > ca. 500 mg kg^{-1}. Using the fitted curve (Figure 3), it was possible to determine a critical B concentration (90% maximum yield) in the YMB for deficiency as 190 mg kg^{-1}, a value considerably higher than the 34 mg kg^{-1} reported in other studies (Blamey et al., 1979; Fernandez et al., 1985; Bergmann, 1992). This may have been due to the marked increase in dry matter production with increasing B supply in Experiment 1.

The critical B concentration in the YMB for toxicity was determined as 1130 mg kg^{-1}, again indicating a high level of tolerance to excess B in solution.

Conclusions

Results of this study indicate that sunflower is tolerant of high B supply in the root environment, and fail to support the existence of a narrow range of supply between B deficiency and toxicity in this crop. On the contrary, sunflower appears tolerant of an extended range in external B supply. Furthermore, the critical concentration in the YMB for toxicity (1130 mg B kg^{-1}) indicates tolerance of high B in plant tissue (based either on the critical concentration for deficiency (34 mg B kg^{-1}) established by Blamey et al. (1979) or on the value of 190 mg B kg^{-1} established in this study). This finding supports that of Chapman et al. (1997) with other crop species.

Although sunflower appears tolerant of high B supply, care is needed in the application of B fertilisers. Other crops in rotation with sunflower may, indeed, be sensitive to high B supply, and there is the potential for the leaching of applied B into the groundwater.

Acknowledgements

We are grateful for the financial support received from the Australian Grains Research and Development Corporation and for the seed supplied by Pacific Seeds, Toowoomba, Australia.

References

Aitken R L and McCallum L E 1988 Boron toxicity in soil solution. Aust. J. Soil Res. 26, 605–610.
Asher C J and Blamey F P C 1987 Experimental control of plant nutrient status using programmed nutrient addition. J. Plant Nutr. 10, 1371–1380.
Bergmann W 1992 Colour Atlas of Nutritional Disorders of Plants: Visual and Analytical Diagnosis. Gustav Fischer, Jena. 386 p.
Blamey F P C 1976 Boron nutrition of sunflowers (Helianthus annuus L.) on an Avalon medium sandy loam. Agrochemophysica 8, 5–10.
Blamey F P C, Edwards D G and Asher C J 1987 Nutritional Disorders of Sunflower. Dept of Agriculture, Univ. of Qld, St. Lucia. 72 p.
Blamey F P C, Mould D and Chapman J 1979 Critical boron concentrations in plant tissues of two sunflower cultivars. Agron. J. 71, 243–247.
Blamey F P C, Zollinger R K and Schneiter A A 1997 Sunflower production and culture. In Sunflower Science and Technology. Ed A A Schneiter. American Society of Agronomy, Madison, WI. (in press.)
Cerda A, Salinas R M and Romero M 1981 Tolerancia del girasol (Helianthus annuus L.) al boro. Anales Edaf. Agro. 40, 2245–2253.
Chapman V J, Edwards D G, Blamey F P C and Asher C J 1997 Challenging the dogma of a narrow supply range between deficiency and toxicity of boron. In Boron in Soils and Plants. Proceedings Eds R W Bell and B Rerkasem, pp 151–155. Kluwer Academic Publ., Dordrecht, the Netherlands.
Eaton F M 1944 Deficiency, toxicity, and accumulation of boron in plants. J. Agric. Res. 69, 237–271.
Fernandez P G, Baudin C G, Esquinas T M and Vara J M M 1985 Boron deficiency in sunflowers in Spain. In Proc. XI Int. Sunflower Conf., Mar Del Plata, Argentina, 10–13 March 1985, pp 1–8 (Printed by Borax Holdings Ltd, London).
Hudson D J 1966 Fitting segmented curves whose join points have to be estimated. J. Am. Stat. Assoc. 61, 1097–1129.
Marschner H 1995 Mineral Nutrition of Higher Plants. Academic Press, London. 889 p.
Schneiter A A and Miller J F 1981 Description of sunflower growth stages. Crop Sci. 21, 901–903.
Schofield C S, Wilcox L V and Blair G Y 1940 Boron absorption by sunflower seedlings. J. Agric. Res. 61, 41–56.
Scott E G 1960 Effect of supra-optimal boron levels on respiration and carbohydrate metabolism of Helianthus annuus. Plant Physiol. 35, 653–661.

R.W. Bell and B. Rerkasem (eds.), Boron in Soils and Plants, 151–155.
© 1997 *Kluwer Academic Publishers.*

Challenging the dogma of a narrow supply range between deficiency and toxicity of boron

V J Chapman, D G Edwards, F P C Blamey & C J Asher
Department of Agriculture, The University of Queensland, Brisbane, Queensland 4072, Australia

Key words: boron, deficiency, lentil, lupin, medic, pea, toxicity, wheat

Abstract

It is generally accepted that there is a narrow range of boron (B) supply between B deficiency and B toxicity in plants. In this study, five temperate crop species and cultivars were grown in flowing solution culture for 28 to 30 days at carefully maintained solution B concentrations from 0.15 to 640 μM. A no added boron (NBA) treatment was also included. Most of the species grew poorly in the NBA treatment, with relative whole plant dry matter yields ranging from 1.8% for barrel medic (*Medicago truncatula*) to 89% for wheat (*Triticum aestivum*). However, a solution B concentration between 0.15 and 2.5 μM was sufficient to produce maximum yield in all species and cultivars. Yield reduction due to B toxicity became apparent at 160 μM, and at 640 μM B relative yield ranged from 35% for barrel medic to 83% for field pea (*Pisum sativum*). Hence, there was a wide range of solution B concentrations over which maximum or near-maximum yield was produced. A lack of sensitivity in internal B concentration to solution B concentration between 0.15 and 40 μM was consistent with the operation of a B absorption regulation mechanism. There was no evidence to support the notion that there is a narrow range of B supply between B deficiency and B toxicity.

Introduction

Boron (B) deficiency has been identified in numerous crops throughout the world in countries including the USA, Canada, Nigeria and England (Gupta et al., 1985). Reports of B deficiency in commercial crops in Australia have largely been restricted to the slopes of the Great Dividing Range from north Queensland to Tasmania. In these areas, B deficiency has been reported in a number of crops (e.g. lucerne (*Medicago sativa*), beetroot (*Beta vulgaris*), celery (*Apium graveolens*) and apple (*Malus domestica*)) (Jackson and Chapman, 1975). Reports of B toxicity and tissue concentrations associated with B toxicity are limited. Boron toxicity may occur due to naturally high concentrations of B in the soil, the overuse of B fertiliser or the continued use of irrigation waters high in soluble salts, including B (Bradford, 1966). Naturally-occurring B toxicity in temperate cereals, particularly barley (*Hordeum vulgare*), has now been recognised in southern Australia (Cartwright et al., 1984; 1986).

Cartwright et al. (1984) reported that grain yield of barley grown in a red-brown earth at Gladstone, South Australia, was depressed by up to 17% by B toxicity.

Work by Eaton (1944), involving a range of plant species grown in sand culture with solution B concentrations ranging from 3 to 2313 μM, suggested that there is only a narrow range of external B concentration between B deficiency and B toxicity. Morrill et al. (1977) concluded that B toxicity and associated yield reduction in field grown peanut occurred at application rates of between 1.7 and 2.2 kg B ha^{-1}. On the basis of these studies and others, the opinion that there is a narrow external B concentration range between B deficiency and B toxicity and, subsequently, a narrow range of B concentration tolerable in the plant tissue, has been perpetuated throughout the literature (e.g. Marschner, 1986; Mengel and Kirkby, 1987).

The B nutrition of many plant species has been studied using soil culture (e.g. Haddad and Kaldor, 1982), sand culture (e.g. Eaton, 1944) and conventional solution culture with nutrient solution replacement after a number of days (e.g. Nable et al., 1990). However, these techniques do not allow careful maintenance of the external B concentration. Flowing solution culture (FSC), on the other hand, does allow this. The relationship between external B concentration, plant growth and plant tissue B concentration for selected temperate crops was examined using FSC. For each B treatment, B concentration was maintained within narrow limits throughout the 4 weeks' experimental period.

Materials and methods

Wheat and selected dicotyledonous species, grown in rotation with temperate cereals in southern Australia, were grown in FSC at maintained solution B concentrations ranging from 0.15 to 640 μM for 28 to 30 days. A no B added (NBA) treatment was also included. The species selected for discussion in this paper are wheat cv. Hartog, lentil (*Lens culinaris*) cv. Callisto, white lupin (*Lupinus albus*) cv. Ultra, barrel medic cv. Jemalong and field pea cv. Bonzer.

A low level of background B present in the basal nutrient solution was the basis for the NBA treatment. A further seven B treatments were established, viz. maintained solution B concentrations of 0.15, 0.625, 2.5, 10, 40, 160 and 640 μM. Solution B concentrations were monitored using an inductively coupled plasma atomic emission spectrophotometer (ICPAES) and adjusted, where necessary, on a daily basis. A concentration step was required to accurately detect solution concentrations $\leq 10 \ \mu M$ B (Chapman et al., 1996).

The basal nutrient solution was designed to simulate soil solution (Gillman and Bell, 1978). The concentrations of nutrients in the basal solution were as follows (μM): N 1100, P 2, K 500, Ca 1200, Mg 100, S 1000, Cl 10, Fe 2, Zn 0.5, Mn 0.15, Cu 0.1, Mo 0.02, Co 0.04, Ni 0.04. The composition of the basal nutrient solution (except N and P) was monitored twice weekly using ICPAES. The solution N concentration was determined using a nitrate-specific electrode. Daily solution analyses for P utilised the malachite green method (Motomizu et al., 1983). The solution within each FSC unit was maintained at pH 5.8 ± 0.1 and at 18° C.

Eight seedlings, with 15 to 20 mm radicles, were planted per pot; the exception was barrel medic, where poor germination limited the number per pot to four. The experiment consisted of four replicates of each treatment. Plants were thinned to four per pot between 5 and 7 days after transplanting. Plants from the thinning harvest (Harvest 1), representative of those that were continuing in the experiment, were dried for 48 h at 70° C and root and shoot dry weights were recorded. The final harvest (Harvest 2) was conducted 28–30 days after transplanting: lupin and field pea on Day 28, barrel medic and wheat on Day 29, and lentil on Day 30. Plants were divided into root, shoot and youngest mature blade (YMB). The roots were rinsed in deionised water three times to remove the B present in the cell wall free space. Plant material was dried for 48 h at 70° C and dry weight determined. Plant material for B analysis was dry ashed at 500° C for 6 h. The ash was dissolved in 10 mL 1 M HCl prior to B analysis using ICPAES.

Results and discussion

Symptom development

In the NBA treatment, B deficiency symptoms appeared on the younger leaves of the dicotyledonous species only. Symptom development on the younger leaves was expected, given that B is required for cell wall growth (Hu and Brown, 1994). Boron deficiency symptoms were not evident on any species at 0.15 μM B. Boron toxicity symptoms developed on the tips and margins of the older leaves of wheat, lentil, white lupin and field pea at 640 μM B. The location of these symptoms was in accordance with the work of Oertli and Kohl (1961) who noted that B toxicity symptoms developed at the termini of the transpiration stream, where B accumulates. Older leaves developed toxicity symptoms first.

Relative yield

All species showed reduced growth in the NBA treatment (Table 1) with relative yield ranging from 89% in wheat to 6.2% in lentil and 1.8% in barrel medic. The relatively better growth of wheat in the NBA treatment is in accordance with reports that monocotyledons generally have a lower external B requirement than dicotyledons (Marschner, 1986), and that wheat is tolerant to B deficiency (Weir and Cresswell, 1994). Plants grown at 0.15 μM B achieved relative yields > 80%, with the only exception being lentil (73%). Increase in solution B concentration to 160 μM generally maintained relative yields in the range 80 to 100%;

Table 1. Relative yield of whole plants of temperate crops grown in flowing solution culture for 28–30 days (Harvest 2) at solution B concentrations ranging from 0.15 μM to 640 μM; a no B added treatment (NBA) was also included.

Species	Cultivar	Solution B concentration (μM)							
		NBA	0.15	0.625	2.5	10	40	160	640
		Relative yield (%)							
Wheat	Hartog	89 (6.0)[a]	99 (6.7)	91 (3.1)	96 (6.2)	100 (4.3)	96 (3.1)	91 (6.9)	63 (3.3)
Lentil	Callisto	6.2 (0.51)	73 (7.6)	90 (7.6)	90 (10.3)	100 (13.8)	96 (11.2)	76 (13.3)	73 (2.4)
White lupin	Ultra	35 (2.5)	88 (3.7)	100 (4.0)	96 (7.1)	94 (0.67)	95 (5.5)	99 (5.7)	65 (2.7)
Barrel medic	Jemalong	1.8 (0.25)	86 (28.3)	94 (20.1)	100 (8.8)	86 (11.3)	80 (16.1)	69 (8.2)	35 (1.9)
Field pea	Bonzer	31 (2.2)	96 (3.7)	100 (3.3)	95 (2.4)	95 (1.2)	92 (6.4)	86 (4.5)	83 (6.9)

[a] Values in brackets are standard errors.

Table 2. Boron concentration in the YMB of temperate crops grown in flowing solution culture for 28–30 days (Harvest 2) at solution B concentrations ranging from 0.15 μM to 640 μM; a no B added treatment (NBA) was also included.

Species	Cultivar	Solution B concentration (μM)							
		NBA	0.15	0.625	2.5	10	40	160	640
		B concentration in YMB (mg kg^{-1})							
Wheat	Hartog	13.1 (0.90)[a]	12.1 (0.78)	13.4 (0.83)	14.2 (1.12)	13.8 (0.81)	24.6 (1.88)	46.2 (1.84)	234 (27.1)
Lentil	Callisto	[b]	29.6 (0.13)	38.3 (1.92)	38.3 (2.10)	40.9 (2.22)	61.1 (0.34)	137 (11.6)	544 (9.65)
White lupin		24.3 (0.35)	50.6 (0.31)	64.9 (4.14)	56.8 (4.42)	64.9 (4.33)	74.8 (5.45)	120 (7.51)	389 (7.35)
Barrel medic	Jemalong	[b]	35.9	50.5 (2.22)	52.9 (1.45)	62.2 (2.24)	60.4 (0.25)	109 (7.31)	391 (2.35)
Field pea	Bonzer	27.7 (3.05)	35.6 (1.82)	38.7 (1.72)	40.3 (6.15)	37.5 (1.35)	43.5 (3.40)	71.1 (1.23)	133 (4.80)

[a] Values in brackets are standard errors.
[b] No data available; insufficient plant material available for B analysis.

the exception was barrel medic, which had a relative yield of 69% at 160 μM B. Further increase in the external B concentration to 640 μM reduced growth in all species. Growth reduction at 640 μM was most pronounced in barrel medic (35% relative yield) and least pronounced in field pea (83% relative yield).

The production of near-maximal (> 90%) yield of all species, except barrel medic, over a wide range of solution B concentrations from 0.15 or 0.625 μM to 40 or 160 μM contrasts markedly with earlier reports (e.g. Eaton, 1944) that have led to the widely accepted belief that plants can tolerate only a narrow range of external B supply. The present finding is supported by the work of Blamey et al. (1997) with sunflower (*Helianthus annuus*).

The present results pertain to plants grown for 28–30 days. A higher minimum solution B concentration may be required when plants are grown to maturity because the B supply needed for seed and grain production is usually higher than that needed for vegetative growth only (Marschner, 1986; Rerkasem et al., 1988).

Boron concentration in the youngest mature blade

In the NBA treatment, the B concentration in the YMB ranged from 13.1 mg kg^{-1} in wheat to 27.7 mg kg^{-1} in field pea (Table 2). The B concentration in the YMB was little affected by increase in solution B concentration from 0.15 to 10 μM in wheat and from 0.15 to 40 μM B in field pea; this lack of response was also observed in white lupin and lentil over the narrower concentration range from 0.625 to 10 μM. Increase in the solution B concentration from 40 to 160 μM resulted in a 1.5- to 2.0-fold increase in B concentration in the YMB, while further increase to 640 μM increased B concentration in the YMB of most species 3- to 4-fold. The B concentration in the YMB of plants grown at 640 μM ranged from 133 mg kg^{-1} in field pea to 544 mg kg^{-1} in lentil.

Boron distribution in the plant

In the NBA treatment, from 49% (white lupin) to 62% (field pea) of the plant B content was present in the shoots (Table 3). The proportion of the plant B present

154

Table 3. Boron distribution index of temperate crops grown in flowing solution culture for 28–30 days (Harvest 2) at solution B concentrations ranging from 0.15 μM to 640 μM; a no B added treatment (NBA) was also included.

Species	Cultivar	Solution B concentration (μM)							
		NBA	0.15	0.625	2.5	10	40	160	640
		B distribution index (%)							
Wheat	Hartog	51 (3.6)[a]	55 (1.0)	57 (1.7)	61 (4.4)	64 (0.6)	72 (1.6)	82 (1.0)	92 (0.6)
Lentil	Callisto	[b]	72 (0.4)	76 (0.1)	77 (0.2)	77 (0.7)	84 (0.4)	93 (0.2)	96 (0.6)
White lupin	Ultra	49 (3.2)	54 (1.0)	71 (1.2)	79 (0.9)	80 (0.3)	82 (2.8)	89 (0.8)	94 (0.1)
Barrel medic	Jemalong	[b]	82	87	90 (0.5)	91 (0.2)	90	93	95
Field pea	Bonzer	62	82 (1.6)	80 (0.2)	82 (1.4)	84 (2.2)	85 (0.07)	90 (0.5)	94 (0.2)

[a] Values in brackets are standard errors.

[b] No data available; insufficient plant material available for B analysis.

Table 4. Mean rate of B absorption per unit of dry weight of roots of temperate crops grown in flowing solution culture for 28–30 days (Harvest 2) at solution B concentrations ranging from 0.15 μM to 640 μM; a no B added treatment (NBA) was also included.

Species	Cultivar	Solution B concentration (μM)							
		NBA	0.15	0.625	2.5	10	40	160	640
		Rate of B absorption (mg kg^{-1} day^{-1})[a]							
Wheat	Hartog	5.7 (0.06)[b]	6.1 (0.81)	6.3 (0.32)	7.1 (0.10)	7.5 (0.10)	11.0 (0.24)	19.9 (1.07)	87.2 (3.64)
Lentil	Callisto	[c]	7.0	9.1	9.6	11.3	19.0	53.5	195
White lupin	Jemalong[c]	12.3	16.1	24.2	22.4	28.7	32.6	64.4	205
Barrel medic	Jemalong[c]								
Field pea	Bonzer	6.7	18.8	22.7	23.7	22.8	29.1	56.2	137

[a] Mean rate of B absorption = $(\ln R_2 - \ln R_1)/(t_2 - t_1)*(M_2 - M_1)/(R_2 - R_1)$ where R_1 and R_2 are the initial (Harvest 1) and final (Harvest 2) root weights, $(t_2 - t_1)$ is the time between harvests and M_1 and M_2 are the initial and final plant B contents

[b] Values in brackets are standard errors.

[c] No data available; insufficient plant material available from Harvest 1 for B analysis or Harvest 1 not conducted.

in the shoots increased with increase in solution B concentration, reaching values ranging from 92 to 96% at 640 μM. At all solution B concentrations ≥ 0.15 μM, barrel medic and field pea had $\geq 80\%$ of the plant B present in the shoots.

Reduced transpiration, caused by the lower shoot production of most species grown in the NBA treatment, at least partially accounted for the greater root retention of B in that treatment. The present findings that root retention of B declined with increasing solution B concentration and that there was little root retention of B at the higher solution concentrations are consistent with those of Nable (1988) for several wheat and barley cultivars grown at solution B concentrations from 15 to 5000 μM. We conclude, as did Nable (1988), that root retention is not a mechanism for plant tolerance to high external B concentrations.

Rate of boron absorption

As expected, the mean rate of B absorption (Table 4) was minimal when plants were grown in the NBA treatment. In field pea, the B absorption rate increased as solution B concentration was increased from 0.15 to 0.625 μM, with further increase to 10 μM having little effect. In the remaining species, the rate of B absorption increased slowly with increase in solution B concentration from 0.15 to 10 μM. Increase in solution B concentration above 40 μM increased the rate of B absorption by all species.

The five species regulated the rate of B absorption through a considerably lowered uptake efficiency with increase in solution B concentration from 0.15 to 10 or 40 μM. This regulation of B absorption was suppressed or lost at solution B concentrations > 160 μM. This finding contrasts with that of Nable (1988), who reported that B uptake by three barley genotypes, grown in 5 L of nutrient solution that was replaced every 5 days, was linearly related to solution B concentration over the range 10 to 6400 μM. In that study, there was minimal focus on the 0.15 to 40 μM range. Over this range, there was little change in B concentration in the YMB and in the rate of B absorption in the present experiment. In fact, a close analysis of the data of Nable (1988) over the solution B concentration from 10 to 30 μM also suggests a lack of sensitivity of B absorption rate to solution B concentration.

Acknowledgments

The authors wish to thank Mr Gil Walters and Ms Janette Mercer for technical assistance. Advice regarding solution and plant tissue analysis was provided by Mr Graham Kerven and Mr John Oweczkin.

References

Blamey F P C, Asher C J and Edwards D G 1997 Boron toxicity in sunflower. *In* Boron in Soil and Plants. Proceedings. Eds R W Bell and B Rerkasem, p 145–149. Kluwer Academic Publishers, The Netherlands.

Bradford G R 1966 Boron. *In* Diagnostic Criteria for Plants and Soils. Ed H D Chapman, pp 33–61. University of California, Division of Agricultural Sciences, Riverside, California, USA.

Cartwright B, Zarcinas B A and Mayfield A H 1984 Toxic concentrations of boron in a red-brown earth at Gladstone, South Australia. Aust. J. Soil Res. 22, 261–272.

Cartwright B, Zarcinas B A and Spouncer L R 1986 Boron toxicity in South Australian barley crops. Aust. J. Agric. Res. 37, 351–359.

Chapman V J, Kerven G L, Edwards D G, Asher C J and Blamey F P C 1996 Determination of low boron concentrations in nutrient solution. Plant and Soil 180, 97–99.

Eaton F M 1944 Deficiency, toxicity, and accumulation of boron in plants. J. Agric. Res. 69, 237–277.

Gillman G P and Bell L C 1978 Soil solution studies on weathered soils from tropical North Queensland. Aust. J. Soil Res. 16, 67–77.

Gupta U C, Jame Y W, Campbell, C A, Leyshon A J and Nicholaichuk W 1985 Boron toxicity and deficiency: A review. Can. J. Soil Sci. 65, 381–409.

Haddad K S and Kaldor C J 1982 Effect of parent material, natural available soil boron, and applied boron and lime on the growth and chemical composition of lucerne on some acidic soils of the Central Tablelands of New South Wales. Aust. J. Exp. Agric. Anim. Husb. 22, 317–323.

Hu H and Brown P H 1994 Localization of boron in cell walls of squash and tobacco and its association with pectin. Plant Physiol. 105, 681–689.

Jackson J F and Chapman K S R 1975 The Role of Boron in Plants. *In* Trace Elements in Soil-Plant-Animal Systems, Eds D J D Nicholas and A R Egan, pp 213–226. Academic Press, New York, USA.

Marschner H 1986. Mineral Nutrition of Higher Plants. Academic Press, London, UK. 674 p.

Mengel K and Kirkby E A 1987. Principles of Plant Nutrition, 2nd ed. pp 559–572. International Potash Institute, Bern.

Morrill L G, Hill W E, Chrudimsky W W, Ashlock, L O, Tripp L D, Tucker, B B and Weatherly L 1977 Boron requirements of Spanish peanuts in Oklahoma: Effects on yield and quality and interaction with other nutrients. Report of the Agricultural Experiment Station, Oklahoma State University, MP-99.

Motomizu S, Wakimoto T and Toei K 1983 Spectrophotometric determination of phosphate in river waters with molybdate and malachite green. Analyst 108, 361–367.

Nable R O 1988 Resistance to boron toxicity amongst several barley and wheat cultivars: A preliminary examination of the resistance mechanism. Plant and Soil 112, 45–52.

Nable R O, Paull J G and Cartwright B 1990 Problems associated with the use of foliar analysis for diagnosing boron toxicity in barley. Plant and Soil 128, 225–232.

Oertli J J and Kohl H C 1961 Some contributions about the tolerance of various plant species to excessive supplies of boron. Soil Sci. 92, 243–247.

Rerkasem B, Netsangtip R, Bell R W, Loneragan J F and Hiranburana N 1988 Comparative species responses to boron on a Typic Tropaqualf in Northern Thailand. Plant and Soil 106, 15–21.

Weir R G and Cresswell G C 1994 Plant Nutrient Disorders: 4. Pastures and Field Crops. Inkata Press, Melbourne, Australia.

R.W. Bell and B. Rerkasem (eds.), Boron in Soils and Plants, 157–160.
© 1997 Kluwer Academic Publishers.

Water supply influences boron uptake by transplanted oilseed rape (*Brassica napus* cv. Eureka) grown in low boron soil

Longbin Huang[1,2], Ke Wang[3] & Richard W. Bell[1]

[1] *School of and Environmental Science, Murdoch University, Murdoch, WA 6150, Australia*
[2] *Division of Forestry and Forest Products, CSIRO, Wembley, WA 6014, Australia*
[3] *Institute of Remote Sensing, College of Natural Resources and Environmental Sciences, Zhejiang Agricultural University, Hangzhou, P.R. China*

Key words: boron (B), B uptake, leaf B concentrations, oilseed rape (*Brassica napus*), water stress

Abstract

Previous studies have shown that low soil water supply depresses boron (B) uptake and may induce B deficiency in crops at low soil B. The present study was conducted to assess the possibility that increases in root/shoot ratio induced by water stress might offset the negative effect of water stress on B absorption in low B soil. Oilseed rape (*Brassica napus* cv. Eureka) plants were subject to a factorial combination of two levels of B: +B (0.45 mg B kg^{-1} air dry soil) and -B (without added B); and three levels of water supply: adequate (Wet), subsoil dry (Wet-Dry) and surface soil dry (Dry). In -B soil, the water stress treatments significantly increased root dry weights, but decreased shoot dry weights, resulting in a decrease in shoot/root ratio. By contrast, water stress had no effect on shoot/root ratios of plants in +B soil. Decreasing water supply in the soil significantly decreased B absorption per unit root mass in -B soil but not in +B soil. The increase in root/shoot ratio was unable to significantly offset the decrease in B uptake induced by dry soil conditions and water stress still strongly depressed B concentrations in young leaves.

Introduction

Boron uptake in plants is determined by the availability of boron, transpiration rates in the leaves (Marschner, 1995; Brown and Hu, 1993) and the growth rate of the plants. By contrast, plants respond to low water supply in soil by decreasing leaf water potential, stomatal conductance and transpiration rates and by increasing root/shoot ratio (Schulze, 1986). In the field, crops may be exposed to both drought and low B supply in the soil. As a result, low water supply in soil may lead to decreases in B availability, leaf transpiration, and eventually the net uptake of B. In particular, when the pool size of available B in soil is small or inadequate for the normal growth of plants, limited water supply could enhance B deficiency through decreasing the mass-flow of B in soil solution to plants (Hobbs and Bertramson, 1949). However, the possibility that increased root/shoot ratio induced by drought may off-

set the effects of drought on B uptake in plants has not been previously considered.

Transplanting which is a common practice by the farmers in China decreases root/shoot ratio. The time for transplanting oilseed rape (*Brassica napus* L.) in southeast China often coincides with the dry season. As a result, the transplanted oilseed rape seedlings are particularly susceptible to B deficiency due to the root injury at transplanting and to water stress due to low soil water. However, the consequences of the water stress on root/shoot ratio, and of transplanting on root injury for B uptake and B responses are not understood.

Water stress treatments in pot-soil are commonly induced by withholding watering for a number of days to allow the water content in the soil to decline to a fixed percentage of its field capacity. To maintain the level of water stress in pot-soil, water is added onto the surface of pot soil, which would simulate a water deficit in the subsoil in the field. Plants in the field may also experience water deficit in the surface soil with high water content in the sub-soil. In the present experiment, these two methods were used to achieve water deficit in the sub-soil and top-soil, respectively, to determine how the water deficit regime interacted with low soil B to affect B uptake by transplanted oilseed rape seedlings.

Material and methods

Oilseed rape (cv. Eureka) plants were subject to a two-way factorial combination of B and water supply treatments in potted-soil. Soil B treatments were +B (soil treated with 0.45 mg B kg^{-1} air-dried soil) and -B (soil without added B). The soil used contained 0.19 mg hot $CaCl_2$-extractable B kg^{-1} air-dry soil. Water supply treatments in the soil were well watered (80% field capacity (FC) – refered to as Wet), Dry (the pot soils started with 67% FC and declined to 30% FC) and Wet-Dry (water contents in soil were 80% FC and allowed to naturally dry to 30% FC). To achieve the Dry treatment, a plastic tube was buried at the bottom of a pot and an appropriate amount of water was added into the tube and allowed to diffuse upwards by capillary rise to maintain 30% FC in the soil throughout the experiment. Water contents in the Dry soil declined from 67% to 30% FC by 16 day after transplanting and were maintained at this water supply level for the remaining 29 days by adding appropriate amounts of water into the base of the pots. For Wet-Dry treatment, soils had 80% FC water content at transplanting and naturally dried to 30% FC after 27 days and then an appropriate amount of water was applied evenly on the surface of the soil to maintain the water content of soils at 30% FC for the remaining 18 days. Water content for each pot was determined by weighing them daily.

A sandy soil was collected from east of Pinjarra in Western Australia, air-dried and sieved through a 4-mm stainless steel screen. Soil was collected from a field site (5–15 cm depth) under pasture which had previously shown seed yield responses to B fertilizer in subterranean clover (Salardini and Robson, personal communication). The soil was mixed with the following amounts of basal nutrients (mg kg^{-1} air-dried soil): K_2SO_4, 348; KH_2PO_4, 160; $CaCl_2.2H_2O$, 75; $MgSO_4.7H_2O$, 24; NH_4NO_3, 240; $Na_2MoO_4.2H_2O$, 0.73; $ZnSO_4.7H_2O$, 10; $CuSO_4.5H_2O$, 5; $MnSO_4.H_2O$, 6.77. Only analytical grade chemicals were used in this experiment. The basal nutrient solutions were purified with Boron-

Table 1. Effect of soil water and boron (B) treatment on total B uptake per unit root weight (mg B plant^{-1} g^{-1} root dry wt) of oilseed rape plants. Values are means of 4 replicates. Means within a column followed by the same letter were not significantly ($p \leq 0.05$) different. The values in the lower table are F-values and their respective levels of significance.

Water treatment	−B	+B
Wet	0.27A	1.34A
Wet-dry	0.15B	1.92A
Dry	0.11B	1.78A

Source of variance	df	F-value
B supply	1	292***
Water supply	2	1.3
Interaction	2	4.0*

*$p \leq 0.05$; ***$p \leq 0.001$.

specific resin (Amberlite, IRA743) to remove any boron contamination and then applied to the soil.

Seeds were sown directly in the soil (without additional boron) in a plastic tray (8 kg soil per tray), on May 7 1994. Soil in the tray was kept at 80% FC and 15–18°C. Seedlings were thinned as required to maintain uniformity.

The seedlings with 3–4 leaves (39 days after sowing) were transplanted into pot-soil with B and water supply treatments as described above. Five kg lots of soil were placed in 5-litre plastic pots lined with polythene bags. The basal nutrients were mixed into the soil, at the same rates used for the seedling-bed soil. The soil in each of the pots was mixed thoroughly. Boric acid solution was mixed into the soil of +B treatments. Pots were arranged in a temperature-controlled water bath set at 18–21°C, in a randomized block design and rearranged daily to minimize positional effects on growth. Water used to irrigate pots was passed through B – specific resin to remove B contamination.

At 5, 23, 28 and 34 days after sowing, the seedlings were top-dressed with 480 mg NH_4NO_3 on each occasion. During plant growth, N, P, K fertilisers were top-dressed onto the surface soil for the treatments of Wet-Dry, and Wet treatments. For the Dry treatment (dry surface soil), solutions of N, P and K were injected into the bottom soil layer.

The plants were harvested 45 days after transplanting (DAT). They were separated into the youngest open leaves (YOL) (Huang et al., 1996), shoot tips (younger

Table 2. Effects of B and water supply on B concentrations (mg B kg^{-1} dry wt) in the youngest open leaves (YOL), shoot tips (<YOL), old leaves and stems after 45 days. Values are means of 4 replicates. For each B treatment, means in a column followed by the same letter were not significantly ($p \leq 0.05$) different.

Boron	Water	Shoot tips	YOL	Old leaves	Stems
−B	Wet	8.5A	4.3A	7.6A	8.4A
	Wet-dry	4.4B	3.0B	7.2AB	7.5B
	Dry	6.0C	3.3C	5.9B	7.3B
+B	Wet	26.2a	26.4a	68a	16.0a
	Wet-dry	24.6b	26.2a	106b	15.7ab
	Dry	22.0b	21.0b	76a	14.3b

Table 3. The dry weights and shoot to root ratios of oilseed rape plants subject to B and water supply treatments in soil for 45 days. The values are means of 4 replicates. Means in a column followed by the same letter were not significantly ($p \leq 0.05$) different.

	Shoot dry weight (g plant^{-1})		Root dry weight (g plant^{-1})		Shoot/root	
	−B	+B	−B	+B	−B	+B
Wet	1.46A	1.51a	0.061A	0.061a	25.3A	25.5a
Wet-dry	1.22B	1.22b	0.074AB	0.048a	16.8B	25.9a
Dry	1.13B	1.35c	0.091B	0.062a	12.4B	22.2a

than YOL), old leaves, and stem. These samples were dried at 70°C and weighed. Milled plant samples were analysed for B by inductively coupled plasma-atomic emission spectroscopy, after digestion in concentrated HNO$_3$ at 140°C (Zarcinas et al., 1987).

Two-way analysis of variance was performed to detect main effects and interactions of B and water supply on dry weights of shoots and roots, shoot to root ratios, and B concentrations in plant parts, using SPSS-X software (SPSS Inc, 1988). One way analysis of variance and Duncan's tests were also carried out to determine water treatment effects within a B treatment group. The data generally conformed to homogeneity requirements ($p \leq 0.05$) for the analysis of variance, but log$_{10}$ transformation of original data was made where it did not. Total B uptake in plants was calculated as the total amount of B in a plant per unit root dry weight.

Results and discussion

The present results showed that low soil water supply decreased total B uptake in plants (Table 1) and B con-

centrations in leaves when plants were grown in low B soil (Table 2). Low water supply in soil interacted with low soil B on plant growth through adjusting assimilate partitioning to achieve a smaller shoot/root ratio, compared to plants with adequate water (Table 3). Root growth was particularly stimulated by the water stress treatments in -B soil, but not in +B soil. Nevertheless, the stimulated root growth in the water stress treatments did not effectively offset the effects of low soil B supply on B concentrations in plant parts. This is supported by the fact that in -B soil, water stress significantly decreased the total B absorption in the plants on the basis of per unit root dry weight, even though water stress had a compensatory effect on root growth (Table 1 and 3).

By comparison to the effect of -B soil, the effect of water stress treatments on B concentrations in plant parts appeared relatively small. In the Wet treatment, B concentrations in the YOL of plants in -B soil were 6 times lower than those in +B soil. In contrast, the water stress treatments decreased B concentrations in the YOL by only 20–30%, compared to those in the Wet treatment regardless of B supply levels in the soil. However, in the case of severe water stress, the effect

160

of soil B addition on plant B status would be minimal (Hobbs and Bertramson, 1949).

Root growth stimulation and adjustment in shoot/root ratios may be an important strategy for plants in moderate water stress to cope with low soil B supply. The alteration in shoot/root ratios in the plants may have resulted in a change in the balance of B uptake capacity in roots and B demand in shoots. In soils with low but not deficient B supply, this strategy may enable the water-stressed plants to take up more B by exploring a larger volume of soil as long as the B supply in soil is not exhausted. However, in B deficient soil like the present -B soil, this strategy may not be effective in achieving a significant increase in B uptake as the B supply is readily exhausted in the limited volume of potted-soil.

Since inhibition of root growth is often the earliest response of plants to B deficiency (Marschner, 1995), the stimulation of root growth in -B soil and Dry treatment may have been induced by water stress before the onset of B deficiency. By the end of experimental period, the oilseed rape plants were B deficient as B concentrations in shoot tips and the YOL were well below $10–14\,mg\,B\,kg^{-1}$ dry the recommended B critical concentration in the YOL (Huang et al., 1996). Judging from the higher B concentrations in leaves older than the YOL+2, the plants in -B soil may have had adequate or marginally adequate B supply from the soil during the early growth when plants were small. As leaf cell expansion is dependent on cell tugour (Schulze, 1986), it is not clear if decreased leaf elongation rates induced by water deficit could lower the B requirements for leaf growth, but this warrants further investigation.

On the basis of the present results, it is difficult to differentiate the effects of the two types of water stresses on plant growth and B concentrations in plant parts. Water stress induced by the two methods differed in both their severity and nature. Soils in the Wet-dry and Dry treatments had different initial water content (80% and 67% FC, respectively) and different periods of time at 30% FC (18 days for the Wet-dry treatment, 29 days for the Dry treatment). As a result, in the early growth phase, plants with adequate B supply (+B soil) in the Wet-dry treatment were able to take up more B than those in the Dry treatment, which was indicated by the higher B concentrations accumulated in the old leaves, compared to those in the Wet treatment. As plants grew bigger and water stress developed in the Wet-dry treatment, water available in surface soil was inadequate and therefore, water stress in plants was more severe in the Dry treatment than the Wet-Dry treatment, as indicated by the differences in plant growth responses to these two water stress treatments. The surface soil water deficit in the Dry treatment may have induced more root growth into the lower portion of the pot-soil than the Wet-dry treatment, as a result of the different distribution of soil water with depth.

In conclusion, when plants were not in severe drought (soil water 30% FC), soil B levels, but not water supply, predominantly determined B supply in the plants. However, when the level of available soil B is very low, decreasing water supply in soil can significantly limit B supply to the new growth of shoots, leading to as much as 30-50% decrease in B concentrations in shoot tips, by comparison to those in plants with adequate water. Therefore, in the dry periods, B deficiency diagnosis should be interpreted, together with soil water contents, in order to make an accurate decision about B fertilizer application in the field.

Acknowledgment

This study is a part of Project 9120 funded by the Australian Centre for International Agricultural Research.

References

Brown P H and Hu H 1993 Boron uptake in sunflower, squash and cultured tobacco cells – Studies with stable isotope and ICP-MS. *In* Plant nutrition – from Genetic Engineering to Field Practice, Ed. N J Barrow, pp 161–164. Kluwer Academic Publishers, Dordrecht, The Netherlands.

Hobbs J A and Bertramson B R 1949 Boron uptake by plants as influenced by soil moisture. Soil Sci. Soc. Amer. Proc. 14, 257–261.

Huang L, Ye Z and Bell R W 1996 The importance of sampling immature leaves for the diagnosis of boron deficiency in oilseed rape (*Brassica napus* cv. Eureka). Plant Soil 183, 187–198.

Marschner H 1995 Mineral Nutrition of Higher Plants. Academic Press, London, UK. 889 p.

Schulze E-D 1986 Whole-plant responses to drought. Aust. J. Plant Physiol. 13, 127–141.

SPSS Inc 1988. SPSS-X User's Guide. 3rd Ed. SPSS Inc, USA. 677 p.

Zarcinas B A, Carwright B and Spouncer L R 1987 Nitric acid digestion and multi-element analysis of plant material by inductively coupled plasma spectrometry. Commun. Soil Sci. Plant Anal. 18, 131–146.

R.W. Bell and B. Rerkasem (eds.), Boron in Soils and Plants, 161–164.
© 1997 *Kluwer Academic Publishers.*

Effect of boron on the growth of mungbean seedlings under aluminium stress

Hongyan Zhang & Yonghua Yang
Department of Biological Sciences and Technology, Nanjing University, Nanjing 210093, P. R. China

Key words: aluminium, boron, cuttings, growth, mungbean, seedling

Abstract

The growth of mungbean (*Vigna radiata* (L.) Wilczek) seedlings and cuttings (without roots) was studied in a growth chamber, to determine the effect of boron (B) on aluminium (Al) toxicity for plant growth by adding two concentrations of B (5, 50 μM) under Al stress. We found that B at 50 μM in the nutrient solutions at 2 or 5 mM Al, increased the height of intact seedlings, and the fresh and dry weights of the seedlings compared to that at 5 μM. It also promoted the elongation of the epicotyl and hypocotyl of intact seedlings under Al stress. Boron at 50 μM increased the content of chlorophyll in the seedlings which suffered from Al toxicity. However, since B had no amelioration effect for the cuttings under Al stress it appears that B alleviation of Al toxicity was related to the roots. The possible relationship between the effects of B and Al toxicity is discussed.

Introduction

Aluminium (Al) toxicity is one of the most important factors limiting plant growth on acid soils (Taylor, 1988a). The major approach to ameliorating Al toxicity has been to precipitate toxic ionic species of Al from the soil solution by liming to raise soil pH (Foy, 1988). Unfortunately, the management of soil acidity by liming is not always economically feasible, especially in strongly acid soil, and when the subsoil is high in Al. While an ideal solution to increasing the agricultural productivity of acid soil is to utilize the plant genetic potential for tolerance to Al stress (Foy, 1988), locally adapted, Al-tolerant varieties for a number of crop species are not yet available.

A possible alternative solution to the Al toxicity problem, has been suggested by several reports that indicated that boron (B) may alleviate the toxic effects of Al. The application of B into Al-toxic subsoil can promote the root growth of alfalfa (LeNoble et al., 1996a). Boron also affected patterns of uptake of Al and nutrient elements in plants (LeNoble et al., 1991). LeNoble et al. (1996b) showed that B protected against Al inhibition of root and shoot growth of squash seedlings in solution culture. Their results showed that more B were required under acidic, high-Al conditions

than under low-Al conditions. They suggested that supplementary B might shift the equilibrium toward continued formation of B-ester complexes (LeNoble et al., 1996b), since the formation of the complexes may be the equilibrium process which enables the controlled creep of expanding cell walls while maintaining the cell wall strength necessary for continued cell growth (Hu and Brown, 1994). Similar results were obtained in a soil experiment except that supplemental B did not protect shoot growth against Al toxicity (LeNoble et al., 1996a). However, Taylor and Macfie (1994) argued that the effect of supplemental B on Al toxicity in root growth must reflect the unique characteristics of soil-based media, and (or) differences among species in internal requirements for B and the effects of Al on B nutrition.

The purpose of the present paper was to investigate the B amelioration of Al toxicity and demonstrate the effect of B on the morphology and physiology of mungbean under Al stress.

Materials and methods

Seeds of mungbean (*Vigna radiata*) were surface-sterilized in 0.1% HgCl$_2$ for 16 min, rinsed thorough-

162

Table 1. Effect of Al and B on the growth of mungbean seedlings after sixteen days treatment. Values are means ± SE of three replicates.

| Treatment Al (mM)+B(μM) | Length (cm) of | | Fresh weight (g 10 plants^{-1}) | Dry weight (mg plant^{-1}) |
	Epicotyl	Hypocotyl		
Al(0)+B(5)	7.2 ± 0.1	3.8 ± 0.1	3.86 ± 0.26	34.7 ± 0.18
Al(2)+B(5)	6.4 ± 0.2	2.9 ± 0.2	2.89 ± 0.19	29.3 ± 0.29
Al(2)+B(50)	6.9 ± 0.2	3.3 ± 0.2	3.34 ± 0.21	31.4 ± 0.41
Al(5)+B(5)	4.3 ± 0.1	2.0 ± 0.2	2.34 ± 0.05	26.1 ± 0.14
Al(5)+B(50)	4.7 ± 0.1	2.5 ± 0.1	2.85 ± 0.27	30.6 ± 0.22

Table 2. Effect of Al and B on height of seedlings at 8 and 16 days after treatment and on chlorophyll content of leaves of seedlings and cuttings of mungbean. Values are means of three replicates.

| Treatment Al (mM)+B(μM) | Height (cm) | | Chlorophyll (mg g^{-1} F.Wt) | |
	8d	16d	Seedlings	Cuttings
Al(0)+B(5)	8	10.8	2.99	1.28
Al(2)+B(5)	6.3	9.6	2.46	1.13
Al(2)+B(50)	7.1	10.3	3.01	1.16
Al(5)+B(5)	3.6	6.3	2.31	1.12
Al(5)+B(50)	3.9	7.1	2.71	1.08
LSD(0.05)	0.6	0.7	0.36	0.40

ly with distilled water, immersed in distilled water overnight, and then germinated in flasks of aerated, distilled water for two days. Seedlings were elongated in sand for four days before being transferred to the treatments. Seedlings of the same size were selected, then the cotyledons were removed. The cuttings which comprised the apical bud, one pair of primary leaves, the epicotyl, and 1.5 cm long segments of the hypocotyl were taken from the above seedlings but had their roots removed. The seedlings or cuttings were transplanted into a container with 50 mL of a nutrient solution containing different concentrations of Al as AlCl$_3$ and B as H$_3$BO$_3$ and grown in a growth chamber. Three levels of Al (0, 2, 5 mM) and two levels of boron (5, 50 μM) were used. The composition of the nutrient solution was as follows (in mM): 0.5 KNO$_3$; 0.5 Ca(NO$_3$)$_2$; 0.2 MgSO$_4$.7H$_2$O; 0.1 KH$_2$PO$_4$; (in μM): 0.9 MnCl$_2$.4H$_2$O; 0.1 ZnSO$_4$.7H$_2$O; 0.03 CuSO$_4$.5H$_2$O; 0.012 Na$_2$MoO$_4$.2H$_2$O; and 4.0 FeEDTA. The pH was adjusted to 4.1.

Each pot contained ten mungbean seedlings or cuttings. All treatments were replaced with fresh treatment solutions every two days. The seedlings or cuttings were harvested after sixteen-days in treatment solutions. The length of the epicotyl and hypocotyl, the height of intact seedlings and cuttings, and the fresh and dry weight of the intact seedlings were measured. A bulk sample of all leaves was analyzed for the content of chlorophyll by the method of Chen and Chen (1984). Each treatment was replicated three times in a complete randomized design.

Results and discussion

High Al concentration (5 mM) has been reported to inhibit the elongation of the epicotyl of mungbean seedlings after six days treatment (Yang and Shen, 1996). The present results showed that 2 mM Al can also inhibit the elongation of the epicotyl and hypocotyl of the intact seedlings, and reduce the height, and the fresh and dry weight of the seedlings after sixteen days treatment (Table 1 and 2).

Boron at 50 μM in solution partly alleviated the depression in seedling growth at both 2 and 5 mM Al. The length of the epicotyls at 50 μM B was increased by 7.8% and 9.3% in contrast to that at 5 μM B under 2 and 5 mM Al stress, respectively. A similar result was observed for the length of hypocotyls (Table 1). The height of intact seedlings in the solution with 5 mM Al and 50 μMB was increased by 12.7% compared to that with 5 mM Al and 5 μMB. By contrast, the

Table 3. Effect of Al and B on the growth of mungbean cuttings after sixteen days treatment. Values are means ± SE of three replicates.

Treatment Al(mM)+B(μM)	Length (cm) of Epicotyl	Fresh weight (g 10 plants^{-1})	Dry weight (mg plant^{-1}
Al(0)+B(5)	6.2 ± 0.2	2.49 ± 0.17	15.5 ± 0.51
Al(2)+B(5)	6.2 ± 0.4	2.24 ± 0.52	13.7 ± 0.40
Al(2)+B(50)	6.4 ± 0.3.	2.42 ± 0.07	14.2 ± 0.30
Al(5)+B(5)	6.3 ± 0.5	2.05 ± 0.10	12.6 ± 0.52
Al(5)+B(50)	6.3 ± 0.5	2.18 ± 0.11	12.8 ± 0.54

seedling height in the solution containing 2 mM Al and 50 μMB was increased only by 7.3% compared to that containing 2 mM Al and 5 μMB. This increase in the height was, in fact, apparent after eight-day treatment (Table 2). In addition, we also found that B increased accumulation of fresh and dry mass under Al stress (Table 1).

Some reports suggested that B may be capable of alleviating the toxic effects of Al, providing improved plant performance on acid soil (LeNoble et al., 1996a; LeNoble et al., 1991). Our observation of a reversal of Al-inhibited elongation of the epicotyl, and hypocotyl, and of height, fresh and dry weight of the seedlings by adding more B supports this hypothesis.

Pan et al. (1989) suggested that one mode of action by which Al may affect shoot growth is by inhibiting the synthesis and subsequent translocation of cytokinin to the meristematic regions of the shoot. Boron may play a role in the balance of plant hormones (Liu et al., 1987). Deficiency of B decreases synthesis of cytokinin (Wagner and Michael, 1971). Thus, a possible mode of action by which B may increase the growth of the seedlings under Al stress is by promoting the synthesis of cytokinin and its transportation to the meristem. Another explanation is that Al toxicity may induce B deficiency, and B deficiency consequently reduced the ascorbate concentration in root apices. The symptoms of Al toxicity may be a consequence of disrupted ascorbate metabolism (LeNoble et al., 1996b; Lukaszewski and Blevins, 1996).

In contrast, in the cuttings there were no significant differences between the treatment of 50 $\mu M$and 5 μMB under 2 or 5 mM Al stress in the length of epicotyl, or in the fresh and dry weight (Table 3). The immobilization of Al by binding in cell walls of the root could reduce uptake into the symplasm to ameliorate Al toxicity in shoots (Taylor, 1988b). The mungbean cuttings would accumulate more Al in the symplasm than that in intact seedlings because of the absence of roots. As a result,

Al toxicity is more severe and B could not alleviate Al toxicity for the cuttings.

Petterson et al. (1985) reported that enhanced degradation of the thylakoid was the most pronounced ultrastructual change induced in *Anabaena cylindrica* by adding Al. Aluminium caused severe injuries to chloroplast membranes of *Spinacea oleracea* (Hampp and Schnabl, 1975). Our results showed that B increased the content of chlorophyll in the intact seedlings under Al stress, but for the cuttings, it seemed that B had no such effect (Table 3).

Previous research has showed that Al could depress the uptake and translocation of other elements, such as Ca, Mg, P (Foy, 1988), K (Olivetti et al., 1995), and B (Poschenrieder et al., 1995), thereby inducing mineral deficiency. The first symptom of B deficiency is the inhibition of root growth. Deficiency of B can also impair the function of cell membranes and then depress the transportation of nutrients and activity of enzymes (Blaser-Grill et al., 1989). Boron deficient plants, may also experience increased damage to the membrane structure of the chloroplast, and a decreased number and thickness of chloroplast grana. Such influences inevitably reduce photosynthesis of the plants. These symptoms due to the B deficiency are similar to those induced by Al stress, with most being associated with membrane function and root growth. However, whether Al exactly increases the B requirement or B improves the metabolism of the mungbean to resist Al stress is unclear. Further work to study the mechanism of B amelioration of Al toxicity in the growth of mungbean seedlings is underway.

References

Blaser-Grill J, Knoppik D, Amberger A and Goldbach H 1989 Influence of boron on the membrane potential in *Elodea densa* and *Helianthus annuus* roots and H proton extrusion of suspension cultured *Daucus carota* cells. Plant Physiol. 90, 280–284.

164

Chen F M and Chen S W 1984 Research of the measurement of chlorophyll by mixed solution method. Commun. Forestry Sci. Technol. 2, 4–8.

Foy C D 1988 Plant adaptation to acid, aluminum-toxic soils. Commun. Soil Sci. Plant Anal. 19, 959–987.

Hampp H and Schnabl H 1975 Effect of aluminum ions on $^{14}CO_2$ fixation and membrane system of isolated spinach chloroplasts. Z. Pflanzenphysiol. 76, 300–306.

Hu H and Brown P H 1994 Localization of boron in cell walls of squash and tobacco and its association with pectin. Plant Physiol. 105, 681–689.

LeNoble N, Blevins D G and Miles R J 1991 Mineral composition of alfalfa roots grown in high aluminum soils with additional boron. Current Topics in Plant Biochem. and Physiol. 10, 230–239.

LeNoble M E, Blevins D G and Miles R J 1996a Prevention of aluminium toxicity with supplemental boron. II. Stimulation of root growth in an acidic, high-aluminium subsoil. Plant Cell Environ. 19, 1143–1148.

LeNoble M E, Blevins D G, Sharp R E and Cumbie B G 1996b Prevention of aluminium toxicity with supplemental boron. I. Maintenance of root elongation and cellular structure. Plant Cell Environ. 19, 1132–1142.

Liu W D, Pi M M and Wu L S 1987 Interaction between boron and potassium in cotton nutrition. Acta Pedologica Sinica. 24, 43–50.

Lukaszewski K M and Blevins D G 1996 Root growth inhibition in boron-deficient or aluminum-stressed squash may be a result of impaired ascorbate metabolism. Plant Physiol. 112, 1135–1140.

Olivetti G P, Cumming J R and Etherton B 1995 Membrane potential depolarization of root cap cells precedes aluminum tolerance in snapbean. Plant Physiol. 109, 123–129.

Pan W L, Hopkins A G and Jackson W A 1989 Aluminum inhibition of shoot lateral branches of *Glycine max* and reversal by exogenous cytokinin. Plant and Soil 120, 1–9.

Petterson A, Haellbom L and Bergman B 1985 Physiological and structural responses of the cyanobacterium *Anabeana cylindrica* to aluminum. Physiol. Plant. 63, 153–158.

Poschenrieder C, Llugany M and Barcele J 1995 Short-term effects of pH and aluminum on mineral nutrition in maize varieties differing in proton and aluminium tolerance. J. Plant Nutr. 18, 1495–1507.

Taylor G J 1988a The physiology of aluminum tolerance. *In* Metal Ions in Biological Systems, Vol. 24: Aluminum and Its Role in Biology. Ed. H Sigel and A Sigel, pp. 165–198. Marcel Dekker, New York.

Taylor G J 1988b The physiology of aluminium tolerance in higher plants. Commun. Soil Sci. Plant Anal. 19, 1179–1194.

Taylor G J and Macfie S M 1994 Modeling the potential for boron amelioration of aluminum toxicity using the Weibull function. Can. J. Bot. 72, 1187–1196.

Wagner H and Michael G 1971 The influence of varied nitrogen supply on the production of cytokinins in sunflower roots. Biochem. Physiol. Pflanz. 162, 147–158.

Yang Y H and Shen Z G 1996 Effect of aluminium and 6-benzylaminopurine on the growth of the epicotyl in mungbean seedlings. J. Plant Nutr. 19, 63–71.

R.W. Bell and B. Rerkasem (eds.), Boron in Soils and Plants, 165–169.
© 1997 *Kluwer Academic Publishers.*

The effect of boron on growth and development in strawberry (*Fragaria ananassa* Duch.) shoot culture

Ante Bisko[1], Tomislav Cosic[1], Milan Poljak[1] & Sibila Jelaska[2]
[1]*Department of Plant Nutrition, Faculty of Agronomy*
[2]*Department of Molecular Biology, Faculty of Science, University of Zagreb, HR-10000 Zagreb, R Croatia*

Key words: Boron, Fragaria sp., in vitro culture, roots, shoots, strawberry

Abstract

The effect of boron was tested on three strawberry (*Fragaria ananassa*) cultivars: Korona, Elsanta and Gorella. Boron was added into the culture medium as boric acid (H_3BO_3) with the concentration ranging from 0.0 to 8.68 mg B L^{-1}, and the medium was solidified with 0.8% agar. A liquid medium without B was tested, also. The tested B concentration did not affect the time of shoot appearance but affected the number of newly developed shoots at the multiplication phase of culture. Regardless of the B quantity added to the medium in the rooting phase, the time of appearance and the number of rootlets did not differ in the same cultivar. Strawberry plantlets in axenic culture were tolerant of high B quantities in a medium, but they also survived in a medium with no boron added. Increase of B concentration to eight times the amount in MS (Murashige and Skoog) nutrient medium did not provoke necrosis nor inhibit shoot growth. The B concentration in plantlets reflected the increase of B concentration in the medium.

Introduction

Studies of the effect of B upon strawberry showed that it is very susceptible to B excess, and that the boundary between the indispensable and excessive concentration was narrow (Purvis and Hanna, 1940; Latimer, 1943; Haydon, 1981; Williams and Gostick, 1981). Investigations of the effect of B toxicity on strawberry have been conducted in field and pot trials, while those of the B effect on strawberry *in vitro* are inconclusive. Jungnickel (1988) refers to Haydon's (1981) investigations and points to the need for great caution with B in strawberry tissue culture.

Boron added to the nutrient medium for strawberry growing *in vitro* culture varies from 0.31 mg H_3BO_3 L^{-1} (Jungnickel and Gliemeroth, 1986) to 6.18 mg H_3BO_3 L^{-1} (Murashige and Skoog, 1962; Linsmaier and Skoog, 1965; Boxus, 1981).

The objective of our investigation was to test the morphogenic response of strawberry grown in axenic culture to the B supply in the nutrient medium (from B-deficient medium to an eight time increase of B amount relative to the standard Murashige and Skoog medium),

as well as to determine the strawberry uptake of B in response to the B concentrations in medium.

Materials and methods

Investigations were carried out on strawberry plants, cvs. Korona, Elsanta and Gorella under *in vitro* conditions. The shoot explants were individually inoculated into test-tubes (23x160 mm, made of B-free glass), on 20 mL of medium. The test-tubes were closed with cotton plugs and aluminium foil.

Initial culture

The basal nutrient medium (Boxus, 1981) supplemented with organic constituents (mg L^{-1}): 100 myo-inositol, 0.1 thiamine-HCl, 0.5 pyridoxine-HCl, 2.0 glycine, 0.5 nicotinic acid, 0.1 N_6 benzyladenine (BA), 1.0 indole-3-butyric acid (IBA), 0.1 gibberellic acid (GA_3) and 40 g glucose L^{-1} was used. Prior to autoclaving, the medium pH was adjusted to 5.6 and the medium was solidified with 8 g agar L^{-1}(A-1296,

166

Sigma^R) and autoclaved at 121° C at 103.4 kP for 15 min.

Cultures grew in a controlled environment room at 24° C and under artificial cool light of fluorescent lamps (400–700 nm), at 17 Wm^{-2} and a photoperiod of 16 h. After a week, the cultures were replanted into a fresh medium of the same composition. The initial culture, as well as the two subsequent subcultures (multiplication and elongation phases) lasted a month each.

Multiplication and elongation phases

Shoot explants grown in the initial culture were individually transferred to the multiplication medium which comprised the basal medium supplemented with 0.5 mg BA L^{-1}, 0.5 mg IBA L^{-1} and 0.1 mg GA$_3$ L^{-1} GA$_3$. Before using the shoots in B experiments, shoots were separated and individually replanted into basal medium with 0.1 mg BA L^{-1}, 0.5 mg IBA L^{-1} and 0.1 mg GA$_3$ L^{-1} (elongation phase).

Boron effect trial

Multiplication phase

Preparation of shoot inoculum, composition of multiplication medium (except for B) and growing conditions were identical to those of the mentioned multiplication phase. When introduced into the trial, microplants of all the three cultivars consisted of the apical bud and four to five leaves.

Boron concentrations used (added as H$_3$BO$_3$) were: 0.00 – liquid nutrient medium (with paper bridges), and agar solidified medium with 0.00, 0.27, 0.54, 1.08, 2.17, 4.34, 6.51 and 8.68 mg L^{-1}. Each treatment had 10 replicates (test-tubes). Cultures were examined every 3–4 days and all changes in morphogenesis and organogenesis were recorded. The multiplication phase lasted for five weeks.

Rooting phase

Part of the uniform shoots, developed during the multiplication phase, were separated, their roots removed, and subcultured into the basal medium with 1.0 mg IBA L^{-1} for rooting. Shoots were transferred to the medium with the same B concentration as in the multiplication phase. The agar medium with 0.27 mg B L^{-1} was not included in the rooting phase, so there were eight treatments in all. Each treatment consist-

ed of seven replicates. The subculture lasted for five weeks.

Determination of total B in plant material

After the multiplication phase, one part of the specimens was separated for B analysis. Rootlets and the basal part of plantlets were cleaned of the nutrient medium and rinsed 4–5 times under a jet of distilled water. After that, specimens were dried on a filter paper at 105° C to constant weight. The same procedure was applied to plantlets after the rooting phase. The total B concentration of plant material was determined by the method of Berger and Truog (1944).

Statistical analysis

The factorial trial (treatment x genotype) was processed using the two-way analysis of variance (MSTATC). Significance was determined at $P<0.05$.

Results

Multiplication phase

To the end of the multiplication phase, cultures showed no signs of B deficiency (specified according to Shorrocks, 1984; Bergman, 1992) on the nutrient media to which no boron was added. Increase of B concentration up to eight times (8.68 mg L^{-1}) relative to the standard concentration of MS medium did not lead to chlorosis or any other visible damage. Boron concentration in the medium affected moderately the number of shoots and the root length. The smallest number of shoots was recorded on liquid B-free medium and agar-solidified medium with 8.68 mg B L^{-1}, while the cultures on nutrient medium with 1.08 mg B L^{-1} had the significantly largest number of shoots. The shortest roots were formed by cultures in the nutrient medium with 8.68 mg B L^{-1}. Significant differences among the cultivars were proved for the root growth; the longest roots being those of cv. Korona. The cultivars were significantly different with respect to the number of shoots, which was highest in Gorella. The effect of the medium B concentration on the strawberry plantlets is shown in Figure 1 A, B, C, D.

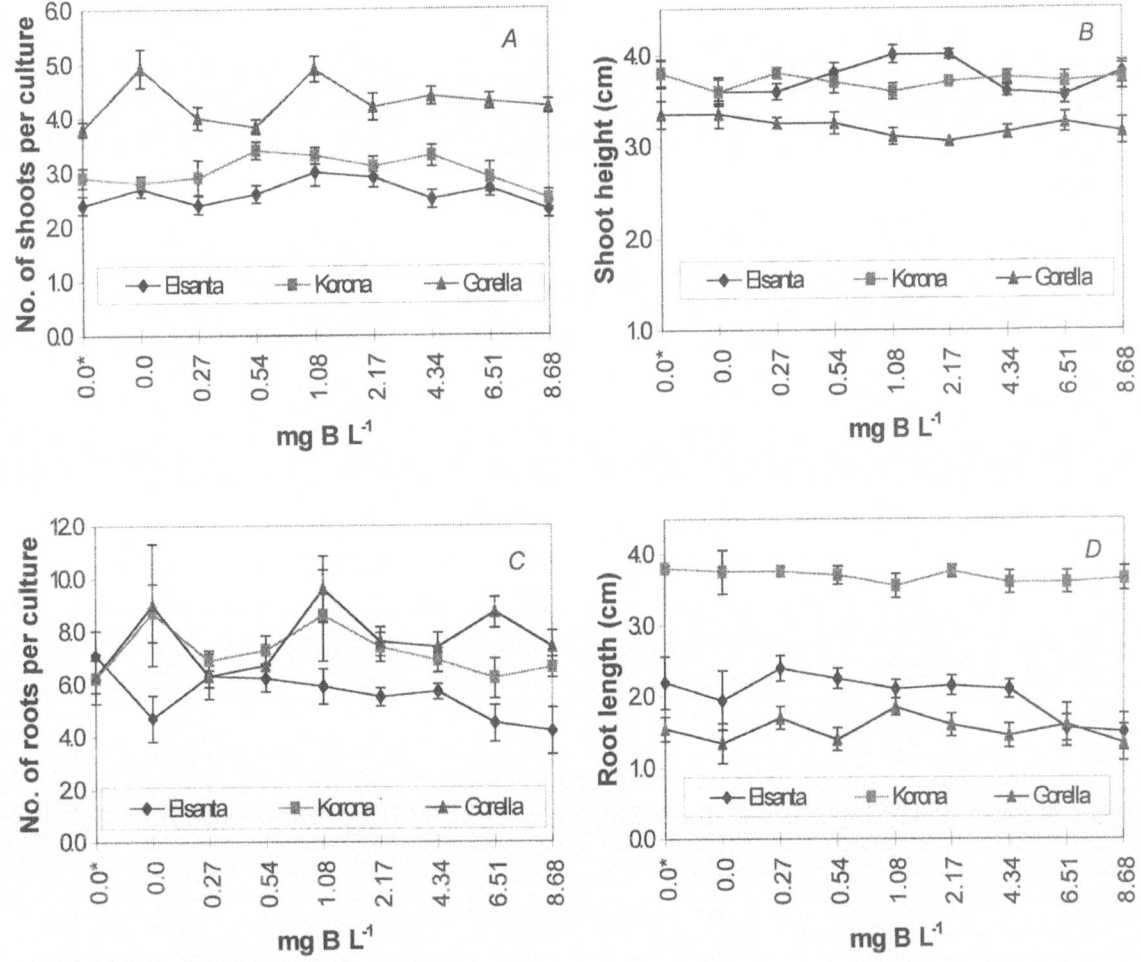

Figure 1. Effect of the boron concentration in nutrient medium on: A) number of strawberry shoots, B) shoot height (cm), C) number of roots and D) length of roots (cm) after five weeks of culture (multiplication phase). Each point on the graph represents an average (n=10). 0.0∗=liquid B-deficient medium. Standard errors of the mean are shown.

Rooting phase

Microscopic examination of apical and root meristems, after completion of the rooting phase, confirmed normal development regardless of the B concentration in which plantlets were grown. Boron addition of 8.68 mg L^{-1} inhibited regeneration of roots in all three cultivars. The height of shoots in the liquid nutrient medium without B was significantly larger, while there were no significant differences between the other treatments (Figure 2A). The cultivars differed significantly in shoot height. The numbers of leaves and roots per shoot showed no statistically significant differences between B treatments but these difference were significantly affected by the cultivar (Figure 2B, C). Roots in the liquid nutrient medium without B were signifi-

cantly longer. Comparison of agar-added media clearly shows that an increase in B concentration above 4.34 mg L^{-1} (cvs. Korona and Gorella) and 2.17 mg L^{-1} (cv. Elsanta) caused a significant inhibition of the root growth (Figure 2D).

Boron content in plant material

Boron uptake and total B content in the strawberry plant material were increased by the B addition in the medium (Figures 3 and 4). Significantly different B contents in plant material were determined between cultures grown on liquid medium and on agar solidified medium (in both cases without addition of B).

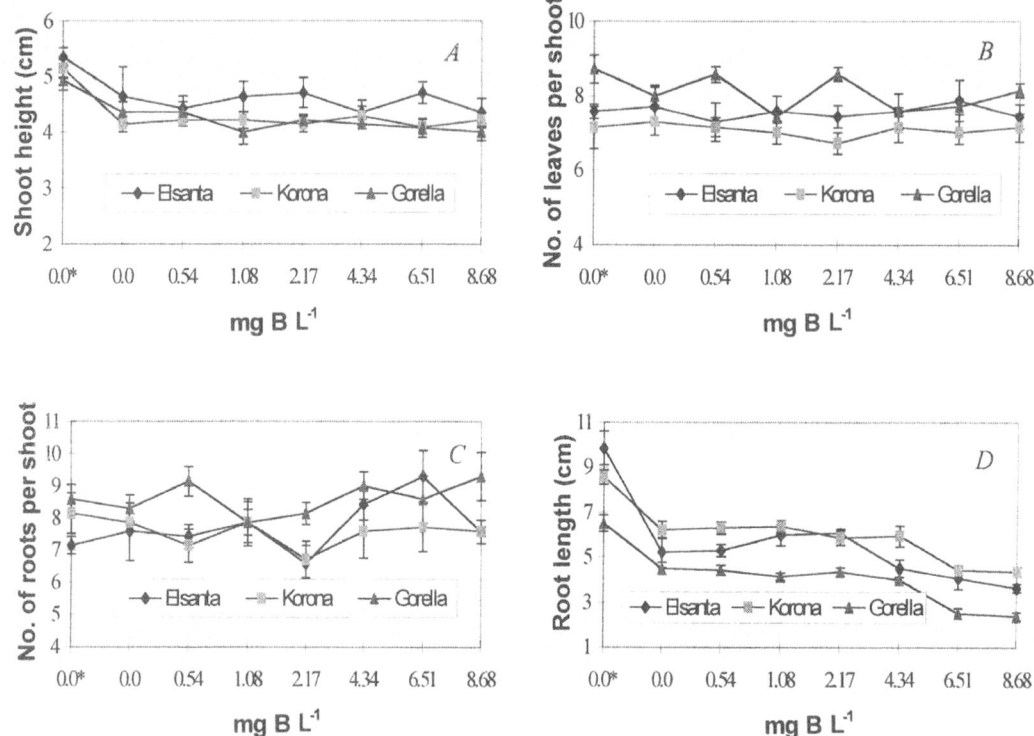

Figure 2. Effect of the boron concentration in nutrient medium on: A) shoot height (cm), B) number of leaves per shoot, C) number of roots and D) length of roots (cm) after five weeks of culture (rooting phase). Each point on the graph represents an average (n=7). Standard errors of the mean are shown.

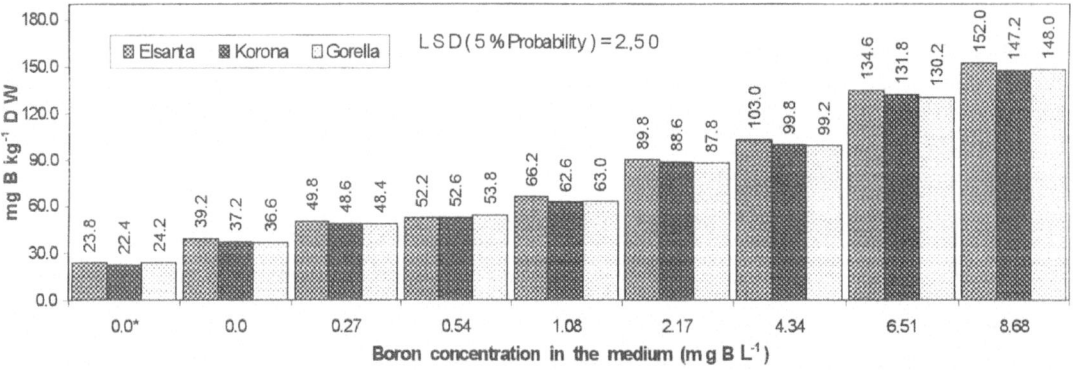

Figure 3. Relation between boron concentration of the medium and boron concentration in strawberry microplant tissues (multiplication phase). Each point on the graph represents an average (n=5). 0.0∗=liquid B-deficient medium.

Discussion

The B treatments had no influence on the time of appearance and height of newly generated shoots, and on the number of roots per culture. Initiation of roots was more related to the cultivar and state of the nutrient medium than by B concentration in the medium. The number of shoots and root length are the two traits

that were significantly related to B. Efficient culture development on liquid medium without addition of B may be explained by the fact that cultures were provided with a certain amount of B via distilled water and partly also through nutrient salts containing B as an impurity. Salinas et al. (1986) report that B is contained in nutrient salts as an impurity in the amount of 0.1–0.15 mg B L^{-1}. Agar is an additional source of

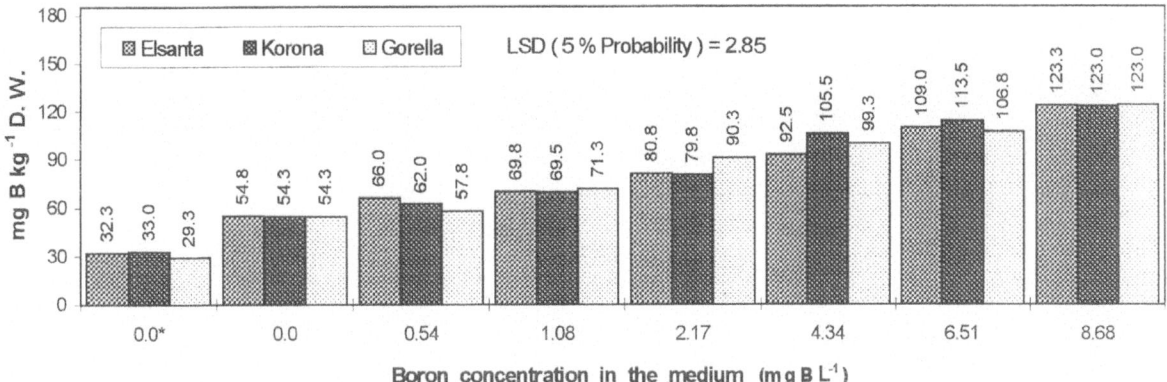

Figure 4. Relation between boron concentration of the medium and boron concentration in strawberry microplant tissues (rooting phase). Each point on the graph represents an average (n=4).

B (Singha and Edwin, 1985). A B increase in nutrient medium up to 6.51 mg L^{-1} did not inhibit morphogenesis and organogenesis of strawberry grown in the *in vitro* culture, which points to the conclusion that strawberry tolerates an increased level of B in nutrient medium.

The number of leaves and the number of roots did not differ significantly whereas root lengths were significantly different in the mentioned B treatments. Liquid nutrient medium yielded the longest roots, which was also the case of their shoots. Higher susceptibility of roots to B than of shoots is emphasised by Davis et al. (1993) in their study involving pearl millet seedlings. A B increase of over 4.34 mg L^{-1} inhibited the strawberry root growth significantly. Inhibition of the root growth at the mentioned B concentrations was noticeable already during culture growth, and this is the only trait in which the effect of raised B concentration in the nutrient medium was clearly manifested.

References

Berger K C and Truog E 1944 Boron tests and determination for soils and plants. Soil Sci. 57, 25–36.

Bergman W 1992 Nutritional Disorders of Plants: Development, Visual and Analytical Diagnosis. Gustav Fischer, Jena. 178 p.

Boxus P 1981 Commercial production of strawberry plants produced by meristem culture and micropropagation. *In* Col Sci. 15, Florales Int., Montreal, pp 310–348.

Davis J G, Hossner L R and Persaud N 1993 Elemental toxicity effects on germination and growth of pearl millet seedlings. J. Plant Nutr. 4, 1957–1968.

Haydon F G 1981 Boron toxicity of strawberry. Commun. Soil Sci. Plant Anal. 12 (11), 1085–1091.

Jungnickel F 1988 Strawberries (*Fragaria spp.*and *Hybrids*) *In* Biotechnology in Agriculture and Forestry 6, Crops II. Ed. Y P S Bajaj, pp 39–103, Springer-Verlag, Berlin.

Jungnickel F and Gliemeroth K 1986 In-vitro-Depots vegetativ vermehrter Blutenpflanzen. Wiss. Z. Fridrich-Schiller-Univ., Jena, Math. Nat. R. 35, 631–638.

Latimer L P 1943 The response of strawberries to boron. J. Am. Soc. Hortic. Sci. 42, 441–443.

Linsmaier E M and Skoog F 1965 Organic growth factor requirements of tobacco tissue cultures. Physiol. Plant. 18, 100–127.

Murashige T and Skoog F 1962 A revised medium for rapid growth and bioassays with tobacco tissue cultures. Physiol. Plant. 15, 473–497.

Purvis E R and Hanna W J 1940 Vegetable crops affected by boron deficiency in eastern Virginia. Va Agric. Exp. Sta. Bull. 105.

Salinas R M, Cerda A, Romero M and Caro M 1986 Boron tolerance of pea (*Pisum sativum*). J. Plant Nutr. 4, 205–217.

Shorrocks V M 1984 Boron deficiency-its prevention and cure. Borax Holdings, London. 30 p.

Singha S and Edwin C T 1985 Mineral nutrient status of crabapple and pear shoots cultured *in vitro* on varying concentrations of three commercial agars. J. Am. Soc. Hortic. Sci. 110, 407–411.

Williams J H and Gostick K 1981 Water quality for crop irrigation: guidelines on chemical criteria. Leaflet 776 of the Ministry of Agric., Fish and Food, London, 1–6.

R.W. Bell and B. Rerkasem (eds.), Boron in Soils and Plants, 171–174.
© 1997 Kluwer Academic Publishers.

Composition of root exudates and root bleeding sap of a boron-efficient and inefficient oilseed rape genotype

Xiangyun Cao, Wuding Liu & Meimei Pi
Department of Agricultural Chemistry, Huazhong Agricultural University, Wuhan 430070, P.R. China

Key words: bleeding sap, boron deficiency, exudates, rape genotypes

Abstract

A solution culture study examined the influence of boron (B) supply on the composition of root exudates and bleeding sap of oilseed rape (*Brassica napus* L.) cultivars which differed in B efficiency. For both cultivars, root exudation of K and NO_3 was not affected by B supply. However, B deficiency reduced soluble sugars in root exudates of the B-inefficient cultivar, but not those of the B-efficient cultivar. In root exudates, the total amount of amino acids was not changed, but its composition changed under B deficient conditions. Among the individual amino acids, exudation of tyrosine (Tyr) was specific for the B efficient cultivar under B-deficient conditions. Under B-deficient condition, the amount of root bleeding sap, and its amino acids and soluble sugars content were significantly depressed in the B-inefficient cultivar, but not in the efficient one. Among the detected amino acids in root bleeding sap, the concentration of glutamine, alanine, serine, and Tyr were more strongly suppressed under B deficient conditions in the B-inefficient cultivar. This further suggests that Tyr has some relationship with B efficiency.

Introduction

More than 70% of oilseed rape (*Brassica napus* L.) crops in China are cultivated in the Yangtse River Valley where the soils are commonly boron (B)- deficient (Liu and Chen, 1986). As rape production and macronutrient fertilization have been expanding, the demand for B fertilizer has increased and in some areas of China is in short supply. It is therefore worthwhile exploring the possibility of utilizing genetic variation in B efficiency in oilseed rape and other crops to decrease the need for B fertilizer and to use existing fertilizer and native soil B more efficiently.

Significant differences exist among rape species and cultivars in response to boron deficiency (Peng et al., 1995; Wang and Lan, 1995a; Yang et al., 1993). A B-efficient and B-inefficient cultivar whose B efficiencies [(-B/+B)x100] for seed yield in pots were 91% and 0, respectively, have been identified among 86 rape cultivars screened by Wang and Lan (1995 b).

The B-inefficient cultivar had deeper coloured, larger and thicker leaves which needed more B for nutritional growth, but its absorption and translocation of B were poor at low B supply (Xiong et al., 1995). Under B-deficient conditions, the root growth and elongation of this cultivar were retarded seriously, and deformities in root anatomy such as missing root cap, misshapen and disorderly root tip cells, reduced root activity and lowered nutrient absorption were noted (Cao, 1996). The suppression of root growth may affect root exudation. Our previous study indicated that the media pH for growth of several crops in solution culture was changed by B supply (Cao et al., 1995); boron affects H^+ in sunflower seedlings roots (Belver et al., 1988) and carrot cells (Blaser-Grill et al., 1989). It is unclear whether the composition of root exudates is related to B deficiency. The present study was undertaken to examine the influence of B on composition of root exudates and root bleeding sap of the two cultivars previously shown to differ in their B-efficiency.

Table 1. Effect of B supply for 15 days on growth (g dry wt 15 plants^{-1}) and concentrations of K, and NO$_3$ and soluble sugars (mg kg^{-1} root dry wt h^{-1}) in exudates from roots of two rape genotypes. Values are means of 3 replicates. For each cultivar, means in a row followed by the same letters were not significantly different at $p<0.05$.

	01		03	
	−B	+B	−B	+B
Shoot dry weight	1.25a	1.74b	1.35y	1.45y
Root dry weight	0.102a	0.166b	0.104y	0.104z
K	302a	360a	397y	493y
NO$_3$	184a	160a	126y	108y
Soluble sugar	14.3a	32.9b	32.3y	32.8y

Table 2. The influence of B supply for 15 days on amino acid exudation of 2 rape genotypes (mg kg^{-1} root dry wt 5 h^{-1}). Values are means of 2 replicates.

Amino acid	01		03		Amino acid	01		03	
	−B	+B	−B	+B		−B	+B	−B	+B
Ala	3.64	0.61	3.97	3.23	Leu	2.95	0.77	3.79	2.93
Asp	4.37	1.22	3.93	1.21	Lys	1.53	1.95	1.15	1.99
Arg	2.17	Tr[1]	0.90	Tr	Phe	11.3	14.7	8.08	14.3
Cys	20.0	21.5	16.6	13.9	Ser	7.69	5.73	12.9	11.3
Glu	27.7	24.4	21.0	25.3	Thr	2.29	0.70	3.34	2.84
Gly	3.39	3.91	3.94	6.58	Tyr	Tr	Tr	3.08	Tr
His	0.97	Tr	1.74	Tr	Val	5.40	11.4	6.13	6.69
Ile	3.04	6.88	3.86	4.22	Total	96.4	93.8	94.4	94.5

[1] Trace amounts

Materials and methods

Two oilseed rape cultivars were used: 01 was a B-inefficient cultivar and 03 an efficient one (Wang and Lan, 1995b).

Rape seeds were soaked in distilled H$_2$O for 6h, then incubated in the dark at 20° C. After germination, the seeds were transferred to nylon fabric placed on darkened boxes filled with distilled water. About one week later, the seedlings were planted in darkened plastic pots filled with 2 L of one quarter-strength Hoagland solution (Hoagland and Arnon, 1950), containing Fe as Fe-EDTA. The strength of the nutrient solution was increased after one week to half- and then after one more week to full-strength. At 4-leaf stage, rape plants were treated with 2 B levels, 0 (-B) and 0.6 mg B L^{-1} (+B), and grown for a further 15 days when samples were collected. Each treatment had 20 replicates.

After 15 days treatment, plant roots were washed with distilled water 3 times, then with sterilized (110° C, 30 min) double distilled water (SDDW) for another 3 times, and transferred to SDDW and incu-

Table 3. Influence of B supply for 15 days on amounts and composition of root bleeding sap collected for 4 h from two rape genotypes. Values are means of two replicates. For each cultivar, means in a row followed by the same letter were not significantly different ($p<0.05$).

	01		03	
	−B	+B	−B	+B
Amount of bleeding sap (g 5 plants^{-1})	2.29b	3.42a	3.28y	3.40y
Concentration (mg kg^{-1} fresh sap)				
Soluble sugar	93b	263a	189y	221y
Ca	391a	428a	392y	439y
Mg	184a	208a	164y	191y
K	902a	994a	741z	955y
NO$_3$	551b	476a	460z	563y
P	95a	97a	106y	127y
Total amount (μg plant^{-1})				
Soluble sugar	53b	179a	111y	147y
Ca	177b	297a	249z	295y
Mg	83b	143a	105z	129y
K	412b	665a	485z	645y
NO$_3$	254a	328a	303y	383y
P	43b	67a	68z	85y

bated for 5 h. The resultant solution containing root exudates was filtered and concentrated to 25 mL at 50° C under vacuum. Each sample contained exudate from about 80 rape plants and 500 mL sterilized water.

Plants were cut at the stem base on which was fitted a plastic tube filled with acid-washed cotton fibre that can easily absorb bleeding sap: 4 h later, the tubes were weighed. From the weight difference, the weight of bleeding sap was calculated. After that, the cotton fibre was washed several times with distilled water which was collected and concentrated to 25 mL.

Amino acids in root exudates and root bleeding sap were determined by an Amino Acid Analyzer, soluble sugar by spectrophotometry using the anthrone reagent; NO_3 by UV spectrophotometry at 220 nm wavelength; K, Ca and Mg by atomic absorption spectrophotometry and; P by spectrophotometry (Nanjing Agricultural University, 1986).

Results

Plant growth

The growth of rape plants was inhibited by withholding B for 15 days (Table 1). For both cultivars, the dry weight of roots was reduced significantly by B deficiency, but only the shoots of the inefficient cultivar were obviously decreased which accorded with the observed B deficiency symptoms in the B-inefficient cultivar.

Root exudates

For both cultivars, K exudation decreased and NO_3^- increased under B-deficient condition, but neither was statistically significant (Table 1). Boron deficiency halved the content of soluble sugars in root exudates of 01 cultivar, but it did not alter that of 03 cultivar. Total amino acid exudation for the two rape cultivars was nearly equal at the 2 B levels, but the composition of amino acids was markedly different between B treatments and between cultivars (Table 2). At B sufficiency, both cultivars secreted 12 kinds of amino acids; under B deficient conditions, both cultivars also exuded histidine (His) and arginine (Arg) whilst cultivar 03 exuded tyrosine (Tyr).

Root bleeding sap

For cultivar 01, the quantity of root bleeding sap decreased from 3.42 g at adequate B supply to 2.29 g at B deficiency; by contrast, for cultivar 03, it remained

Table 4. Influence of B supply on amino acids in root bleeding sap of two rape genotypes (μg plant^{-1} h^{-1}). Values are means of two replicates.

Amino acid	01		03	
	−B	+B	−B	+B
Ala	4.41	10.8	10.5	12.6
Arg	2.17	5.14	3.84	10.8
Cys	0.86	0.86	0.86	0.96
Glu	5.75	13.8	10.4	13.2
Gly	0.71	1.01	0.62	1.01
His	1.02	1.45	1.12	2.68
Ile	1.20	2.71	2.00	3.72
Leu	0.41	1.60	0.68	1.48
Lys	0.26	2.94	2.05	4.71
Met	Tr[1]	0.36	Tr	0.40
Phe	1.40	0.97	1.17	0.98
Pro	0.78	7.68	0.48	3.06
Ser	3.88	8.16	6.10	7.47
Thr	45.7	61.9	57.2	63.3
Tyr	Tr	0.92	0.69	1.21
Val	1.78	3.39	2.34	4.51
Total	70.3	124	100	132

[1] Trace amounts

unchanged with changing B supply (Table 3). Under B deficient conditions, the concentration of inorganic ions (Ca, Mg, K, P and NO_3) in root bleeding sap of both cultivars decreased to a small extent (NO_3^- in 01 cultivar was an exception), but their total amount declined remarkably, probably because of depressed dry matter. Both the concentration and total amount of soluble sugar were reduced by B deficiency in 01 cultivar but not in 03 cultivar (Table 3). The total amount of amino acids in bleeding sap of 01 and 03 cultivar were lessened by 43% and 24%, respectively, by B deficiency (Table 4). In B deficient plants, methionine (Met) was not detected for both cultivars, and Tyr was not detected for 01 cultivar, resulting in reduced numbers of amino acids in root bleeding sap under B deficiency. The contents of glycine (Gly), cysteine, valine, isoleucine, Arg and proline were reduced to similar extents by B deficiency for two cultivars; by contrast the decrement of threonine, serine, glutanine, alanine, leucine and lysine in 01 cultivar was more than that in 03, and that of His was lower in 01. The concentration of phenylalanine was increased by B deficiency in both cultivars. Whereas asparagine failed to be tested out, asparagine levels were relatively high especially with adequate B supply.

174

Discussion

The effect of B on inorganic ions such as K, NO_3^- was much different between root exudates and root bleeding sap (Table 1 and Table 3), indicating that ion secretion from the roots and transport to the shoots in rape plant are two independent processes, and the influence of B on ion absorption and transportation is complex. In a previous study, it was found that the cmin of K absorption was increased in 03 cultivar, but decreased in 01 cultivar (Cao, 1996). It could be concluded that the contents of inorganic ions in both root exudates and bleeding sap have no relationship with B-efficiency of rape. By contrast, the present results showed that the organic compounds were closely related with B deficiency susceptibility. The soluble sugars in both root exudates and root bleeding sap were obviously depressed by B deficiency in the inefficient cultivar (01), but not in the efficient one (03) (Table 1 and Table 3). The composition of sugars in exudates and root bleeding sap needs further investigating.

To some extent, the influence of B on amino acids differed in root exudates and bleeding sap (Table 2 and Table 4). In root exudates, no effect was observed on total amount, but their composition was affected by B deficiency. In addition, the B-inefficient cultivar (01) failed to secrete Tyr, but for the B-efficient one (03), much Tyr was detected in B deficient condition. Though the total amount and components of amino acids in bleeding sap responded differently to B compared to root exudates, there was a similarity in that Tyr seemed to be specific to the B efficient cultivar in B-deficient solutions (Table 4). At B deficiency, the Tyr in root bleeding sap in the B-inefficient cultivar (01) was not detected, however it only reduced in concentration in the efficient cultivar (03). Perhaps its content was too low to be determined, or it had specific relation with B efficiency.

Previous investigation showed that B deficiency was closely related with ethylene release in cotton plants (Wu, 1995), but this response seemed to not appear in rape plants since ethylene release was very low in both B adequate and deficient conditions (data not shown). Methionine which was the precursor for ethylene synthesis, in this experiment was not detected in bleeding sap in either B deficient solutions. Whether the low ethylene release in oilseed rape was related to low free Met is an interesting subject and warrants further research.

Acknowledgments

This work was supported by National Natural Science Foundation of China and the Doctoral Scientific Foundation of the National Education Committee of China. The senior author is grateful to the Australian Agency for International Development for financial assistance under its International Seminar Support Scheme.

References

Belver A, Roldan M, Rodriguez-Rosales M P and Donaire J P 1988 Plant senescence in relation to boron. Plant Physiol. Biochem. 26, 383–388.

Blaser-Grill J, Knoppik J D, Amberger A and Goldbach H 1989 Influence of boron on membrane potential in *Elodea densa* and *Helianthus annuus* and H^+ extrusion of suspension cultured *Daucus carota* cells. Plant Physiol. 90, 280–284.

Cao X 1996 The research on root physiology of two rape genotypes responding differently to boron deficiency. Ph D thesis. Huazhong Agric. University, Wuhan, PR China. (In Chinese with English abstract).

Cao X, Fu F and Liu W 1995 The effect of boron on the pH of nutrient solution in several crops. J. Huazhong Agricultural University. Sup. Sum. 21, 28–30. (In Chinese with English abstract).

Hoagland D R and Arnon D I 1950 The water-culture method for growing plants without soil. *In* California Agricultural Experiment Station Circular 347. The College of California, Berkeley, CA.

Liu C and Chen Z 1986 Boron nutrition of oilseed crops. *In* Research and Application of Micronutrient Fertilizers, Ed. Agriculture Bureau of the Ministry of Agriculture. pp 112–134. Hubei Science and Technology Press, Wuhan. (In Chinese).

Nanjing Agricultural University 1986 The Analysis of Soils, Plants and Fertilizers. Chinese Agriculture Press, Beijing. pp 214–254. (In Chinese).

Peng Q, Pi M and Liu W 1995 Screening of rape cultivars in response differently to boron deficiency. J. Huazhong Agricultural University. Sup. Sum. 21, 92–97. (In Chinese with English abstract).

Wang Y and Lan L 1995a A study on boron efficiency of rape (*Brassica napus* L.) (I). J. Huazhong Agricultural University. Sup. Sum. 21, 71–78. (In Chinese with English abstract).

Wang Y and Lan L 1995b A study on boron efficiency of rape (*Brassica napus* L.) (II). J. Huazhong Agricultural University. Sup. Sum. 21, 79–82. (In Chinese with English abstract).

Wu L 1995 The study on interaction of boron and plant hormone. J. Huazhong Agricultural University. Sup. Sum. 21, 18–20. (In Chinese with English abstract).

Xiong S, Wu L and Wang Y 1995 Absorption and distribution of boron in different varieties of rape (*Brassica napus* L.). J. Huazhong Agricultural University. Sup. Sum. 21, 85–91. (In Chinese with English abstract).

Yang Y, Xue J, Ye Z and Wang K 1993 Responses of rape genotypes to boron application. Plant and Soil 155/156, 321–324.

R.W. Bell and B. Rerkasem (eds.), Boron in Soils and Plants, 175–177.
© *1997 Kluwer Academic Publishers.*

Role of boron in the plasmalemma turbo reductase activity

E. Cseh & F. Fodor

Eötvös University, Department of Plant Physiology, P.O.B. 330. H-1445, Budapest, Hungary

Key words: boron, microelement composition, nutrient solution, plasmalemma, turbo reductase

Abstract

Turbo reductase activity of the roots of 8-day-old, iron-deficient cucumber (*Cucumis sativus*) plants was investigated by FeCN-reduction assay. The turbo reductase activity was induced only in the plants grown in nutrient solution containing boron (B). Boron either present in the assay solution or applied during pre-treatment for 4 hours had no effect. Therefore the plasmalemma turbo reductase cannot be the direct target of B. It is suggested that the small amount of B entering the cytoplasm and tightly bound there facilitates the induction of turbo reductase of iron-deficient (strategy I) plants via an effect on the metabolism. Manganase also stimulated ferricyanide reduction but its simultaneous supply with B did not result in an additive effect.

Introduction

Boron (B) is required for membrane integrity and function influencing ion influx (Pollard et al., 1977) and efflux (Cakmak et al., 1995) suggesting a role for B in the plasmalemma. Ion influx is stimulated by adding B within 20 minutes showing a direct effect. In root and leaf cells and in cells growing in suspension cultures, the presence of B causes hyperpolarization and increases H^+ efflux (Blaser-Grill et al., 1989; Goldbach et al., 1990; Schon et al., 1990).

The essentiality of B in plant growth is apparently related to its role in the cell wall structure, but the function of the unremovable portion in the cytoplasm remained obscure (Brown and Hu, 1994). Since B hyperpolarizes, and its lack depolarizes, the membrane, it seems important to investigate its effect on the transmembrane electron transport e.g. the reductase activity.

Plant species that acquire Fe according to strategy I (Bienfait, 1988) have a standard and a so-called turbo reductase in the plasmalemma (Römheld, 1987). Although the activity of the two reductase enzymes are different in magnitude, the turbo reductase has not been isolated, while the standard reductase is quite well characterised (Askerlund et al., 1991; Bagnaresi and Basso, 1996; Fredlund, 1996; Fredlund et al.,

1996; Holden et al., 1994). On the basis of preliminary experiments with cucumber (*Cucumis sativa*) and wheat (*Triticum aestivum*) we supposed that the failure in isolating the turbo reductase occurred because it is induced only in strictly determined conditions. Apart from the effects of plant species and iron deficiency, nitrogen-forms, nutrient supply and especially microelement composition in the culture medium are also important in the induction of the turbo reductase. The purpose of the present paper was to prove the requirement for B in the induction of the turbo reductase.

Materials and methods

The experiments were conducted with 8-day-old, iron-deficient cucumber (*cv.* Budai korai). Seeds were germinated on two layers of wet filter paper in a dark thermostat at 26° C for 36 hours. Ten plantlets with about 5 mm radicles were placed on Netlon discs between rings of polystyrene foam floating over 0.5 mM CaSO$_4$ solution. The plantlets were kept in the dark for an additional day then transferred to nutrient solution and light. The nutrient solution contained 1.25 mM KNO$_3$, 1.25 mM Ca(NO$_3$)$_2$.4H$_2$O, 0.5 mM MgSO$_4$.7H$_2$O, 0.25 mM K(H$_2$PO$_4$). Microelements were supplied in different mixtures according to the aim of the experiment: 11.6 μM H$_3$BO$_3$, 4.6 μM MnCl$_2$.4H$_2$O, 0.19 μM ZnSO$_4$.7H$_2$O, 0.12 μM Na$_2$MoO$_4$.2H$_2$O, 0.08 μM CuSO$_4$.5H$_2$O. Iron was not supplied at all. The culture solution was renewed every second day. The plants were irradiated with 75 W m^{-2} light intensity and with 10/14 h night/day photoperiod. The temperature was 20/25°C night/day and the relative humidity of the air was 55-60%.

Reducing capacity of the roots was measured as FeCN reduction by the method of Avron and Shavit (1963). Ten intact plantlets grown together in the same container were placed in incubation solution containing 2.5 mM KNO$_3$, 2.5 mM Ca(NO$_3$)$_2$.4H$_2$O, 1 mM MgSO$_4$.7H$_2$O, 0.5 mM KH$_2$PO$_4$ and 0.1 mM FeCN. After 30-min continuous shaking (120 rpm), 8 mL samples were taken from the solution and added to buffer (1.8 mL) containing 1 mL 3 M Na-acetate, 0.5 mL 0.4 M citrate and 0.3 mL distilled water. 0.1 mL 0.01 M FeCl$_3$ and 0.1 mL 0.02 M BPDS were added to form a stable complex with Fe(II). FeCN is reduced by the roots to ferrocyanide which in turn reduces equivalent FeCl$_3$. The amount of Fe(BPDS)$_3$ was measured at 535 nm with a spectrophotometer. Five replicate samples were used in all measurements.

Results

In the experiment shown in Table 1, B was supplied to the plants only during the incubation period, 4 hours preceding the measurement or during the whole growth period. Clearly the turbo reductase activity can only be measured if B was supplied together with macroelements during the whole growth period.

In order to find out whether B is the only responsible microelement for the formation of the turbo reductase activity, two separate experiments were done. In the first experiment single microelements were successively omitted from the nutrient solution (Figure 1) while in the second one the additive effect of different microelements was tested (Figure 2). The omission of B prevented the formation of the reductase activity inducible by iron deficiency (Figure 1). In the nutrient solutions containing Cu, Mo, Zn and Mn in different mixtures, the supply of Mn significantly increased the reducing capacity. However, the combined effect of B and Mn was not additive.

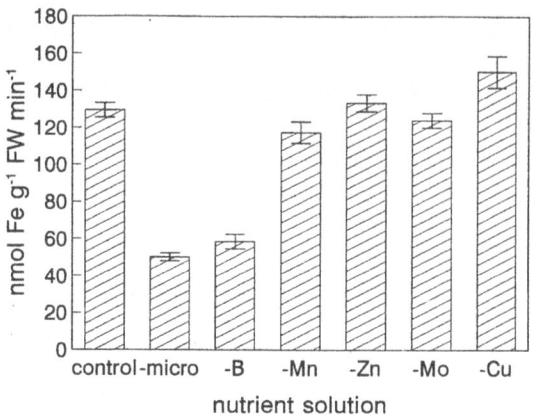

Figure 1. Reducing capacity of 8-day-old iron deficient cucumber (*Cucumis sativus*) roots supplied with nutrient solution containing all the necessary microelements (control), no microelements (-micro) or missing one microelement during the whole growth period. Values are presented as means of 5 replicates ±SE.

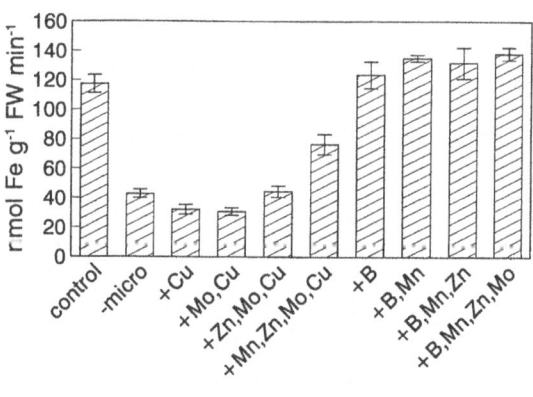

Figure 2. Reducing capacity of 8-day-old iron deficient cucumber roots supplied with nutrient solution containing all the necessary microelements (control), no microelements (-micro) or different mixtures of microelements during the whole growth period. Values are presented as means of 5 replicates ±SE.

Discussion

It has long been known from experiments with intact plants following strategy I in Fe acquisition that so-called standard and turbo redox systems are discernible (Bienfait, 1985, 1988). In the latter case, the enhanced H$^+$ and electron release from the root cell plasmalemma is induced by iron deficiency. The standard system may reduce electron acceptors with high redox potential (ferricyanide, O$_2$) while the turbo system reduces Fe-chelates and ferricyanide (Moog and Brüggemann, 1994).

Table 1. Short- and long-term effect of boron (B) on the reducing capacity of iron-deficient cucumber roots. Values are presented as means of five replicates $\pm$ SE.

$-B$[1]	+B (incubation)[2] (nmol Fe g^{-1} FW min^{-1})	+B (pretreatment)[3]	+B (growth period)[4]
35.2 ± 2.2	46.4 ± 4.3	39.4 ± 2.4	127.1 ± 5.9

Boron was supplied:
[1] neither in the growth period nor during the experiment
[2] during the incubation in the experiment only
[3] 4 hours before the experiment but not during the incubation
[4] during the whole growth period

The results of this paper show that the presence of B in the nutrient solution is necessary for turbo reductase activity. Short-term application of B causes fast membrane hyperpolarization and ion uptake stimulation, but does not affect reducing capacity even in a 4-hour pre-treatment period. Only continuous supply of B during measurement of the turbo reductase activity is effective. We suggest that the small amount of B entering the cytoplasm facilitates the induction of turbo reductase of iron deficient (strategy I) plants via an effect on the metabolism.

References

Askerlund P, Laurent P, Nakagawa H and Kader J C 1991 NADH-ferricyanide reductase of leaf plasma membranes. Partial purification and immunological relation to potato tuber microsomal NADH-ferricyanide reductase and spinach leaf NADH-nitrate reductase. Plant Physiol. 95, 6–13.

Avron M and Shavit N 1963 A sensitive and simple method for determination of ferrocyanide. Anal. Biochem. 6, 549–554.

Bagnaresi P and Basso B 1996 Soluble maize root NADH ferric-chelate reductase. J. Plant Nutr. 19, 1171–1177.

Bienfait H F 1985 Regulated redox processes at the plasmalemma of plant root cells and their function in iron uptake. J. Bioenerg. Biomembr. 17, 73–83.

Bienfait H F 1988 Mechanisms in Fe-efficiency reactions of higher plants. J. Plant Nutr. 11, 605–629.

Blaser-Grill J, Knoppik D, Amberger A and Goldbach H 1989 Influence of boron on the membrane potential in *Elodea densa* and *Helianthus annuus* roots and H$^+$ extrusion of suspension cultured *Daucus carota* cells. Plant Physiol. 90, 280–284.

Brown P H and Hu H 1994 Boron uptake by sunflower, squash and cultured tobacco cells. Physiol. Plant. 91, 435–441.

Cakmak I, Kurz H and Marschner H 1995 Short-term effects of boron, germanium and high light intensity on membrane permeability in boron deficient leaves of sunflower. Physiol. Plant. 95, 11–18.

Fredlund K M 1996 NAD(P)H dehydrogenases in plant membranes. Doctoral dissertation, Section of Plant Physiology, Lund University, Lund.

Fredlund K M, Widell S, Struglics A, Askerlund P, Kader J, Bérczi A and Møller I M 1996 NADH-ferricyanide reductases in plant membranes. *In* Plant Membrane Biology. Eds I M Møller and P Brodelius, pp 143–151. Clarendon Press, Oxford.

Goldbach H E, Hartmann D and Rotzer T 1990 Boron is required for the stimulation of ferricyanide-induced proton release by auxins in suspension-cultured cells of *Daucus carota* and *Lycopersicon esculentum*. Physiol. Plant. 80, 114–118.

Holden M J, Luster D G and Chaney R L 1994 Enzymatic iron reduction at the root plasma membrane: Partial purification of the NADH-Fe chelate reductase. *In* Biochemistry of Metal Micronutrient in the Rhizosphere. Eds J A Manthey, D E Crowley, D G Luster, pp 285–294. CRC Press, Boca Raton, Florida.

Moog P R and Brüggemann W 1994 Iron reductase systems on plant plasma membrane. A review. Plant and Soil 165, 241–260.

Pollard A S, Parr A J and Loughman B C 1977 Boron in relation to membrane function in higher plants. J. Exp. Bot. 28, 831–841.

Römheld V 1987 Different strategies for iron acquisition in higher plants. Physiol. Plant. 70, 231–234.

Schon M K, Novacky A and Blevins D G 1990 Boron induces hyperpolarization of sunflower root cell membranes and increases membrane permeability to K$^+$. Plant Physiol. 93, 566–571.

R.W. Bell and B. Rerkasem (eds.), Boron in Soils and Plants, 179–182.
© *1997 Kluwer Academic Publishers.*

Growth of lateral buds versus changes of endogenous indoleacetic acid and zeatin/zeatin riboside content in pea plants grown under boron deficiency

Chunjian Li, Hongyu Yuan, Yigong Zhang & Fusuo Zhang
Department of Plant Nutrition, China Agricultural University, Beijing 100094, P.R. China

Key words: Apical dominance, boron, deficiency, indoleacetic acid, pea, zeatin/zeatin riboside

Abstract

The concentration changes of endogenous indoleacetic acid (IAA) and zeatin/zeatin riboside (Z/ZR) in different pea (*Pisum sativum*) organs and their possible relation to the growth of lateral buds were investigated under boron (B)- deficient treatment. The results showed that the level of endogenous IAA in the plants continuously decreased from the outset of the B starvation, while the content of Z/ZR rapidly increased and reached its maximum on the eleventh day after the commencement of B treatment. It was just at this time that the elongation of lateral buds was visibly observed. These results suggest that the content changes of endogenous IAA and Z/ZR might be responsible for the growth of lateral buds in pea plants grown under B deficiency.

Introduction

It is well known that boron (B) is an essential micronutrient for vascular plants, diatoms and some species of green algae (Loomis and Durst, 1992). Under B deficiency one of the rapid visible responses is the inhibition or cessation of root and shoot elongation (Bohnsack and Albert, 1977; Kouchi, 1977). Another morphological change under the same conditions is the growth of lateral buds in some plants (Li et al., 1996). This is puzzling since the growth of lateral buds is normally inhibited in the presence of the apex, and this is most likely mediated by the polar auxin transport system (Bangerth, 1989; Sachs and Thimann, 1967). Lateral buds can grow when plants lose their apical dominance, for example, when they are decapitated (Hall and Hillman, 1975; Thomas, 1983). In this case, the source of IAA is removed, and therefore cytokinin (CYT) concentration in plants increases, which results in growth of lateral buds. This may imply that indoleacetic acid (IAA) originating from apex maintains CYT at low concentration (Bangerth, 1994; Li et al., 1995). However, the relationships between B nutrition and auxin levels in plants are still not clear. In B-deficient plants, auxin levels may increase (Coke and Whittington, 1968), remain

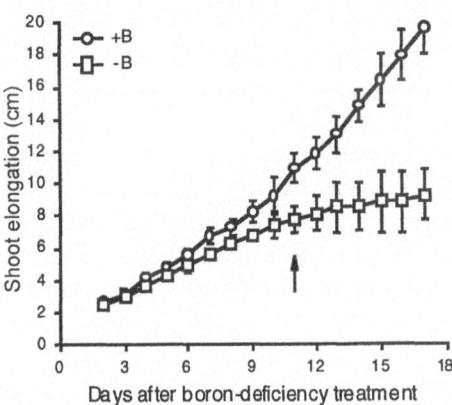

Figure 1. Effect of boron deficiency on shoot elongation of pea plants. Arrow indicates the time when the lateral buds started to elongate.

unchanged (Hirsch et al., 1982) or decrease (Fackler et al., 1985). The present research was conducted to measure the changes of endogenous IAA and Z/ZR concentrations. Their possible relation to the growth of lateral buds in pea plants grown under B deficiency condition are discussed in this paper.

Materials and methods

Seeds of pea (*Pisum sativum* L.) were surface sterilized with 10% (v/v) H_2O_2 for 5 min., washed with deionized water, and then germinated on wet sterile sand at 25°C in the dark. The seedlings were transferred 5 days later into an aerated nutrient solution containing (mM): K_2SO_4 0.75; $MgSO_4$ 0.65; KH_2PO_4 0.25; KCl 1.0; $Ca(NO_3)_2$ 2.0; FeEDTA 0.1; $MnSO_4$ 1×10^{-3}; $ZnSO_4$ 1×10^{-3}; $CuSO_4$ 5×10^{-4}; $(NH_4)_6Mo_7O_{24}$ 5×10^{-3}. There were two treatments, i.e., with $10 \mu M$ H_3BO_3 or without B supply.

Deionized water was used for the nutrient solution after being passed through an B-specific resin (Amberlite IRA -743) to remove B present as a contaminant. The pH of the solution was adjusted to 6.2.

The plants grew up to 17 days in a growth room (photoperiod 14/10h; 25/20°C; light intensity 220 μmol m^{-2} s^{-1}). Each pot had 1.5 L nutrient solution and contained 6 seedlings. The nutrient solution was replaced every three days. Lengths of 10 shoots were measured on the second day after the treatment commenced and repeated every day until the harvest. The number and fresh weight of the lateral buds in 10 plants was recorded at harvest.

The apex, stem and roots were separately excised at the different times after transferring the plants into the nutrient solution (see figures). Tissues of 30 plants from five separate pots were used each time to quantify the hormones. They were immediately frozen in liquid N_2 and stored at $-20°$ C until being analyzed for endogenous IAA and Z/ZR. The frozen tissues were homogenized in ice-cold methanol (80%, v/v) with polyvinylpyrrolidone (100 mg g^{-1} fresh wt.) and extracted overnight at 4°C. The supernatant was purified by passing it through a C-18 Sep Pak cartridge after centrifugation (40000 g, 20 min). Having been further washed with 80% methanol the samples were dried under vacuum. They are then dissolved in 0.1 M phosphate buffer (pH 7.5) and used for determining IAA and Z/ZR by ELISA (Zhang et al., 1991a,b). There were three analytical replicates of each sample.

Results

Omitting B from solutions inhibited shoot elongation of pea plants, especially on the eleventh or twelfth day after the treatment commenced (Figure 1). By contrast, the lateral buds started to elongate just as the elongation of the main shoot ceased (Figure 1). At harvest, the

Table 1. Growth of lateral buds in pea plants grown under B-deficient condition.

Node Number	Number of lateral buds per node		Fresh weight of lateral buds (g 6 buds^{-1})	
	+B	−B	+B	−B
First	0	0.55 ± 0.50	0	0.08
Second	0	1.18 ± 0.39	0	0.48
Third	0	1 ± 0	0	0.27
Fourth	0	1 ± 0	0	0.16
Fifth	0	0.82 ± 0.39	0	0.02

plants grown without added B had five nodes and there was a lateral bud in each node, although the fresh weight of the lateral buds was greatest at nodes two and three. By contrast no lateral buds grew on the control plants (Figure 1, Table 1).

A decrease of endogenous IAA in the apex, stem and roots was observed one day after the B-free treatment commenced. They continued to decline over the following 14 days. A steady increase of IAA concentration in the tissues of the control plants was evident during the same period (Figure 2A). In contrast, the concentration of endogenous Z/ZR increased significantly in the tissues of the plants grown without added B. It reached its maximum in roots and stem on the eleventh day after the treatment, and then declined in all of these tissues (Figure 2B).

Discussion

Boron deficiency resulted in growth of lateral buds in pea plants (Table 1, Figure 1), which can also be observed when a plant is decapitated (Thomas, 1983; Li et al., 1995), or CYT is applied to a node of an intact pea plant (Sachs and Thimann, 1967; Li and Bangerth, 1992). It is well accepted that IAA inhibits the growth of lateral buds and CYT has the opposite effect (Bangerth, 1989; Cline, 1994). IAA originating from the apex maintains a low CYT concentration in the shoot of intact pea plants, consequently, lateral buds can not grow out (Li et al., 1995). Based on this concept it is proposed that B deficient results in a decrease of endogenous IAA and an increase of Z/ZR. The changes of content of these hormones have to occur before the growth of lateral buds is visibly observed. This hypothesis is strongly supported by the present results (Figure 1, 2). These results suggested that the growth of lateral buds is due to the loss of apical dominance

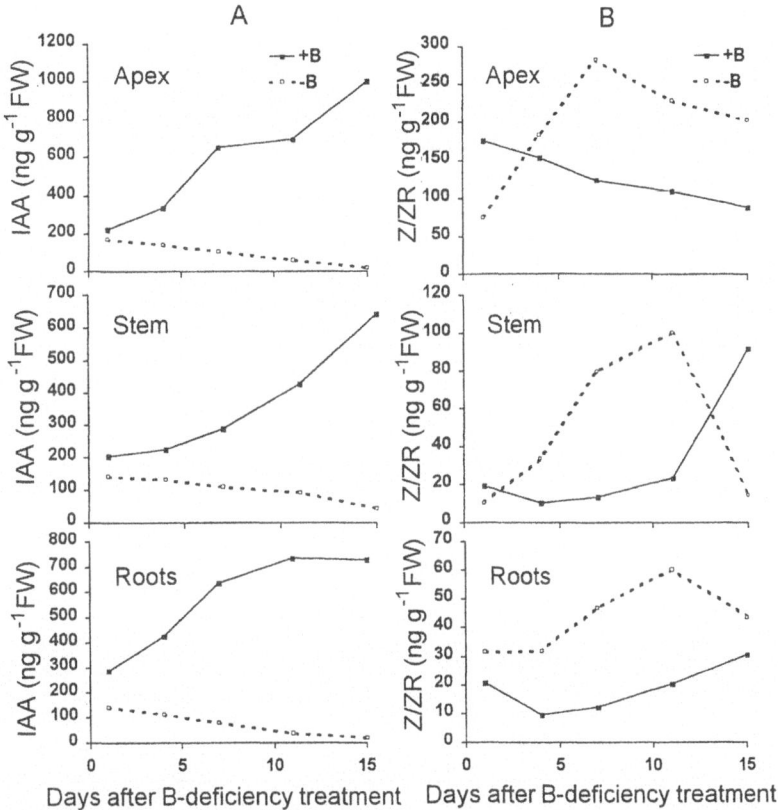

Figure 2. Changes of concentration of (A) indoleacetic acid and (B) zeatin/zeatin riboside (Z/ZR) in different tissues of pea plants grown with (+B) and without boron (-B) in solution for 15 days. 30 plants were used per analysis. There were no biological but three analytical replications. Variation of analytical replicates from the mean did not exceed ±3.5%.

in pea plants grown under B deficient conditions. The changes of concentration of endogenous IAA and Z/ZR might be responsible for it.

The results demonstrated clearly that B deficiency resulted in the decline of endogenous IAA (Figure 2A). The reason for it, however, is not clear. Since B is necessary for the biosynthesis of cell walls (Hu and Brown, 1994), young tissues, for example, tips of roots and shoots may cease elongating, and even die under the B deficiency (Kouchi, 1977). This may be one reason for the decline of endogenous IAA, because IAA is mainly synthesized in shoot apex and young leaves in plants. Another possibility is the destruction by IAA, since IAA oxidase activity increases dramatically soon after the B supply is cut off, and falls rapidly when B is resupplied (Bohnsack and Albert, 1977). In addition, from the synthesis site IAA can reach its active site by means of polar transport. Under B deficiency the IAA transport is reduced because of the impaired membrane integrity (Tang and de la Fuente, 1986). This may also be a possible reason for the decreased IAA concentra-

tion in stems and roots (Figure 2A) and therefore the increased Z/ZR content (Figure 2B).

Another unresolved question concerns the origin of B required for lateral shoot growth. Boron is necessary for the biosynthesis of cell walls, and therefore needed for the growth of lateral buds. Boron starvation inhibited the main shoot elongation but resulted in growth of lateral buds (Table 1). Withholding B supply began as the plants were transferred into the nutrient solution but the lateral bud growth took place eleven or twelve days later. Boron used for lateral bud growth clearly could not be supplied by the roots since apical shoot growth was inhibited by lack of B. It is unlikely that lateral buds stored sufficient B before their elongation commenced to supply the needs for growth. Whilst B is generally considered to be immobile in species like pea, the possibility of some retranslocation of B into lateral shoots should not be discounted.

182

Acknowledgments

The authors thank Dr L B Huang, School of Environmental Science, Murdoch University, Australia, for providing boron-specific resin, Dr L B Yan, Department of Natural Resource and Environmental Sciences, University of Illinois, U.S.A and Dr X L Yan, Department of Natural Resources and Environmental Sciences, South Agricultural University, China, for correcting the text, and the National Natural Science Foundation of China for providing financial support.

References

Bangerth F 1989 Dominance among fruits/sinks and the search for a correlative signal. Physiol. Plant. 76, 608–614.

Bangerth F 1994 Response of cytokinin concentration in the xylem exudate of bean (*Phaseolus vulgaris* L.) plants to decapitation and auxin treatment, and relationship to apical dominance. Planta 194, 439–442.

Bohnsack C W and Albert L S 1977 Early effects of boron deficiency on indoleacetic acid oxidase levels of squash root tips. Plant Physiol. 59, 1047–1050.

Cline M G 1994 The role of hormones in apical dominance. New approaches to an old problem in plant development. Physiol. Plant. 90, 230–237.

Coke L and Whittington W J 1968 The role of boron in plant growth, IV. Interrelationships between boron and indol-3-yl acetic acid in the metabolism of bean radicals. J. Exp. Bot. 19, 295–308.

Fackler U, Goldbach H, Weiler E W and Amberger A 1985 Influence of boron deficiency on indol-3yl-acetic acid and abscisic acid levels in root and shoot tips. J. Plant Physiol. 119, 295–299.

Hall S M and Hillman J R 1975 Correlative inhibtion of lateral bud growth in *Phaseolus vulgaris* L.: Timing of bud growth following decapitation. Planta 123, 137–143.

Hirsch A M, Pengelly W L and Torrey J G 1982 Endogenous IAA levels in boron-deficient and control root tips of sunflower. Bot. Gaz. (Chicago) 143, 15–19.

Hu H N and Brown P H 1994 Localization of boron in cell wall of Squash and Tobacco and its association with pectin. Plant Physiol. 105, 681–689.

Kouchi H 1977 Rapid cessation of mitosis and elongation of root tip cells of *Vicia faba* as affected by boron deficiency. Soil Sci. Plant Nutr. (Tokyo) 23, 113–119.

Li C J and Bangerth F 1992 The possible role of cytokinins, ethylene and indoleacetic acid in apical dominance. *In* Progress in Plant Growth Regulation, Eds. C M Karssen, L C van Loon and D Vreugdenhil, pp. 431–436. Kluwer Academic Publishers, Dordrecht, the Netherlands.

Li C J, Guevara E, Herrera J and Bangerth F 1995 Effect of apex excision and replacement by 1-naphthylacetic acid on cytokinin concentration and apical dominance in pea plants. Physiol. Plant. 94, 465–469.

Li C J, Tang Y L, Zhang F S and Cui J Y 1996 Effect of boron-deficiency on growth of shoot and roots and concentration of potassium in different plants. Journal of China Agricultural University 1(1), 17–21.

Loomis W D and Durst R W 1992 Chemistry and biology of boron. BioFactors 3, 229–239.

Sachs T and Thimann K V 1967 The role of auxins and cytokinins in the release of buds from dominance. Am. J. Bot. 54, 136–144.

Tang P M and de la Fuente R K 1986 Boron and calcium sites involved in indole-3-acetic acid transport in sunflower hypocotyl segments. Plant Physiol. 81, 651–655.

Thomas T H 1983 Effects of decapitation, defoliation and stem girdling on axillary bud development in brussels sprouts. Scientia Horticulturae 20, 45–51.

Zhang J, Han B W, Wu Q and He Z P 1991a Development of an indirect enzyme-linked immunosorbent assays for indole-3-acetic acid. Acta Agriculturae Universitatis Pekinensis (Supplement) 17, 139–144.

Zhang J, Han B W, Wu Q and He Z P 1991b Development of an indirect enzyme-linked immunosorbent assays for zeatin riboside. Acta Agriculturae Universitatis Pekinensis (Supplement) 17, 145–148.

R.W. Bell and B. Rerkasem (eds.), Boron in Soils and Plants, 183–186.
© 1997 *Kluwer Academic Publishers.*

Effect of pH and boron concentration in the nutrient solution on translocation of boron in the xylem of sunflower

Frank Dannel, Heidrun Pfeffer & Volker Römheld

Institut für Pflanzenernährung (330), Universität Hohenheim, 70593 Stuttgart, Germany

Key words: boron, *Helianthus annuus*, pH, translocation, uptake

Abstract

The influence of the pH in the nutrient solution (pH 3 to 10) on uptake and translocation of boron (B) in the xylem of young sunflower plants was investigated over a three hour treatment. Two B concentrations were supplied during the whole growth period of seven days: a marginal (1 μM) and a sufficient supply (100 μM). Boron concentrations in the xylem exudate responded both to B and pH treatments. When B supply was sufficient, B concentration in the xylem exudate was the same or lower than that in the nutrient solution and inversely related to the exudation rate. In contrast at marginal supply, B concentration in the xylem exudate was up to 25-fold higher than in the nutrient solution and decreased in direct proportion to the pH dependent concentration of undissociated boric acid. In conclusion, the results indicate that there might be two different mechanisms for uptake and translocation of B in sunflower. At sufficient supply, it seems that a passive diffusion process of boric acid, partly restricted by permeability of membranes, is responsible for uptake. At marginal supply, a concentration mechanism for boric acid builds up a gradient against the nutrient solution, obviously to satisfy the B demand of the plant.

Introduction

There have been several studies investigating the uptake mechanisms for boron (B), which are summarized in the reviews of Raven (1980) and Kochian (1991). Raven concluded from theoretical considerations, that B uptake is likely to be a passive process, but also pointed out, that active uptake or accumulation of B in the cell by ester formation with *cis*-diols cannot be ruled out. Kochian, basically in agreement with Raven, suggested the possibility that B uptake is a combination of active transport and passive diffusion of undissociated boric acid. The view that B uptake is a passive process is supported by the work of Oertli and Grgurevic (1975), who showed, that the amount of B taken up by excised barley roots is not dependent on total B concentration, but dependent on the concentration of free boric acid, which they varied by the pH of the uptake solutions. The aim of this study was to investigate the effect of pH in the nutrient solution not only on B uptake into the roots, as Oertli and Grgurevic did, but also on translocation of B in the xylem. For

this reason decapitated plants with an intact root system were used instead of excised roots and B translocation was monitored by analyzing xylem exudate.

Materials and methods

Plant culture

Sunflower seeds (*Helianthus annuus* cv. Frankasol) were germinated under controlled conditions at 25° C in quartz sand, irrigated with saturated $CaSO_4$ solution. Four day old seedlings were transferred to continuously aerated nutrient solutions containing (in mM): 0.7 K_2SO_4, 0.1 KCl, 2.0 $Ca(NO_3)_2$, 0.5 $MgSO_4$, 0.1 KH_2PO_4, 0.5×10^{-3} $MnSO_4$, 0.5×10^{-3} $ZnSO_4$, 0.2×10^{-3} $CuSO_4$, 0.01×10^{-3} $(NH_4)_6Mo_7O_{24}$, 0.02 Na Fe(III)-EDTA. Boron was added as H_3BO_3 in concentrations of 1 μM (B 1; marginal supply) and 100 μM (B 100; sufficient supply). The plants were grown under controlled environmental conditions with a light/dark regime of 16/8 h, air temperature of 25° C, photon

Table 1. pH in nutrient solutions after short term pH treatments and harvest of sunflower plants, precultured with different boron supply for seven days. Data represent the mean $\pm$ SD of 3 replicates. Boron supply during preculture and short term treatments was $1\mu M$ (B1) or $100 \mu M$ (B100).

pH at start of treatment	pH after collection of xylem exudate	
	B1	B100
3	3.08 ± 0.06	3.07 ± 0.01
4	4.10 ± 0.02	4.06 ± 0.07
5	5.20 ± 0.07	5.23 ± 0.10
6	5.98 ± 0.01	5.92 ± 0.04
7	7.00 ± 0.02	6.96 ± 0.02
8	7.95 ± 0.02	7.91 ± 0.03
9	8.63 ± 0.05	8.52 ± 0.08
10	8.95 ± 0.04	9.11 ± 0.08

flux density of about 450 μmol m^{-2}s^{-1} and a relative humidity of about 70%. During preculture, the nutrient solutions were renewed once after three days.

Short term treatment and harvesting

After seven days of preculture in nutrient solution, plants were transferred for two hours to nutrient solutions containing 5 mM (N-[2-Hydroxyethyl]-piperazine-N'-[2-ethanesulfonic acid]) (HEPES) and adjusted with H_2SO_4 or KOH to pH 3, 4, 5, 6, 7, 8, 9, 10. Boron concentration was the same as during preculture (1 or 100 μM, respectively). Then plants were decapitated at the hypocotyl and xylem exudate was collected over a 1 h period. Thereafter as a check, pH in the nutrient solutions was measured.

Analysis of xylem exudate

Boron was determined by the azomethine-H method (modified after Lohse, 1982) after concentrating the samples by drying under vacuum and dissolving in a small volume of 0.5 N H_2SO_4. The concentration of potassium was determined by flame photometry without further sample preparation other than dilution with double distilled water.

Statistics

Mean values, standard deviations and number of replicates are indicated in tables and figures. The data were statistically analysed by *t*-testing.

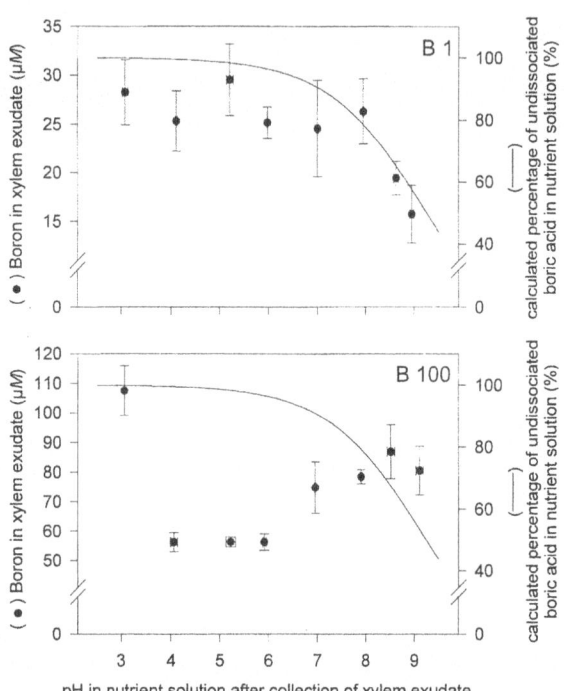

Figure 1. Boron concentrations in the xylem exudate of sunflower plants, precultured with different boron supply for seven days, in response to short term pH treatments. Boron supply during preculture and short term pH treatments was 1 μM (B 1) or 100 μM (B 100). Dots represent the mean of three replicates and vertical and horizontal bars indicate the standard deviation of boron concentration and pH values, respectively. The solid line represents the pH dependent percentage of undissociated boric acid in the nutrient solutions, calculated from boric acid – borate equilibrium.

Results

Despite buffering the nutrient solutions for the short term treatment with HEPES, the pH of the treatments with high pH (pH 9 and 10) was markedly lower after collection of xylem exudate (Table 1) and not affected by the B concentration in the nutrient solutions. Boron concentrations in the xylem exudate responded both to B and pH treatments (Figure 1). In the B 1 treatment, B concentrations in the xylem exudate were constant from pH 3 to 8 and decreased subsequently at higher pH. This decrease corresponds to the calculated decrease in concentration of undissociated boric acid in the nutrient solution. In the B 100 treatment (Figure 1), the highest B concentration in xylem exudate was observed at pH 3, decreasing to about 50% at pH 4 to 6 and increasing at higher pH to about 80%. There is obviously no correlation between B concentration in xylem exudate and the calculated concentration of undissociated boric acid in the nutri-

Table 2. Response to short term pH treatments of exudation rate and concentration of potassium in xylem exudate of sunflower plants, precultured with different boron supply for seven days. Data represent the mean $\pm$ SD of 3 replicates. Boron supply during preculture and short term treatments was 1 μM (B1) or 100 μM (B100).

| | Exudation rate | | Potassium concentration | |
| | (μ L h^{-1} [g root fresh weight]$^{-1}$) | | (mM) | |
pH	B1	B100	B1	B100
3	28.7 $\pm$ 7.8	23.7 $\pm$ 1.4	19.3 $\pm$ 1.2	18.4 $\pm$ 0.5
4	109 $\pm$ 5.8	86.5 $\pm$ 8.5	17.4 $\pm$ 1.8	16.9 $\pm$ 0.5
5	98.8 $\pm$ 6.5	94.5 $\pm$ 2.7	20.4 $\pm$ 1.2	19.1 $\pm$ 1.1
6	90.7 $\pm$ 6.2	93.3 $\pm$ 7.3	18.3 $\pm$ 1.2	16.7 $\pm$ 1.0
7	80.5 $\pm$ 4.1	58.4 $\pm$ 14.3	17.6 $\pm$ 1.4	16.7 $\pm$ 1.2
8	53.8 $\pm$ 1.6	32.2 $\pm$ 3.7	17.0 $\pm$ 1.2	16.0 $\pm$ 1.1
9	48.1 $\pm$ 7.0	24.3 $\pm$ 4.7	17.5 $\pm$ 0.6	16.0 $\pm$ 2.5
10	42.1 $\pm$ 1.5	27.8 $\pm$ 8.3	15.5 $\pm$ 0.8	14.6 $\pm$ 0.8

ent solution. The exudation rate was strongly affected by the pH treatment, whereas the B concentration in the nutrient solution played a minor role (Table 2). At pH 3, the exudation rate was only about 25% of the exudation rate at pH 4 to 6 and decreased at higher pH. In the B 1 treatment, the exudation rate at pH 10 was about 40% and in the B 100 treatment about 25% of the maximum. The B concentration in the xylem exudate of the B 100 treatment was inversely related to the exudation rate. Potassium concentration in the xylem exudate was determined for comparison. It ranged from 15 to 20mM and did not respond to pH in the nutrient solution and B supply (Table 2).

Discussion

The pH of the nutrient solutions adjusted to 9 and 10 declined after the short term treatment and collection of xylem exudate (Table 1). The reason for this may be the secretion of protons from the plant roots, which are not fully buffered by HEPES (pK$_a$ = 7.5, i.e. useful pH range 6.8 to 8.2). Irrespective of pH in the nutrient solution potassium concentration in the xylem exudate was nearly the same in all treatments (Table 2), supporting the concept that uptake and loading of potassium to the xylem is the driving force of exudation. The low exudation rate at pH 3 (Table 2) is likely to be caused by the decrease of the electropotential of root cells due to enhanced influx of protons from the apoplasm into the cells. In general, this can lead to an inhibition of cation uptake (Marschner, 1995). The decrease in exudation rates in treatments with pH$\geq$7 (Table 2) could result

from a lack of protons for proton-anion cotransport from the apoplasm into the cytoplasm. Thus, anion uptake and in consequence cation and water uptake may be inhibited (Marschner, 1995).

The findings, that the B concentrations in the xylem exudate differed not only in relation to B treatment, but also in relation to the pH of the outer medium (Figure 1) suggest that the mechanisms for uptake and translocation of B might differ between the B supplies of 1 and 100μM.

At a B supply of 100 μM, the results can be explained by passive diffusion processes, which Raven (1980) proposed as one possibility for B distribution. These theoretical considerations were partly based on the results of Bingham et al. (1970), who showed, that B diffuses passively into excised barley roots and that treatments with metabolic inhibitors or lower temperature in nutrient solution have no effect on B uptake. Brown and Hu (1994) demonstrated, that B uptake in cultured tobacco cells increased linearly with increasing B levels (25 μM to 10 000 μM) and Garnett et al. (1993) showed that the flux of B into root tissue of wheat is fast and reversible and obviously caused by the passive diffusion of B between the root symplasm, the apoplasm and the external solution. As shown in Figure 1, at high exudation rates (pH 4 to 6) the B concentration in the xylem exudate of plants grown at 100 μM B was lower than in the nutrient solution. This suggests that diffusion of B into the xylem was too slow to allow equilibrium between the nutrient solution and xylem exudate. In contrast, at pH 3 the B concentrations in xylem exudate and nutrient solution were equal, suggesting that the exudation rate was low enough to allow full adjustment of diffusion equilibrium. Although the exudation rates at pH $\geq$ 8 were as low as at pH 3, the B concentrations in the xylem exudate reached only about 80% of the B concentration in the nutrient solution. This might be due to the increasing percentage of borate, which in contrast to the undissociated boric acid, diffuses at a much lower rate across membranes. In conclusion the results suggest, that the flux of water into the xylem is less restricted than the flux of B. This is in accordance with Nissen (1974) who concluded from literature that at higher B concentration the plasma membrane may be an apparent barrier to free diffusion. Nable (1988) likewise suggested that root plasma membranes of wheat and barley may be a barrier to B uptake, governing the different accumulation of B amongst genotypes with differences in tolerance to B toxicity. These conclusions seem to be contradictory to the findings of Garnett et al. (1993)

186

that the B influx into wheat roots is quite rapid. But, recalculating from Garnetts influx data shows that B concentration in the roots did not exceed 50% of the B concentration in the external solution. This is similar to the relation between B concentration in xylem exudate and external solution (pH 4 to 6) calculated from the results presented here. This indicates that there must be some restriction for the influx of B in Garnetts experiments, too.

In contrast to the sufficient B supply, uptake and translocation of B at 1 μM B supply seemed to be caused by another mechanism. As shown in Figure 1, B concentration in the xylem exudate and the percentage of undissociated boric acid in the nutrient solutions, calculated from boric acid – borate equilibrium, decrease in a similar way at pH $\geq$ 8. This indicates clearly that boric acid and not borate is mainly taken up and translocated in the xylem of sunflower plants supplied with 1 μM B. This is in agreement with Oertli and Grgurevic (1975), who concluded from the pH dependent uptake of B in excised barley roots, that undissociated boric acid was taken up. Surprisingly, the B concentration in the xylem exudate is 25-fold higher than the B concentration in the nutrient solution (Figure 1). Thus it is obvious that at marginal B supply there is a concentration mechanism for B, which is able to build up a B gradient against the nutrient solution. It can not be concluded from the experiments presented here whether this concentration mechanism is acting at the level of uptake or at the level of xylem loading. However, the findings of Pfeffer et al. (1997) that the B concentration in xylem exudate of sunflower plants was always nearly the same as in root cell sap suggest that the key step of the concentration mechanism is the uptake of B from the nutrient solution into the symplasm of the root cells, followed by passive loading of the xylem.

In summary the results suggest, that there are two mechanisms for uptake and translocation of B in sunflower. At sufficient supply it seems, that passive diffusion of boric acid, partly restricted by permeability of membranes, is responsible, whereas at marginal supply a concentration mechanism for boric acid builds up a gradient against the nutrient solution to satisfy the B demand of the plant.

Acknowledgements

The authors like to dedicate this work to the late Prof. Dr. Drs. h. c. H. Marschner who originated and supervised the experiments presented here. When writing this manuscript, we missed his guidance and his constructive comments.

We are grateful to Borax Consolidated Limited, United Kingdom and the Deutsche Forschungsgemeinschaft for financial support and to Ernest A. Kirkby (University of Leeds) for critical evaluation and correction of the manuscript.

References

Bingham F T, Elseewi A and Oertli J J 1970 Characteristics of boron absorption by excised barley roots. Soil Sci. Soc. Amer. Proc. 34, 613–617.

Brown P H and Hu H 1994 Boron uptake by sunflower, squash and cultured tobacco cells. Physiol. Plant. 91, 435–441.

Garnett T P, Tester M A and Nable R O 1994 The control of boron accumulation by two genotypes of wheat. Plant and Soil 155/156, 305–308.

Kochian L V 1991 Mechanisms of micronutrient uptake and translocation in plants. In Micronutrients in Agriculture, 2nd ed. Ed J J Mortvedt, F R Cox, L M Shuman and R M Welch, pp 229–296. Soil Science Society America Book Series No.4.

Lohse G 1982 Microanalytical azomethine-H method for boron determination in plant tissue. Commun. Soil Sci. Plant Anal. 13, 127–134.

Marschner H 1995 Mineral Nutrition of Higher Plants, 2nd ed. Academic Press, London. 889 p.

Nable R O 1988 Resistance to boron toxicity amongst several barley and wheat cultivars: A preliminary examination of the resistance mechanism. Plant and Soil 112, 45–52.

Nissen P 1974 Uptake mechanisms: Inorganic and organic. Ann. Rev. Plant Physiol. 25, 53–79.

Oertli J J and Grgurevic E 1975 Effect of pH on the absorption of boron by excised barley roots. Agron. J. 67, 278–280.

Pfeffer H, Dannel F and Römheld V 1997 Compartmentation of boron in roots and its translocation to the shoot of sunflower as affected by short term changes in boron supply. In Boron in Soils and Plants. Proceedings Eds R W Bell and B Rerkasem. pp 203–207. Kluwer Academic Publishers, the Netherlands.

Raven J A 1980 Short- and long-distance transport of boric acid in plants. New Phytol. 84, 231–249.

R.W. Bell and B. Rerkasem (eds.), Boron in Soils and Plants, 187–190.
© 1997 *Kluwer Academic Publishers.*

187

Boron and calcium distribution in B and Ca deficient pea plants

E. Esteban, R. O. Carpena Ruiz & A. Gárate
Departamento de Química Agrícola. Universidad Autónoma de Madrid. Campus de Cantoblanco 28049 Madrid. Spain.

Key words: boron, calcium, mobility, pea

Abstract

The aim of this work is to study B and Ca distribution in pea (*Pisum sativum*) plants, and the effect of the deficiencies of both elements on this distribution. Pea plants were grown in a complete aerated nutrient solution (9 μM B and 2.0 mM Ca) for two weeks. Then, the youngest part of the plant was tagged, and four treatments were applied for two more weeks: control (complete), -B (without B), 1/5 Ca (with 0.4 mM Ca) and -B+2Ca (without boron and with 4.0 mM Ca). B concentration decreased significantly in the young shoot of both B-deficient treatments, while no effect was observed in roots. Calcium concentration in root, and to a lesser extent in young shoot, decreased for the 1/5 Ca treatment and increased for the -B+2Ca. Shoots were more sensitive to B nutrition and roots more sensitive to Ca supply. Boron distribution in -B plants suggests a degree of B remobilization from old to young shoots.

Introduction

Boron and calcium are considered nutrients of a relatively low mobility in plants (Marschner, 1995). Nevertheless, some reports show evidence of a certain degree of B retranslocation in different species: Shelp (1988) in broccoli, Hanson (1991) and Brown et al. (1992) in fruit trees, Shu et al. (1993) in peaches.

The relationship between B and Ca in higher plants has been studied by many workers. Reeve and Shive (1944) reported that B deficiency symptoms in tomato plants became more pronounced as the Ca concentration in culture solution was increased. Boron uptake and translocation were not affected by Ca supply (Yamanouchi, 1976), but B deficiency increased Ca uptake and decreased Ca translocation to the upper part of the tomato plant (Yamauchi, 1986a; Ramón et al., 1990). Blossom-end-rot in tomato fruits was observed when plants were cultivated with a deficient B supply (Carpena Ruiz et al., 1990).

In previous work pea plants showed a low B requirement (9 μM B; Gárate et al., 1993). The aim of this work was to study B and Ca distribution in pea

plants and the effects of deficiencies of both B and Ca on their distribution in pea plant.

Materials and methods

Plant growth

Sterilized pea seeds (*Pisum sativum* cv. Argona) were germinated at 28° C in darkness. Three-day-old seedlings were placed in a growth cabinet (16 h day-length, 25/15° C, RH 60/65% day/night), in plastic containers filled with one-tenth strength complete nutrient solution with aeration (Gárate et al., 1993). After five days, plants were transferred to pots with complete, full strength nutrient solution (9 μM B and 2.0 mM Ca). Two weeks later, the youngest part of the plant (third internode from the apex of the plant) was tagged, and treatments applied: -B (nutrient solution without B and with 0.1 g of Amberlite IRA-743, to avoid B contamination), 1/5 Ca (with 0.4 mM of Ca), -B+2Ca (without B, with 0.1g of Amberlite IRA-743 and with 4 mM of Ca) and control (complete nutrient solution). Four replicate pots were established for each treatment. Nutrient solutions were renewed weekly. Plants were harvested after two weeks, and separated into young shoot (above the mark), old shoot (below the mark) and root. Roots were washed in demineralized water and dried with tissue paper before processing. Plant material was dried (80° C), and 0.50 g of each plant tissue were oven-ashed and acid-digested at 70° C for 30 min in 6 mL 1N HCl, taking them to a final volume of 25 mL. B concentration was measured in all nutrient solutions. Extreme care was taken in using demineralized water (in which B concentration was also measured) and plastic material for preparing the nutrient solutions. B levels in water and -B nutrient solutions were always under the detection limit of our method (4.5 μM B). pH values of nutrient solutions were adjusted to 6.0 before adding to the plants.

Analytical determinations

Boron was measured in the extracts obtained after mineralization of the tissues using azomethine H as the colorimetric reagent (Wolf, 1974). Ca was determined by atomic absorption spectrophotometry (Perkin Elmer 4000). Statistical analysis of the results obtained (ANOVA, $p=0.05$) was carried out using the statistical package SAS.

Results

Plants did not present visual symptoms of Ca deficiency. Visual symptoms of B deficiency were not observed in shoots, but roots from both B-deficient treatments were shorter and with a higher number of secondary roots than control roots. Nevertheless, at the end of the experiment, differences for plant dry or fresh weight between each of the treatments and the control were not statistically significant (Table 1).

Despite this lack of differences in plant weight, B concentration of shoots reflects clearly the treatments applied (Table 2). Boron levels in both -B young shoots, which had grown without B added to the nutrient solution, were significantly lower than B concentrations of the control young shoot. The B content in -B and -B+2Ca old shoots was significantly lower than in control old shoots. However, B concentration in roots was very similar for all the treatments. Ca deficiency did not seem to have any significant effect on B concentrations in the different plant parts, as also reported Yamauchi et al. (1986) for tomato plants.

The effects of low solution Ca were also observed in Ca concentrations in plants. Shortage of Ca significantly decreased Ca concentration in pea shoots, but the greater reduction was observed in roots (Table 2). Withholding B from nutrient solutions did not cause significant alterations in Ca concentration in pea plants. The higher Ca concentration in the nutrient solution of the -B+2Ca treatment significantly increased Ca concentration in roots and, to a lesser extent, in young shoots.

Discussion

Results show that B deficient pea plants managed to grow much like the control plants (Table 1) for two weeks, although the young shoot, which developed without added B in the nutrient solution, had significantly lower B levels (Table 2) than control young shoots. There was negligible external B supplied for the plants, as B was not added to the -B nutrient solution and Amberlite was used to adsorb residual B contamination, so the B content of young shoots of -B plants could probably come from the old shoots. The significantly lower B concentration in -B old shoots than in control old shoots might be also attributed to plant growth during the two weeks without B. We consider this factor of lesser importance since the growth of old shoot – if it occurred – was small in comparison with growth of young shoots (Table 1). Therefore, the lower B concentration of -B old shoots than in control old shoots of the same age (Table 2) is partly due to translocation of B from old shoots to young shoots, and possibly also to roots, although a dilution effect can not be discarded. Oertli (1993) found retranslocation of B from old shoots to roots of tomato plant, but not to the younger shoot. Differences between tomato and pea plant in relation to the pattern of growth under B deficiency can be observed comparing the results for dry weight obtained by Oertli (1994) and ours. After removing B from the nutrient solution, tomato grew very little in the following 21 days, while we obtained a growth similar to the control for pea after 14 days.

Boron transport from tissues with an average B concentration to growing points under a defficient B supply has also been reported in other plant species (McIlrath, 1965; Shelp, 1998; Shelp et al., 1995).

Boron concentration in young shoot was low when B was not included in the growth medium, while B concentration in roots maintain a similar value in -B and in control. Boron level in pea shoot increased as

Table 1. Effect of B deficiency and Ca shortage on fresh weight (FW) and dry weight (DW) of pea plants after two weeks of growth in control solution and two more weeks under B and Ca treatments. Values in the same column followed by the same superscript were not significantly different (*p*=0.05).

| | Plant weight (g plant^{-1}) | | | | | |
| | Young shoot | | Old shoot | | Root | |
Treatments	FW	DW	FW	DW	FW	DW
Control	2.4^{ab}	0.30^{a}	1.1^{a}	0.15^{a}	1.6^{ab}	0.10^{a}
−B	2.3^{ab}	0.31^{a}	1.2^{a}	0.15^{a}	1.6^{ab}	0.10^{a}
1/5 Ca	2.6^{a}	0.31^{a}	1.1^{a}	0.13^{a}	1.8^{a}	0.10^{a}
−B+2Ca	1.9^{b}	0.26^{a}	1.2^{a}	0.15^{a}	1.4^{b}	0.09^{a}

Table 2. Boron and Ca concentrations in young shoots, old shoots and roots of pea plants grown for two weeks in control nutrient solution and for two more weeks with B and Ca treatments. Young shoots developed during the two week treatment period. Values in each column followed by the same superscript were not significantly different (*p*=0.05).

| | B (mg kg^{-1} dry wt) | | | Ca (g kg^{-1} dry wt) | | |
| | Young | Old | Root | Young | Old | Root |
Treatments	shoot	shoot		shoot	shoot	
Control	17.2^{a}	21.6^{a}	9.7^{a}	9.5^{b}	22.8^{a}	15.6^{b}
−B	2.6^{b}	13.7^{c}	9.2^{a}	10.6^{ab}	19.3^{ab}	13.3^{bc}
1/5 Ca	15.7^{a}	18.4^{a}	10.4^{a}	5.6^{c}	17.1^{b}	2.6^{c}
−B+2Ca	3.6^{b}	14.4^{bc}	6.7^{a}	12.1^{a}	19.5^{ab}	29.5^{a}

B concentration in the nutrient solution was increased (Esteban et al., 1996), but B concentration in roots did not change significantly. Taken as a whole, these results suggest that the shoots of pea plant are more sensitive to B treatments, while roots attempt to maintain constant B levels despite of the level of B in the culture solutions.

Calcium concentration in young shoot of the 1/5 Ca treatment was 41% lower than in the control (Table 2). The Ca supply in the 1/5 Ca nutrient solution was only 20% of Ca concentration in control nutrient solution, and Ca level in old shoots is 25% lower than Ca concentration in old shoots of control plants. Although the small effect of dilution because of old shoot growth may be considered, a certain remobilization of Ca from old to young shoots might also have occurred. The effects of Ca deficiency are more evident in roots, as Ca concentration in 1/5 Ca root decreases by 84% with respect to the control. When Ca level in the growth medium is increased (-B+2Ca treatment) a significant increase in Ca concentration in young shoot (27%) and an increase in Ca concentration in root of 89% with respect to the control were observed. Pea roots appear to be more sensitive to the Ca levels in the growth medium than pea shoots.

The major effects of B deficiency and Ca shortage appear to take place in different plant parts; roots are more sensitive to Ca shortage, and young shoots to B deficiency. Although both B and Ca have been considered as nutrients of low mobility, we have observed a relatively higher mobility for B than for Ca in pea plant. Despite the potential redistribution of B observed in this experiment, it would be unlikely to be adequate to maintain plant growth for a longer period of time. Indeed, pea plants of the same cultivar grown in the same culture conditions (Gárate et al., 1993) showed clear B deficiency symptoms when grown without B from the very first day of establishment in hydroponic culture.

Acknowledgments

This work was supported by the Spanish Comisión de Investigación en Ciencia y Tecnología, project number

190

PB92/1097. Dr. Philip Barak's revision of the manuscript is greatly appreciated.

References

Brown H P, Picchioni G, Jenkin M and Hu H 1992 Use of ICP-MS and [10]B to trace the movement of boron in plants and soil. Commun. Soil Sci. Plant Anal. 23, 2781–2807.

Carpena Ruiz R O, Ramón A M and Gárate A 1990 Podredumbre apical en frutos de *Lycopersicon esculentum* Mill. inducida por la ligera dificiencia de boro. Actas Hortic. 2, 341–346.

Esteban E, Gárate A and Carpena Ruiz R O 1996 N2 fixation in nodulated pea plants: optimization of boron requirements. *In* Transactions of the IX Nitrogen Workshop, pp 1–4. University of Brawnschweig, Brawnschweig.

Gárate A, Esteban E, Lucena J J, Carpena R O and Bonilla I 1993 Effect of boron on nodulated and non-nodulated pea plants. *In* Plant Nutrition-from Genetic Engineering to Field Practice. Ed N J Barrow, pp 417–420. Kluwer Academic Publishers, the Netherlands.

Hanson E J 1991 Boron requirements and mobility in tree fruit species. Current Topics in Plant Biochemistry and Physiol. 10, 149–158.

Marschner H 1995 Mineral Nutrition of Higher Plants. Academic Press, London. 889 p.

McIlrath W J 1965 Mobility of boron in several dicotyledoneous species. Botan. Gaz. 126, 27–30.

Oertli J J 1993 The mobility of boron in plants. *In* Plant Nutrition-from Genetic Engineering to Field Practice. Ed N J Barrow, pp 393–396. Kluwer Academic Publishers, the Netherlands.

Oertli J J 1994 Non homogeneity of boron distribution in plants and consequences for foliar diagnosis. Commun. Soil Sci. Plant Anal. 25, 1133–1147.

Reeve E and Shive J W 1944 Potassium-boron and calcium-boron relationship in plant nutrition. Soil Sci. 57, 1–4.

Shelp B J 1988 Boron mobility and nutrition in broccoli. Ann. Bot. 61, 83–91.

Shelp B J, Marentes E, Kitheka A M and Vivekanandan P 1995 Boron mobility in plants. Physiol. Plant. 94, 356–361.

Shu Z H, Oberly G H and Carry E E 1993 Time course study on the mobility and pattern of distribution of foliar-applied boron in peaches. J. Plant Nutr. 16, 1661–1673.

Wolf B 1974 Improvement in the Azomethine H method for determination of boron. Commun. Soil. Sci. Plant. Anal. 5, 39–44.

Yamanouchi M 1976 The effect of pH and calcium on the absorption and translocation of boron in plants. J. Sci. Soil Manure, 49–57.

Yamauchi T, Hara T and Sonoda Y 1986 Effects of boron deficiency and calcium supply on the calcium metabolism in tomato plant. Plant and Soil 93, 223–230.

R.W. Bell and B. Rerkasem (eds.), Boron in Soils and Plants, 191–195.
© 1997 *Kluwer Academic Publishers.*

Comparative study of foliar uptake and translocation of boron (^{10}B) supplied as four different products in radish plants

Françoise Kerrien[1], André Chamel[1] & Jean-Louis Imbert[2]
[1] *Laboratoire Transferts dans les Systèmes Végétaux, DBMS, CEA/Grenoble, 17 rue des Martyrs, 38054 Grenoble cedex 9, France*
[2] *Service Central d'Analyse, CNRS, Echangeur de Solaize, BP 22, 69390, Vernaison, France*

Key words: ^{10}B, boron, foliar absorption, ICP-MS, radish, translocation

Abstract

The uptake and translocation of ^{10}B by radish plants (*Raphanus sativus*) were determined 48 hours and 12 days after a foliar application of boron (B) in the form of four different ^{10}B enriched products: boric acid, borax decahydrate, high-solubility concentrate of sodium borate (Solubor) and an organic solution containing B (Cultibor). No significant effect of the type of product was found on the B uptake which reached about 10% of the applied amount after 48 hours but only 12.2% after 12 days. The ^{10}B translocation to non-treated parts including aerial parts and roots, increased by a factor 3 to 6 between the two intervals according to the type of product. A greater amount of ^{10}B was translocated 48 hours after foliar application when it was applied as Solubor. Moreover, results showed generally a better homogeneity of re-translocation with Solubor than other B products. In spite of this effect on translocation the source of B fertilizer had a relatively limited effect on the availability of foliar applied B.

Introduction

Boron (B), an essential micronutrient for higher plants is applied frequently by foliar spraying either to alleviate deficiencies or to supply directly an additional amount of this element to crops in order to improve their production, especially in the case of fruit trees, sugar beet and leguminous plants (Askew and Chittenden, 1936; Martini and Thellier, 1980; Schon and Blevins, 1990; Peryea and Drake, 1991). Most investigations on the uptake and the translocation of nutrients applied to leaves have been carried out with the help of radioactive isotopes (Chamel, 1988); however, for elements such as B, the lack of a convenient radioactive isotope has precluded such studies. In the case of foliar absorption, it is necessary to use methods which distinguish between the element applied to leaves from the same element already present in the plant or simultaneously absorbed by roots. Distribution of B after foliar application has already been studied by Thellier (1963) and Martini (1989) with the help of an enriched stable isotope and a (n, α) nuclear reaction, and by Chamel et

al. (1981) using spark-source mass spectrometry and laser-probe mass spectrography. The advent of inductively coupled plasma mass spectrometry (ICP-MS), a highly sensitive technique, made possible the discrimination between ^{10}B and ^{11}B and by working with ^{10}B boron it is possible to distinguish between applied and resident B (Brown et al., 1992; Shu et al., 1993, 1994; Brown and Hu, 1994). It is well-known that foliar uptake and translocation depend on many parameters related to the plant species, climatic conditions and characteristics of the applied solution (Chamel, 1988). The effects of such parameters were scarcely considered in the case of B. The aim of the present investigation was to compare the foliar uptake and translocation of B when applied as four different ^{10}B enriched boron products on radish plants (*Raphanus sativus*) using ICP-MS as the analytical technique.

192

Material and methods

Plant species and growth conditions

The experiments were performed on radish (variety ≪ Fluo ≫ de Vilmorin). Seeds selected were three-way F1 hybrids. They were first left to swell in Petri dishes then left on water-soaked filter paper for 48 h at 20–22° C. After germination, they were planted in pots (10 seeds pot^{-1}) containing compost (sedge and sphagnum peat) in a controlled climate chamber with the following day/night conditions: temperature, 22/15°C; humidity, 70/90%; photoperiod, 16/8h.

Boron products

The solutions were enriched with ^{10}B (90.5%) and made up to a B concentration of 20 mM (0.202 g L^{-1}). The characteristics of solutions were as follows:

– H_3BO_3 industrial-grade boric acid containing 0.225 g B L^{-1}, solution pH: 6.1

– $Na_2B_4O_7.10H_2O$ technical-grade borax decahydrate (borax) containing 0.238 g B L^{-1}, solution pH: 9.3

– Solubor: high-solubility concentrate of sodium borate containing 0.232 g B L^{-1}, solution pH: 8.8

– Cultibor: organic solution containing 0.229 g B L^{-1}, solution pH: 8.9.

Foliar treatment

Plants were treated when they were 20 days old and the second pair of leaves was developed. Two 20 μL droplets were deposited on the midrib of one of the pair of first leaves. The drying time varied from 2 to 2.5 hours. Each 40 μL deposit corresponded to 0.8 μmol of B or 8.08 μg of total B deposited per plant treated. Each B solution was applied to 12 plants. Plants (1 to 4) in a single pot were treated with the same product and the various pots were distributed randomly inside the climate chamber. Eight plants distributed in two pots were treated with distilled water and were used as controls for the determination of B content in plants. The treated leaves were excised either 48 hours (1st experiment) or 12 days (2nd experiment) after the foliar application. They were washed twice for a total of 60 seconds in 1 liter distilled water and then non-treated aerial parts and roots were separated.

Preparation of plant samples for analysis

Plant samples were dried at 50° C for 72 hours then wet ashed with 12 N nitric acid and 15 N perchloric acid (50/50 v/v) in special Teflon beakers at 100° C. The extracts were diluted in double distilled water (1/10 v/v) and were stored in plastic bottles until they were analyzed.

Instrumentation

The ICP-MS used was the VG Elemental PQ 2+. A Meinhard C3 concentric glass nebulizer with a Scott-type double pass spray chamber and a VG Elemental torch were used for this study. A peristaltic pump (Gilson Minipulse) was used for solution uptake. For determinations of total B in plant samples, beryllium was used as the internal standard. Three measurements were carried out for each isotopic ratio and coefficients of variation were always lower than 1%. The sensitivity of the method could reasonably be estimated to less than 5 μg B L^{-1}.

Calculations

The enrichment of each part (in μg) was calculated from the isotopic ratios $^{10}B/^{11}B$ and the initial content of B determined using ICP-MS. Individual values for isotopic ratios varied from 0.367 to 1.621, and from 0.234 to 0.292 for treated leaves and non-treated parts, respectively, for 48 hours absorption. Corresponding values for plants harvested 12 days after foliar application varied from 0.291 to 1.213 and from 0.234 to 0.302. Boron concentration of plants before the foliar treatment, obtained from control plants, was 20 mg B kg^{-1} dry wt. Foliar uptake was expressed in μg and as a percentage of the amount of B applied. Distribution of B in non-treated plant parts was expressed in μg and as a percentage of the amount absorbed by the treated leaf.

Statistics

In each experiment a particular treatment was applied to 12 plants. Mean values calculated for foliar uptake and translocation are given with their variation coefficients. Variance analysis and Newman-Keuls tests were used for statistical analysis using the STAT-ITCF program (STAT-ITCF, Manuel d'utilisation, 1991, ITCF, 8 Avenue du Président Wilson, 75116 Paris).

Table 1. Foliar uptake of [^{10}B] boron, 48 hours and 12 days after its application as four different products on the first leaf of radish (*Raphanus sativus*) plants.

Treatment	Foliar uptake	
	μ g	% of the amount applied
48 hours		
Boric acid	0.828 (47.1)	10.3
Borax decahydrate	0.651 (76.8)	8.1
Solubor	0.976 (35.9)	12.1
Cultibor	0.801 (35.0)	9.9
Significance level	NS	NS
12 days		
Boric acid	1.237 (48.1)	15.3
Borax decahydrate	0.739 (50.8)	9.2
Solubor	0.768 (32.9)	9.5
Cultibor	1.206 (55.8)	14.9
Significance level	NS	NS

Coefficients of variation are given in the brackets; NS: not significant.

Results

Mean values for foliar uptake when expressed as percentages of the amount deposited are 10.1% and 12.2% after 48 hours and 12 days, respectively; and no significant effect appeared between the treatments (Table 1). However, in the two experiments the lowest uptake percentage was observed for borax solution.

Results concerning B distribution in non-treated parts are reported in Table 2. They show that there is a significant treatment effect, after 48 hours, on the amount of B translocated towards the aerial parts and the total non-treated parts. The translocation was greater with Solubor but did not differ significantly between the three other treatments. Similar variations are observed, especially for aerial parts, when results are expressed as percentages of the B absorbed by the treated leaf. It must also be noted that results concerning Solubor exhibited generally lower variability. Translocation rates increased greatly from 48 hours to 12 days, by a factor varying from 3 to 6 according to the treatment to reach values comprised between 40.7 and 58.2% of foliar absorbed B. No significant effect appeared between the four treatments for B translocation after 12 days.

Discussion

Results of the present investigation using ICP-MS confirm the penetration and translocation of foliar applied B as previously shown with other analytical techniques (Martini and Thellier, 1980; Chamel et al., 1981; Martini, 1989). In our experimental conditions the penetration rates were relatively limited as foliar uptake reached only 10%, 48 hours after the foliar treatment, and 12.2% at 12 days post treatment. There was generally high variability which differed greatly between the series of experimental values. It was lower for plants treated with Solubor. However, no significant effect of different treatments on foliar uptake was revealed in spite of large differences in the B source and pH. Data previously reported for foliar B uptake show a great variability depending on plant species. Foliar uptake of ^{10}B, supplied as boric acid, by apple, pear, prune and sweet cherry shoot leaves was 88% to 96% complete within 24 hours of application (Picchioni et al., 1995). However, only 0.2–0.3% of the total foliar-applied ^{10}B, as boric acid, was absorbed by peach tree leaves (Shu et al., 1994).

The translocation of foliar-applied B towards non-treated aerial parts and roots increased greatly from 48 hours to 12 days. A major fraction of the amount absorbed remained in the treated leaf, about 88% 2 days after the foliar application but only 53% after 12 days. There was a significant effect of the chemical composition of the solution applied to leaves on the B dis-

Table 2. Distribution of [^{10}B] boron in untreated parts 24 hours and 12 days after its application as four different products on the first leaf of radish plants. Coefficients of variation are given in parentheses. Values in the same column not followed by the same letter are significantly different.

| | Boron distribution | | | | | |
| | Aerial parts | | Roots | | Total | |
	(μg)	(%)[a]	(μg)	(%)	(μg)	(%)
48 hours						
Boric acid	0.032 (94)[b]	3.9	0.023 (57)	2.8	0.055 (78)[b]	6.7
Borax decahydrate	0.054 (94)[b]	8.3	0.037 (65)	5.7	0.091 (68)[b]	14.0
Solubor	0.113 (38)[a]	11.6	0.039 (74)	4.0	0.152 (46)[a]	15.6
Cultibor	0.057 (88)[b]	7.1	0.049 (59)	6.1	0.106 (63)[ab]	13.2
Significance level	***		NS		***	
12 days						
Boric acid					0.504 (72)	40.7
Borax decahydrate					0.430 (64)	58.2
Solubor					0.365 (53)	47.5
Cultibor					0.525 (96)	43.5
Significance level					NS	

[a] % of the amount absorbed

***, significant at $p<0.001$; NS, not significant ($p=0.05$).

tribution in non-treated parts, 48 hours after the foliar application. The amount of B recovered in non-treated aerial parts was greater when B was applied to leaves with Solubor. However, this effect was temporary as there was no significant difference between treatments 12 days after the foliar application. Experimental values for translocation suggested a lower variability for plants treated with Solubor as observed above for foliar uptake. Our data confirm the high mobility of foliar applied B from leaves, though B translocation was not so rapid as reported in some studies (Picchioni et al., 1995). The question of B translocation from leaves and its phloem mobility has been largely discussed in literature (Gupta, 1993; Shelp et al., 1995). This translocation depends probably on leaf age and plant B supply as observed for mobile elements (Chamel, 1988). Recent studies on the translocation of foliar applied isotopically enriched ^{10}B in six tree species (Brown and Hu, 1996) suggest that B can only move out of a leaf as a stable complex with sorbitol, or possibly with other sugar alcohols such as mannitol. Further research is still necessary in this way to explain the variations observed between plant species and to clarify the nature of soluble B complexes involved in the phloem mobility of this element.

For foliar B practise, it is clear that uptake tended to be most uniform with Solubor and more rapid translocation to aerial parts was observed 2 days after the foliar application with this formulation. These results suggest an advantage to using Solubor to rapidly alleviate deficiency symptoms. Nevertheless, further investigation is still necessary in order to obtain a better efficiency of uptake of B from foliar treatments. It would be useful to develop formulations with appropriate additives to facilitate B uptake and to consider the influence of other parameters related to climatic conditions and to the plant species on uptake and retranslocation (Chamel, 1988). New formulations and application practices can be evaluated using ICP-MS which has proved to be very suitable for the investigation of foliar absorption of B using an enriched stable isotope.

Acknowledgments

This work, including a grant to F. K., was financially supported by Borax Europe Limited. We would like to thank Mr Pascal Cèbe and Mr Martin Phillips for their help in completing the project and paper.

References

Askew H O and Chittenden E 1936 The use of borax in the control of 'internal cork' of apples. III Effect of borax sprays on the boron status of fruit and incidence of 'internal cork' in apples. J. Pomol. Hortic. Sci. 14, 242–245.

Brown P H and Hu H N 1994 Boron uptake by sunflower, squash and cultured tobacco cells. Physiol. Plant. 91, 435–441.

Brown P H and Hu H N 1996 Phloem mobility of boron is species dependent: Evidence for phloem mobility in sorbitol-rich species. Ann. Bot. 77, 497–507.

Brown P H, Picchioni G, Jenkin M and Mu H 1992 Use of ICP-MS and [10]B to trace the movement of boron in plants and soil. Commun. Soil Sci. Plant Anal. 23, 2781–2807.

Chamel A 1988 Foliar uptake of chemicals studied with whole plants and isolated cuticles. *In* Plant Growth and Leaf-Applied Chemicals. Ed P M Neumann, pp 27–50. CRC Press, Boca Raton, Florida.

Chamel A, Andréani A M and Eloy J F 1981 Distribution of foliar-applied boron measured by spark-source mass spectrometry and laser-probe mass spectrography. Plant Physiol. 67, 457–459.

Martini F 1989 Imagerie de la distribution du bore dans les plantes. Thèse de doctorat, Faculté des Sciences et des Techniques de Rouen.

Martini F and Thellier M 1980 Use of an (n, α) nuclear reaction to study the long-distance transport of boron in *Trifolium repens* after foliar application. Planta 150, 197–205.

Peryea F J and Drake S R 1991 Influence of mid-summer boron sprays on boron content and quality indices of 'Delicious' apple. J. Plant Nutr. 14, 825–840.

Picchioni G A, Weinbaum S A and Brown P H 1995 Retention and kinetics of uptake and export of foliage-applied, labeled boron by apple, pear, prune, and sweet cherry leaves. J. Am. Soc. Hort. Sci. 120, 28–35.

Schon M K and Blevins D G 1990 Foliar boron applications increase the final number of branches and pods on branches of field-grown soybeans. Plant Physiol. 92, 602–607.

Shelp B J 1993 Physiology and biochemistry of boron in plants. *In* Boron and Its Role in Crop Production. Ed U C Gupta, pp 53–85. CRC Press, Boca Raton, Florida.

Shelp B J, Marentes E, Kitheka A M and Vivekanandan P 1995 Boron mobility in plants. Physiol. Plant. 94, 356–361.

Shu Z H, Oberly G H and Cary E E 1993 Time course study on the mobility and pattern of distribution of foliar-applied boron in peaches. J. Plant Nutr. 16, 1661–1673.

Shu Z H, Oberly G H and Cary E E 1994 Mobility of foliar-applied boron in one-year-old peaches as affected by environmental factors. J. Plant Nutr. 17, 1243–1255.

Thellier M 1963 Contribution à l'étude de la nutrition en bore des végétaux. Thèse de Doctorat d'Etat. Université de Paris.

R.W. Bell and B. Rerkasem (eds.), Boron in Soils and Plants, 197–201.
© *1997 Kluwer Academic Publishers.*

Absorption and distribution of boron in *Arabidopsis thaliana*

Kyotaro Noguchi*, Toru Fujiwara, Mitsuo Chino, Toshiro Matsunaga[1] &
Hisao Watanabe-Oda[1]
Department of Applied Biological Chemistry, Graduate School of Agricultural and Life Sciences, The University of Tokyo, Bunkyo-ku, Tokyo 113, Japan
[1] *National Institute of Agro-Environmental Sciences, Tsukuba-City, Ibaraki 305, Japan*
*Corresponding author

Key words: Arabidopsis thaliana, ^{10}B, inductively coupled plasma-mass spectrometry, tracer, translocation, transpiration stream, uptake.

Abstract

We investigated the absorption and distribution of boron (B) by tracer experiments in *Arabidopsis thaliana*, a model plant for molecular genetic studies. Plants were grown hydroponically in the presence of B at 12, 30 and $100\mu M$ (^{10}B/^{11}B = 19.9/80.1 atom%) for 4 weeks. Then, they were transferred to solutions containing $30\mu M$ of ^{10}B enriched boric acid (^{10}B/^{11}B = 95.9/4.1 atom%) and incubated for 48 hours (pulse labeling). Old rosette leaves (leaves 1–4), young rosette leaves (leaves 5-10) and shoot apices were harvested before and after the labeling. The ^{10}B and ^{11}B contents were determined by inductively coupled plasma mass spectrometry (ICP-MS).

Regardless of the levels of B supply, the B concentration was lower at the top of the shoot than the bottom. The ratio of B concentration in shoot apices to that in old leaves increased as B supply became restricted, suggesting that B translocation is regulated not only by transpiration stream but also by other mechanisms. Percent boron derived from tracer (%BDFT) ranged from 48.0 to 57.2% and from 6.8 to 14.9% in shoot apices and in old leaves, respectively. The amounts of B contained in shoot apices before the labeling were estimated to be 31.6 to 39.6% of those after the labeling, suggesting that most of the increase in B amount during the period of the labeling can be explained by tracer derived B.

These findings suggest that translocation and distribution of B in *A. thaliana* do not contradict to general patterns of B distribution in higher plants. Thus molecular genetic studies on B nutrition using *A. thaliana* are likely to provide us with further understanding of B functions in higher plants.

Introduction

Boron is one of the least understood essential elements for higher plants. Since Warington (1923) found that B is an essential micronutrient for the broad bean, various studies have been carried out to identify the physiological functions of B. Recently, more than 98% of total B was found to be localized in cell walls under B deficiency (Hu and Brown, 1994; Matoh et al., 1992). Furthermore, B in the cell wall formed ester bonds between two molecules of rhamnogalacturonan II, a polysaccharide constituent of pectin, in radish (*Raphanus sativus*) (Kobayashi et al., 1996) and in

sugar beet (*Beta vulgaris*) (Ishii and Matsunaga, 1996). These findings indicate that B plays an important role for the cell wall structure.

A number of studies were also conducted to clarify the mechanisms of B absorption and translocation in higher plants (see review by Shelp et al., 1995). At the cellular level, a linear increase of B absorption with increasing B supply was observed in cultured tobacco (*Nicotiana tabacum*) cells and it was not affected by metabolic inhibitors, suggesting that B is absorbed by simple and facilitated diffusion (Brown and Hu, 1994). On the other hand, a controversy remains over phloem mobility of B. Several reports suggested that

B is phloem immobile. Hu and Brown (1994) reported that squash plants transferred to B-free medium from B-sufficient medium showed severe growth inhibition in young leaves but not in mature leaves. Oertli (1994) also showed that following withdrawal of B, B deficiency was induced in young leaves whilst the same plants maintained high B concentrations in old leaves where toxicity symptoms were evident. In contrast, several reports suggested that B was phloem mobile. Liu et al. (1993) showed that B concentration was greater in the top of the shoot than in the bottom in field grown broccoli and that the B concentration was higher in phloem sap than in xylem sap. In addition, ^{10}B applied to spur leaves was transported to their adjacent fruit tissues in apple (Picchioni et al., 1995) and ^{10}B applied to leaves of peaches was transported to other plant parts such as fine roots and other leaves (Shu et al., 1994). Furthermore, foliar-applied B was freely mobile and translocated to adjacent fruit tissues in sorbitol-rich species such as almond (*Prunus dulcis*), apple (*Malus domestica*) and nectarine (*Prunus persica* var *nucipersica*), whereas it was largely immobile in sorbitol-poor species such as fig (*Ficus carica*), pistachio (*Pistacia vera*) and walnut (*Juglans regia*) (Brown and Hu, 1996).

The use of *A. thaliana* as an experimental plant has the potential to introduce molecular genetic approaches into the field of B nutrition and could lead to a better understanding of B nutrition. Here, we report the absorption and translocation of B in *A. thaliana* under various levels of B supply using ^{10}B enriched tracer and ICP-MS.

Materials and methods

Plant growth

Arabidopsis thaliana (L.) Heynh. ecotype Columbia, was grown hydroponically in a greenhouse at $22\pm1°$ C under natural light. The concentration of B in the hydroponic culture solution (Fujiwara et al., 1992) was modified to 12, 30 or 100 μM. Four weeks after sowing, when the inflorescence shoot elongated to about 10 cm in height, 6 to 8 plants were harvested from the total of 12 to 16 plants in each treatment. The other plants were transferred to the tracer medium, *i.e.*, culture solution containing 30 μM of ^{10}B enriched boric acid (95.9 atom% ^{10}B), then, harvested after 48 hours of the labeling. Old rosette leaves (leaves 1–4), young rosette leaves (leaves 5–10) and shoot apices were har-

vested separately. Before the labeling, woollen yarn was tied to the stem between the most newly opened flowers and the most developed flower buds. The portion of the shoot above woollen yarn was defined as the 'shoot apex.' It contained 6 to 8 flowers after the labeling, whereas it contained only buds before the labeling.

Sample preparation for ICP-MS

Harvested samples were air-dried at 60° C for more than 60 hours. The samples (0.3–10 mg) were weighed and transferred to previously weighed teflon tubes, and were digested with 0.3–3 mL of concentrated nitric acid on a hot plate at 120–130° C. Digested samples were left to complete dryness and were dissolved in 0.08 N nitric acid. Then, those tubes containing sample solutions were weighed. These samples were diluted with 0.08 N nitric acid to the appropriate B concentration (1–100 μg L^{-1}) for analysis. Beryllium was added to 1 μg L^{-1} and used as an internal standard. Then the samples were subjected to analysis with ICP-MS (SII SPQ8000A, Seiko Instruments Inc., Chiba, Japan). Standard solutions were measured every 6 to 10 samples.

Estimation of B absorption and translocation

Atom%, ^{10}B excess and percent B derived from tracer (%BDFT) were calculated from the isotopic ratio of the labeled samples following the procedure of Rennie and Rennie (1983);

atom% ^{10}B excess = atom% ^{10}B (treated sample) - atom% ^{10}B (control)

%BDFT (%) = (atom% ^{10}B excess in plant / atom% ^{10}B excess in tracer) x 100

Atom% ^{10}B of control (mean isotopic ratio of B in samples harvested before the labeling) was defined as 19.9% in this study.

Results

B contents in Arabidopsis thaliana grown under various levels of B

After 4 weeks of growth, no obvious difference was observed in plant appearance or dry weights among B treatment at 12, 30 and 100 μM B (data not shown). At all levels of B supply, the B concentration showed a decreasing gradient from the bottom to the top of the

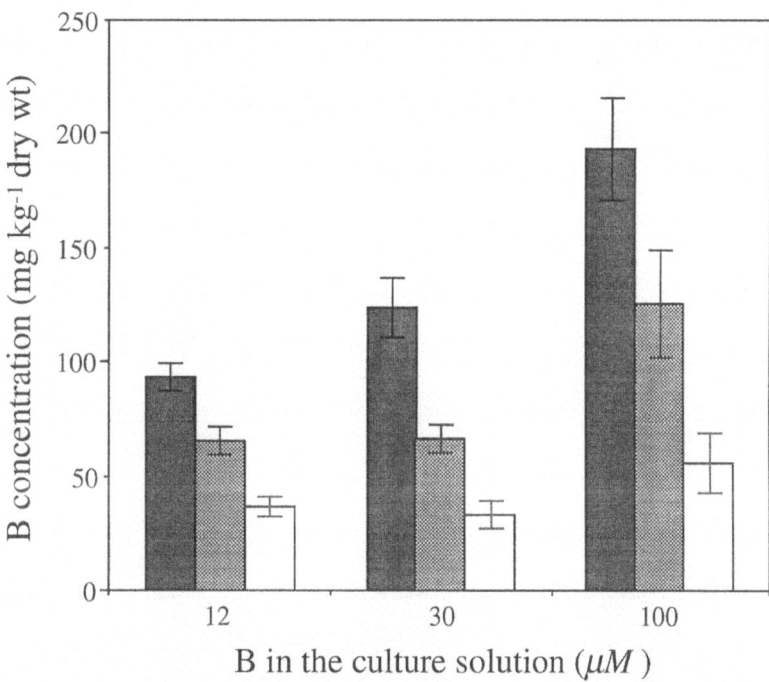

Figure 1. Boron concentrations in *A. thaliana* grown hydroponically at 12, 30 or 100 μM B for 4 weeks. Each column represents old leaves (■), young leaves(▨), and shoot apices(□). Each value is mean ± standard deviation of five replicates.

shoots (old leaves > young leaves > shoot apices) (Figure 1). However, the pattern of B distribution changed with the B supply. The ratio of B concentrations in shoot apices to that in old leaves was 0.40 at 12 μM B, whereas it was 0.27 and 0.29 at 30 and 100 μM B, respectively (Figure 1). Boron concentrations in shoot apices and young leaves at 12 μM B were approximately equal to those at 30 μM B. However, in old leaves, the B concentration at 12 μM B was significantly lower than that at 30 μM B (t-test, $p<0.005$) (Figure 1). At 100 μM B, the B concentration in every organ was significantly higher than those at 30 and 12 μM B (*t*-test, $p<0.01$) (Figure 1).

Distribution of newly absorbed B

After 48 hours of the labeling with ^{10}B, ^{10}B/^{11}B (w/w) value was increased in all samples, whereas it was 0.22 to 0.23 (mean value 0.226, equivalent to 19.9% by atom%) before the labeling (data not shown). This suggests that newly absorbed B was distributed to all plant organs. Atom% ^{10}B excess and %BDFT were calculated as described in materials and methods. In all treatment groups, %BDFT was higher in young organs than in old organs (old leaves < young leaves < shoot apices, Figure 2). The highest %BDFT was

57.2% in shoot apices at 30 μM B and the lowest was 6.8% in old rosette leaves at 100 μM B (Figure 2).

The highest %BDFT in shoot apices can probably be attributed to the greatest growth rate during the period of the labeling among plant parts. The total amounts of B in shoot apices before the labeling were roughly estimated at 31.6 to 39.6% of those after the labeling (Figure 3).

Discussion

In general, B deficiency first affects growing points, such as root tips and apical meristems (Loomis and Durst, 1992; Marschner, 1995). Higher B accumulation in mature organs than in immature organs was often observed (Shelp et al., 1995). These observations suggest that B is translocated mainly by the transpiration stream.

Overall patterns of B accumulation in *A. thaliana* observed in this study were similar to the patterns observed in other plant species (Figure 1). However, the detailed patterns of distribution depended on the levels of B supply. The B concentration in shoot apices at 12 μM B was approximately equal to that at 30 μM B, whereas, in old leaves, it was significantly

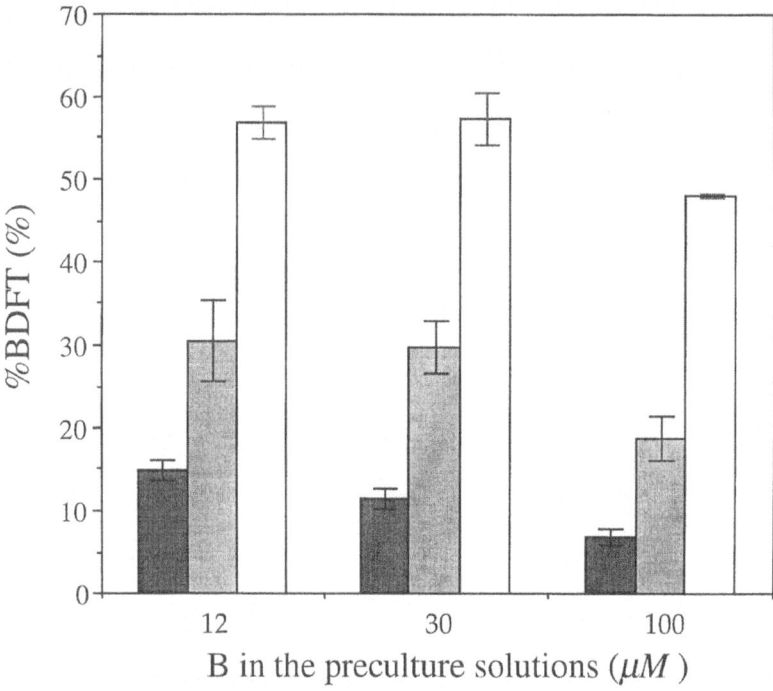

Figure 2. Percent boron (B) derived from tracer (%BDFT) in various tissues of *A. thaliana* supplied with [10]B for 48 hours. Each column represents old leaves (■), young leaves (▨) or shoot apices (□). Each value is mean ± standard deviation of 5 replicates.

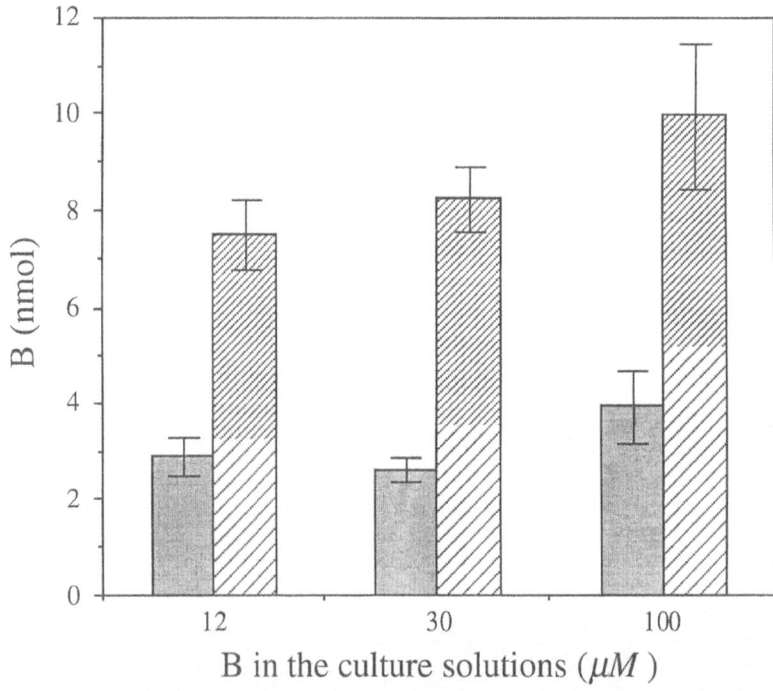

Figure 3. The amounts of boron (B) in shoot apices of *A. thaliana* before and after the labeling. Each column represents the shoot apices harvested before the labeling (▨) or after the labeling (▨). Shadows (▨) indicate the amounts of B derived from tracer estimated by the %BDFT values shown in Figure 2. Each value is mean ± standard deviation of 5 replicates.

lower at 12 μM B than that at 30 μM B (Figure 1). This suggests that increase in the percentage of B allocation to immature organ occurs as the B supply becomes restricted.

As reported previously (Brown and Hu, 1994), B is absorbed by simple and facilitated diffusion at the cellular level, which is mainly governed by the process of B complexation especially to cell walls. Since the shoot apex is the most actively growing organ, it creates a continuously growing sink for B. The B concentration in shoot apices may be determined mainly by sink capacity within a certain range of B supply, which was in agreement with our observation that the B concentration in shoot apices at 12 and 30 μM B were similar (Figure 1). In old leaves, transpiration mass flow determines the levels of B accumulation since B binding sites may already be saturated, which resulted in a significantly lower level of B accumulation at 12 μM B than at 30 μM B (Figure 1).

Tracer experiments were conducted to examine where the newly absorbed B was distributed. At all levels of B supply in the precultures, %BDFT was higher in young organs than in old organs (Figure 2), suggesting that growing organs exert a higher sink strength for B. The %BDFT and total amounts of B suggested that most of the B incorporated into shoot apices during the period of the labeling had derived from tracer B translocated from the medium (Figure 3).

The change in distribution pattern observed in our study (Figure 1) was similar to that observed by Liu et al. (1993) in field grown broccoli which showed increased ratios of B distribution to immature organs when B was not supplemented. Liu et al (1993) attributed this change in the pattern of B distribution to phloem retranslocation since the concentration of B was higher in phloem sap than in xylem sap. However, little or no retranslocation of B from mature organs to immature organs was observed in our study (Figure 3).

Several mutants of *A. thaliana* have been already isolated which relate to mineral nutrients, such as nitrate and phosphate, which led to great progress in understanding the functions of these nutrients at the molecular level (Meyerowitz and Ma, 1994). The present findings suggest that *A. thaliana* has the typical patterns of B absorption and translocation found in higher plants. Thus, the use of genetic approaches using *A. thaliana* is expected to provide us with further understanding of B nutrition.

Acknowledgment

This work is supported in part by a grant from the Ministry of Education, Science, Sports and Culture of Japan to TF (08760055).

References

Brown P H and Hu H 1994 Boron uptake by sunflower, squash and cultured tobacco cells. Physiol. Plant. 91, 435–441.

Brown P H and Hu H 1996 Phloem mobility of boron is species dependent: Evidence for phloem mobility in sorbitol-rich species. Ann. Bot. 77, 497–505.

Fujiwara T, Hirai M Y, Chino M, Komeda Y and Naito S 1992 Effects of sulfur nutrition on the soybean seed storage protein genes in transgenic petunia. Plant Physiol. 99, 263–268.

Hu H and Brown P H 1994 Localization of boron in cell walls of squash and tobacco and its association with pectin. Plant Physiol. 105, 681–689.

Ishii T and Matsunaga T 1996 Isolation and characterization of a boron-rhamnogalacturonan II complex from cell walls of sugar beet pulp. Carbohydr. Res. 284, 1–9.

Kobayashi M, Matoh T and Azuma J 1996 Two chains of rhamnogalacturonan II are cross-linked by borate-diol ester bonds in higher plant cell walls. Plant Physiol. 110, 1017–1020.

Liu L, Shelp B J and Spiers G A 1993 Boron distribution and retranslocation in field-grown broccoli (*Brassica oleracea var. italica*). Can. J. Plant Sci. 73, 587–600.

Loomis W D and Durst R W 1992 Chemistry and biology of boron. Biofactors 3, 229–239.

Marschner H 1995 Mineral Nutrition of Higher Plants, 2nd ed. Academic Press, London. 889 p.

Matoh T, Ishigaki K, Mizutani M, Matsunaga W and Takabe K 1992 Boron nutrition of cultured tobacco BY-2 cells. I. Requirement for and intracellular localization of boron and selection of cells that tolerate low levels of boron. Plant Cell Physiol. 33, 1135–1141.

Meyerowitz E M and Ma H 1994 Appendix B: Genetic variations of *Arabidopsis thaliana*. *In Arabidopsis*. Eds E M Meyerowitz and C R Somerville, pp 1161–1268. Cold Spring Harbor Laboratory Press.

Oertli J J 1994 Non-homogeneity of boron distribution in plants and consequences for foliar diagnosis. Commun. Soil Sci. Plant Anal. 25, 1133–1147.

Picchioni G A, Weinbaum S A and Brown P H 1995 Retention and the kinetics of uptake and export of foliage-applied, labeled B by apple, pear, prune, and sweet cherry leaves. J. Amer. Soc. Hort. Sci. 120, 28–35.

Rennie R J and Rennie D A 1983 Techniques for quantifying N_2 fixation in association with nonlegumes under field and greenhouse conditions. Can. J. Microbiol. 29, 1022–1035.

Shelp B J, Marentes E, Kitheka A M and Vivekanandan P 1995 Boron mobility in plants. Physiol. Plant. 94, 356–361.

Shu Z -H, Oberly G H and Cary E E 1994 Mobility of foliar-applied boron in one-year-old peaches as affected by environmental factors. J. Plant Nutr. 17, 1243–1255.

Warington K 1923 The effect of boric acid and borax on the broad bean and certain other plants. Ann. Bot. 27, 630–672.

R.W. Bell and B. Rerkasem (eds.), Boron in Soils and Plants, 203–207.
© 1997 *Kluwer Academic Publishers.*

Compartmentation of boron in roots and its translocation to the shoot of sunflower as affected by short term changes in boron supply

Heidrun Pfeffer, Frank Dannel & Volker Römheld
Institut für Pflanzenernährung (330), Universität Hohenheim,70593 Stuttgart, Germany

Key words: boron, compartmentation, concentration mechanism, *Helianthus annuus*, translocation, uptake

Abstract

The effect of short term changes in boron (B) supply on B uptake and compartmentation in the roots and its translocation to the shoot was studied using sunflower plants, precultured for seven days with a sufficient (100 μM) and a marginal (1 μM) B supply. Boron translocation to the shoot was determined by analysis of xylem exudate. For compartmental analysis of the roots, B was determined in preparations of cell sap (soluble B in the symplasm) and water insoluble residue (cell wall bound B). Boron supply during preculture and short term treatments influenced both the B concentration in the water insoluble residue and in the cell sap, the latter to a larger extent. Boron bound in the cell wall of the roots during the preculture with sufficient B supply could not be removed or mobilized after short term change to marginal B supply. From this it can be concluded, that this bound B plays a structural role in the cell wall and does not function as a storage pool for further use in the plant. The response of the B concentrations in root cell sap and xylem exudate to the short term treatments suggests, that B uptake at sufficient supply is achieved by passive diffusion, whereas at low supply there is a mechanism to concentrate B against the nutrient solution, which helps to satisfy the B demand of the plants. Additionally, it can be concluded that the key step of the concentration mechanism is the uptake of B from the nutrient solution into the symplasm of the root cells, followed by a passive loading of the xylem. The following three concentration mechanisms are discussed: anion trapping, active energy dependent uptake, and concentration in the symplasm by formation of B complexes.

Introduction

Relatively little work has been carried out on the compartmentation of boron (B) in plants. According to Hu and Brown (1994) a substantial amount of B is bound in the cell wall in particular at low B supply. When supply is sufficient, about 35% of the B is to be found in a soluble form. Thellier et al. (1979) concluded from exchange studies, that B is present in the three classical compartments, free space, cytoplasm and vacuole, and in an additional fourth compartment, which they interpreted as corresponding to borate diesters in the cell wall. Loomis and Durst (1992) proposed that B plays an important role in cross-linking cell wall polysaccharides. Hu and Brown (1994) presented evidence that B is associated with pectin in the cell wall and in later work (Hu et al., 1996), that the variability in B requirement between plant species is correlated with their content of cell wall pectin. This is support-

ed by the work of Matoh et al. (1993), who isolated a B-polysaccharide complex from radish roots which was identified as two chains of rhamnogalacturonan II linked together by boric acid (Kobayashi et al., 1996). Despite the progress in elucidating the role of B in cell walls, knowledge on B compartmentation is still poor. The aim of this study was to investigate the effect of short term changes in B supply of sunflower plants on B compartmentation in the roots and the fluxes of B to the shoots.

Materials and methods

Plant culture

Sunflower plants (*Helianthus annuus* cv. Frankasol) were cultivated in nutrient solution as described elsewhere (Dannel et al. 1997) with a B supply of 1 μM

Table 1. Fresh and dry weight of sunflower plants, precultured with different boron supply for seven days.

		Boron supply during preculture (μM)	
		1	100
Fresh weight (g plant^{-1})	roots	2.39 ± 0.24	2.58 ± 0.24
	shoot	2.99 ± 0.18	3.11 ± 0.23
Dry weight (mg plant^{-1})	roots	94 ± 8	95 ± 10
	shoot	256 ± 21	254 ± 21

Data represent the mean ± SD of 12 replicates.

(B 1; marginal supply) and 100 μM (B 100; sufficient supply).

Short term treatments and harvesting

After seven days of preculture in nutrient solution, plants precultured with a supply of 1 or 100 μM B were transferred to nutrient solutions containing 1 or 100 μM B. After two hours, plants were decapitated at the hypocotyl and xylem exudate was collected over a 1 h period, which was found to be the optimal period of time in preliminary experiments. Then roots were harvested for compartmental analysis. The resulting four treatments are named B 1/1, B 1/100, B 100/1, B 100/100, whereby the first numbers indicate the B concentrations (in μM) in the nutrient solution during preculture and the second numbers during the short term treatment.

Compartmental analysis

After collection of xylem exudate, about 10 g of fresh roots were rinsed for a few seconds with double distilled water and frozen at −18° C for at least 24 h to rupture the cells. Cell sap was prepared by squeezing the roots with a hydraulic press after thawing. The remaining residue was homogenized with 40 mL of double distilled water, the homogenate centrifuged at 3500 × g and the supernatant discarded. The residue was washed three times with 30 mL of double distilled water by repeating the homogenization/centrifugation procedure. Subsequently, the residue was dried at 60° C and termed water insoluble residue (WIR).

Boron analysis

Cell sap was centrifuged at 4000 × g to remove suspended particles and B was determined by inductively coupled plasma-optical emission spectroscopy

Table 2. Boron concentrations (μM) in xylem exudate and cell sap of roots of sunflower plants, harvested after collection of xylem exudate. Plant were precultured with different boron supply for seven days and then transferred to treatment solutions. After 2 hours xylem exudate was collected over a 1 h period.

Nutrient solution (Preculture/Treatment)	Cell sap	Xylem exudate
B 1/1	23.0 ± 2.1	21.8 ± 1.7
B 1/100	149 ± 12.0	156 ± 6.0
B 100/100	67 ± 7.4	61.4 ± 3.2
B 100/1	13.9 ± 2.1	18.4 ± 2.0

Data represent the mean ± SD of 3 replicates.

(ICP-OES). When very low concentrations of B were present, cell sap samples were dried under vacuum, ashed at 500° C, dissolved in a small volume of 0.5 N H_2SO_4 and B was determined by the azomethine-H method (modified after Lohse, 1982). WIR was ashed at 500° C, dissolved in 0.5 N H_2SO_4 and B was determined by the azomethine-H method. Boron in xylem exudate was determined as described elsewhere (Dannel et al. 1997).

Statistics

Mean values, standard deviations and number of replicates are indicated in tables and figures. The data were statistically analyzed by t-testing.

Results

Plants precultured with 1 μM B were not significantly different in fresh and dry weight from plants precultured with 100 μM B (Table 1) and showed no visible symptoms of B deficiency. Figure 1 shows, that B in sunflower roots was present in two different com-

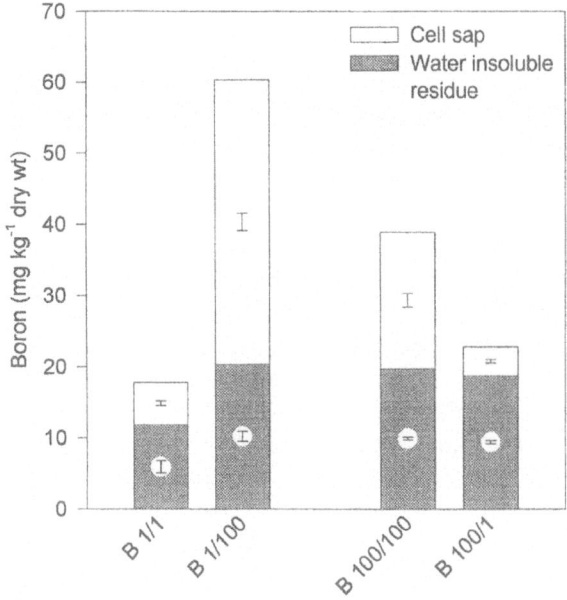

Figure 1. Boron compartmentation in roots of sunflower plants, precultured with 1 or 100 μM boron in nutrient solution for seven days, in response to short term changes in boron supply. First numbers in the names of the four treatments indicate the boron concentrations (in μM) in the nutrient solution during plant preculture and the second numbers during the short term treatment. Bars represent the mean and error bars the standard deviation of three replicates.

partments and that the B concentrations in these compartments were significantly influenced by B supply during preculture and short term treatments. The B concentration in the WIR of plants supplied continuously with 1 μM B (B 1/1) was about 65% of that in the WIR of plants supplied continuously with 100 μM B (B 100/100). Transfer from 1 to 100 μM B (B 1/100) raised the B concentration in the WIR to the level of B 100/100 plants, whereas transfer to lower B supply (B 100/1) did not alter the B concentration in the WIR compared with the corresponding variant (B 100/100). As shown in Figure 1, the B concentration in the cell sap of plants supplied continuously with 1 μM B (B 1/1) was about 34% of the B concentration in the cell sap of plants supplied continuously with 100 μM B (B 100/100). Transfer from 1 to 100 μM B (B 1/100) raised the B concentration in the cell sap to the double of the concentration in sufficient supplied plants (B 100/100). Transfer from 100 μM to 1 μM B (B 100/1) decreased the B concentration to a level, lower than that present in plants with marginal B supply (B 1/1). Strikingly, the concentrations of B in cell sap and xylem exudate (Table 2) were nearly the same in each variant and were strongly influenced by the interaction of preculture and

short term treatment. In plants supplied continuously with 100 μM B (B 100/100) the internal B concentration (cell sap and xylem exudate) was only about 65% of the B in the nutrient solution. Transfer to 1 μM B (B 100/1) decreased the internal B concentration to about 16 μM. In plants supplied continuously with 1 μM B (B 1/1) the internal B concentration was about twenty times higher than in the nutrient solution. Transfer to 100 μM B (B 1/100) increased the internal B concentration to about 150 μM, a concentration two times higher than in the B 100/100 variant.

Discussion

In order to avoid secondary effects due to B deficiency, a B supply was chosen which was just high enough to prevent growth inhibition. This was achieved by preculture of plants with 1 μM B, which had no visible symptoms of B deficiency and the fresh weights and dry weights obtained were not significantly different from the respective weights of plants precultured with 100 μM (Table 1).

For compartmental analysis, roots were rinsed and cell sap was prepared by squeezing the roots with a hydraulic press after a freeze-thaw-cycle. The remaining residue was then homogenized and washed to yield the water insoluble residue. According to Hu et al. (1996) the water insoluble B roughly represents the cell wall bound B. Cell sap is a mixture of cytoplasmic, vacuolar and apoplasmic fluids. As selective removal of apoplasmic B by a rinsing procedure is extremely difficult because of the high membrane permeability of free boric acid, rinsing of the roots was carried out for a few seconds just to remove the remaining nutrient solution from the root surface. Because the free space of 18 d old sunflower roots is only about 8% of the root volume (Göring et al., 1974), the apoplasmic fluid does not influence the composition of the cell sap to a large extent. Thus the cell sap B mainly represents the soluble B in the symplasm (cytoplasm and vacuole). The results of the compartmental analysis (Figure 1) show, that the B supply during preculture and short term treatments influenced the B concentration in both the WIR and the cell sap, the latter to a larger extent. Although, the supply of plants precultured with 1 μMB was sufficient for regular growth, there were free binding sites for B in the WIR, which could be loaded by a short term treatment with sufficient B supply (B 1/100). In contrast it was not possible to decrease the B concentration in the WIR by short term treatments with lower

206

B supply (B 100/1). This suggests, that B once bound in the cell wall fraction (WIR) of the root can not be easily removed or mobilized. Therefore, it is likely that B in the WIR plays a structural role and does not seem to be stored for further use in the plant, a conclusion supported by Hu et al. (1996). This is consistent with the function of B as a cross-link between two pectin chains that has been demonstrated by Kobayashi et al. (1996).

In contrast to the structural B in the WIR, the concentration of B in the symplasm responds very strongly to changes in supplied B concentration (Table 2). Comparing the B concentrations in the cell sap of the B 100/100 and the B 100/1 plants it can be deduced, that about 80% of the B can be mobilized from the symplasm of root cells within a relatively short time. As in the work of Wildes and Neales (1971), who demonstrated that about 90% of the B present in discs of carrot and red beet storage tissues can be desorbed within a period of four hours, it is likely that in our experiment B leaked from the root cells into the nutrient solution. Beside this, translocation to the shoot may play a role, too. In plants supplied continuously with 1 μMB (B 1/1) the B concentration in cell sap and xylem exudate is more than 20-fold higher than in the nutrient solution, suggesting that there is a mechanism to concentrate B against the nutrient solution, which helps to satisfy the B demand of the plants. This conclusion is supported by the results of Dannel et al. (1997) who showed that B concentration in the xylem exudate of sunflower plants cultivated with a B supply of 1 μM was about 25-fold higher than in the nutrient solution. Treating plants, precultured with marginal B supply, with sufficient B (B 1/100) does not seem to deactivate this mechanism in the short term. Thus, the B concentration in the cell sap and xylem exudate was more than the double that in plants supplied continuously with 100 μM B (B 100/100), in which the concentration mechanism was obviously not activated because B supply was sufficient. Despite the great differences in response to the B treatments, it is remarkable that the B concentration in xylem exudate was always nearly the same as in root cell sap (Table 2). Because the short term changes were made in the nutrient solution, it can be concluded that the key step of the concentration mechanism is the uptake of B from the nutrient solution into the symplasm of the root cells, followed by passive loading of the xylem.

The concentration mechanism for B is still unknown. The possibility that the relatively alkaline cytoplasm acts as an anion trap for the weak boric acid can be ruled out by a simple calculation. To achieve a B concentration of 20 μM in the cell sap against a nutrient solution (pH = 6) containing 1 μM B (B 1/1), a symplasmic pH at least > 12 is required, which is non-physiological. A second possibility is that uptake of B is by an active, energy dependent mechanism, especially at low B supply (Wildes and Neales, 1971; Nissen, 1974). But because of the high permeability of membranes for the undissociated boric acid, Raven (1980) pointed out that an active transport of boric acid to maintain B distribution across a membrane away from thermodynamic equilibrium is likely to be energetically expensive. As a third possibility, B complexing compounds may also be responsible for the concentration gradient between nutrient solution and the root symplasm (Kochian, 1991). If such compounds are present in the symplasm the concentration of free boric acid would decrease and B be delivered from the nutrient solution along a concentration gradient until the equilibrium between free B in the symplasm and the nutrient solution is achieved. Total B in the symplasm would then be a sum of free boric acid and complexed B, dependent on the stability constant of the B complex. A number of plant compounds with the appropriate cis-diol structure are able to form mono- and diesters with boric acid, and some of them have relevance in physiological processes for example fructose (Yokota and Konishi, 1990), ribose (Loomis and Durst, 1992) and sorbitol (Brown and Hu, 1996). Recently Hu et al. (1997) presented evidence for the occurence of soluble B polyol complexes (mannitol, sorbitol and fructose) in the phloem of plant species showing high phloem mobility of B. As yet B complexes have not been found in sunflower and it remains an open question for further ongoing research, whether they exist and play the proposed role in B uptake.

Acknowledgements

The authors like to dedicate this work to the late Prof. Dr. Drs. h. c. H. Marschner who originated and supervised the experiments presented here. When writing this manuscript, we missed his guidance and his constructive comments.

We are grateful to Borax Consolidated Limited, United Kingdom and the Deutsche Forschungsgemeinschaft for financial support and to Ernest A. Kirkby (University of Leeds) for critical evaluation and correction of the manuscript.

References

Brown P H and Hu H 1996 Phloem mobility of boron is species dependent: Evidence for phloem mobility in sorbitol-rich species. Ann. Bot. 77, 497–505.

Dannel F, Pfeffer H and Römheld V 1997 Effect of pH and boron concentration in the nutrient solution on translocation of boron in the xylem of sunflower. *In* Boron in Soils and Plants. Proceedings Ed R W Bell and B Rerkasem, pp 183–186. Kluwer Academic Publishers, The Netherlands.

Göering H, Ehwald R and Sammler P 1974 Bestimmung des 'freien Raumes' pflanzlicher Gewebe una seine Bedeutung für die Stoffaufnahme. Archiv Acker Pflanzenbau Bodenkunde 18, 223–232.

Hu H and Brown P H 1994 Localization of boron in cell walls of squash and tobacco and its association with pectin – Evidence for a structural role of boron in the cell wall. Plant Physiol. 105, 681–689.

Hu H, Brown P H and Labavitch J M 1996 Species variability in boron requirement is correlated with cell wall pectin. J. Exp. Bot. 47, 227–232.

Hu H, Penn S G, Lebrilla C B and Brown P H 1997 Isolation and characterization of soluble boron complexes in higher plants. Plant Physiol. 113, 649–655.

Kobayashi M, Matoh T and Azuma J 1996 Two chains of rhamnagalacturonan II are cross-linked by borate-diol ester bonds in higher plant cell walls. Plant Physiol. 110, 1017–1020.

Kochian L V 1991 Mechanisms of micronutrient uptake and translocation in plants. *In* Micronutrients in Agriculture. Eds J J Mortvedt, F R Cox, L. M. Shuman and R M Welch, pp 229–296. Soil Science Society of America, Book Series No 4.

Lohse G 1982 Microanalytical Azomethine-H method for boron determination in plant tissue. Commun. Soil Sci. Plant Anal. 13, 127–134.

Loomis W D and Durst R W 1992 Chemistry and biology of boron. Bio Factors 3, 229–239.

Matoh T, Ishigaki K, Ohno K and Azuma J 1993 Isolation and characterization of a boron-polysaccharide complex from radish roots. Plant Cell Physiol. 34, 639–342.

Nissen P 1974 Uptake mechanisms: Inorganic and organic. Ann. Rev. Plant Physiol. 25, 53–79.

Raven J A 1980 Short- and long-distance transport of boric acid in plants. New Phytol. 84, 231–249.

Thellier M, Duval Y and Demarty M 1979 Borate exchange of *Lemna minor* as studied with the help of the enriched stable isotopes and of a (n,α) nuclear reaction. Plant Physiol. 63, 283–288.

Wildes R A and Neales T F 1971 The absorption of boron by disks of plant storage tissues. Aust. J. Biol. Sci. 24, 873–884.

Yokota H and Konishi S 1990 Effect of the formation of a sugar-borate-complex on the growth inhibition of pollen tubes of *Camellia sinensis* and cultured cells of *Nicotiana tabacum* by toxic levels of borate. Soil Sci. Plant Nutr. 36, 275–281.

R.W. Bell and B. Rerkasem (eds.), Boron in Soils and Plants, 209–212.
© *1997 Kluwer Academic Publishers.*

Absorption, movement and distribution of boron applied to peach (*Prunus persica* L. Batsch) fruits

Z.-H. Shu[1], G. H. Oberly[2] & E. E. Cary[3]

[1] *Department of Plant Industry, National Pingtung Polytechnic Institute, Pingtung, 912, Taiwan, ROC*
[2] *Pomology Department, Cornell University*
[3] *U. S. Plant, Soil and Nutrition Laboratory, Ithaca, NY 14853, USA*

Key words: boron, distribution, fruit, movement, *Prunus persica* L., uptake, sink

Abstract

Six to eight-year-old 'Reliance' peach *(Prunus persica)* trees grown at the Ludlowville Orchard, Cornell University were selected and used in this study. Enriched-^{10}B boric acid solutions were locally applied to fruits to estimate the uptake, movement and distribution in peaches. After harvesting, the plant parts were analyzed with inductively coupled plasma (ICP) - atomic emission spectrometry and ICP mass spectrometry for total boron and ^{10}B/^{11}B ratio. The ^{10}B content was used as the primary measure of B uptake and mobility. In 24 hours, the ^{10}B taken up by the fruit could be translocated to non-treated leaves. The highest value of ^{10}B uptake by peach fruits, as a percent of total ^{10}B applied, was 0.3% with an average of 0.24%. In general, the treated tissue had the highest ^{10}B content, followed by the non-treated fruit tissue and the leaf. Shoot-stems had the lowest ^{10}B contents. The ^{10}B exported from the treated fruit part as a percent of the ^{10}B absorbed ranged from 49.8% (1 day) to 89% (7 days). It is suggested that in peaches, even B in strong sinks like fruit is freely mobile in the phloem.

Introduction

Foliar-absorbed mineral nutrients were classified by Bukovac and Wittwer (1957) into three groups: mobile, partially mobile or immobile. Due to the difficulty in finding a suitable tracer, very little is known about the mobility of foliar- or fruit-absorbed boron (B). In a study using conventional B and measured by spark source mass spectrometry and laser-probe mass spectrography, Chamel et al. (1981) found that 24 hr after foliar application of conventional B (as H_3BO_3) to radish (*Raphanus sativus*), 78 to 98% of the absorbed B was still present in the treated leaf, 4.5 to 7% had migrated to the epicotyl, and 2.5 to 17.7% had migrated to the hypocotyl. Shu et al. (1993) found that the movement of ^{10}B in peach (*Prunus persica*) trees was rapid, but the amount of ^{10}B taken up by leaves was very small.

Sink organs are net importers of assimilate. In terms of assimilate transport, the ability of an organ to import assimilate is the sink strength (Ho, 1988). Being rated as a strong sink, can fruits take up more B than leaves? What is the fate of the absorbed B by the fruit? Will B remain in the fruit or be retranslocated out of the treated tissue? The present study tries to answer these questions.

Materials and methods

Seven 6 to 8-year-old mature fruiting 'Reliance' peach trees grown at the Ludlowville Orchard, Cornell University were selected and used in this study. Fifteen branches of about the same size, angle and vigour having only one fruit about 6 cm long at stage II (Tukey, 1936) were chosen to be treated from the seven trees. Enriched ^{10}B-boric acid which contains 94.7% ^{10}B and a ^{10}B/^{11}B ratio of 17.86 obtained from Blue Eagle Inc, Quapaw, OK. was used in this study. The concentration of the enriched ^{10}B-boric acid solution used was 600 mg L^{-1}, with 0.02% Charger-E surfactant (Agway, Inc.) and pH adjusted to 3. Thirty mL of enriched ^{10}B-boric acid solution was applied as a spot treatment of about 1 cm diameter to the fruit skin with a repeating pipette. Water was applied to the control fruits to serve as checks. On the 2nd, 3rd, 5th and 8th day after the application, the branches were harvested. The branches were divided into leaves, shoot stems and each fruit were subdivided into 2 samples. The site of application was collected by removing a section about 2.5 cm in diameter and 0.5 cm thick. The remaining fruit was analyzed as untreated fruit samples.

All samples were washed, freeze-dried, ground, dry-ashed and determined using inductively coupled plasma (ICP)-atomic emission spectrometry, for total B and ICP- mass spectrometry (Elan model 250) for $^{10}B/^{11}B$ ratio. The procedure and calculation was the same as described in Shu et al. (1993). When calculating, it was assumed that little or none of the absorbed ^{10}B was lost from the branches.

Results

Table 1a shows ^{10}B enrichment was found 1 day after the application in the treated and untreated parts of the fruit and the leaf, but not in the shoot stem. In the treated tissue, ^{10}B enrichment was highest in day 2 after the treatment and declined to the lowest level on the day 8. The patterns of ^{10}B enrichment in other plant parts were slightly different from the treated tissue with the maximum reached in day 3, dropping to the minimum on day 8. The total absorption as a percent of the total ^{10}B applied ranged from 0.3% on day 3 to 0.1% on day 8 with an average of 0.24% (Table 1a). The absorbed ^{10}B exported out of the treated tissue increased from 49.8% on day 2 to 89% on day 8 with an average of 77.8% (Table 1a). About 63% of the ^{10}B absorbed by the whole fruit was subsequently retranslocated to other plant parts (calculation not shown).

Table 1b shows percent distribution of the ^{10}B transported from the treated tissue to other plant parts. More than 79.3% of the ^{10}B absorbed was retranslocated to the untreated fruit tissue: 18.8% or less of the ^{10}B absorbed was retranslocated to the leaf. The shoot stem received the least amount of ^{10}B from the treated tissue (Table 1b).

Discussion

Since the branches used in this study were not closed systems, the loss of the absorbed ^{10}B from the branches could not be avoided. The continuous decrease of total ^{10}B content in all plant parts, especially from days 5 to day 8 (Table 1a) suggests that branches were indeed an open system. This could also account for the average percent of the total absorbed ^{10}B being limited to only 0.24%. However, disregarding the loss from the system, the highest percent absorption of the applied ^{10}B reached 0.3% which was comparable to that of the leaf (Shu et al., 1993) on day 3. Moreover, the absorption area of the fruit (about 3.1 cm^2) was smaller than that of the leaf (about 16.5 cm^2, Shu et al. 1993). It is thus suggested that fruits are capable of absorbing at least as much and probably more ^{10}B than the leaf. The relatively high uptake may be due to the different cuticle composition that is more permeable to ^{10}B than in the leaf.

The fact that more than 90% of the ^{10}B absorbed by the fruit could be translocated out of the treated tissue in 7 days has two implications (Table 1b). Firstly, B moves rapidly in peach trees. Secondly, despite the fact that the fruit is a strong sink, B was not retained in the fruit.

Boron has usually been rated as phloem immobile in higher plants (Mengel and Kirby, 1982). However, there has been findings showing that B is mobile in some species (Shelp, 1988; Hanson, 1991; Shu et al., 1993. Recently, Brown and Hu (1996) and Hu et al. (1996) reported that B mobility in the phloem was related to the formation of B-sorbitol complexes and was species dependent. The peach is one of the species using sorbitol as the primary translocated photosynthate and this explains the findings that B is phloem mobile in peaches in the previous and present studies.

The fruit, being a strong sink, was not anticipated to translocate such a high proportion of B. There is not an obvious explanation for the rapid and substantial translocation of ^{10}B from the fruit to the other plant parts. However, it may be pertinent to note that the amount of ^{10}B absorbed by the fruit was not large if the amount absorbed is divided by the weight of the whole fruit.

Table 1 also shows that it is difficult to detect B uptake and/or movement when using conventional B determination. For example, total B concentrations in the leaf on day 2 and in the untreated fruit tissue on day 5 were about the same (Table 1c), but their ^{10}B contents had a 11-fold difference with the untreated fruit tissue greater than the leaf (Table 1a). It is thus suggested that ^{10}B content is essential for tracing the uptake and movement of the B, especially for those

Table 1. Absorption and distribution of enriched-[10]B boric acid applied to the skin of peach fruits[z].

Time after treatment	Leaf	Shoot stem	Fruit Treated part	Fruit Untreated part	[10]B absorbed[y]	[10]B exported[x]
	(a)[10]B (ng)				%	%
Control	0.0b[w]	0.1b	–[v]	0.3b	–	–
Day 2	3.5ab	0.3b	24.9a	20.9ab	0.28	49.8
Day 3	8.8a	0.9a	7.3b	37.1a	0.30	86.5
Day 5	4.1ab	1.5a	7.4b	39.1a	0.29	85.8
Day 8	0.0b	0.2b	1.9b	15.2ab	0.10	89.0
Mean					0.24	77.8
	(b) % distribution of [10]B transported from treated tissue[u]					
Control	–[v]	–	–	–		
Day 2	14.2ab	1.2a	–	84.6ab		
Day 3	18.8a	1.9a	–	79.3b		
Day 5	9.2ab	3.4a	–	87.5ab		
Day 8	0.0b	1.3a	-	98.7a		
	(c) Total B ([10]B + [11]B; mg B kg^{-1} dry wt)					
Control	30.6a	23.9b	–	30.1cd		
Day 2	32.7a	24.3b	74.0a	34.8ab		
Day 3	31.1a	36.1ab	31.8b	36.9a		
Day 5	19.2a	59.2a	28.7b	32.6bc		
Day 8	34.2a	20.1b	32.2b	28.3d		

[z] 30 ml, 600 mg L^{-1} (18μg), enriched-[10]B boric acid was applied.
[y] [10]B absorbed as % of the [10]B applied, calculated from (total [10]B absorbed/total [10]B applied) $\times$ 100.
[x] % exported of the total [10]B absorbed by the treated tissue, calculated from 100$\times$(total [10]B absorbed $-$ [10]B absorbed by treated tissue)/total [10]B absorbed.
[w] Mean separation within columns by Duncan's multiple range test, $p \leq 0.05$.
[v] Data not available.
[u] Calculated from ([10]B in plant part/(total [10]B absorbed $-$ [10]B in treated tissue)) $\times$ 100.

species, like peaches, which absorb small amounts of B and translocate it rapidly.

To summarize, peach fruits are more efficient in taking up and translocating [10]B than found with the leaf in previous investigations. The uptake amount as a percent of the total absorbed [10]B was greater than 0.3% per 3 cm^2. About 78% and 63% translocation of the [10]B out of the treated tissue and the whole fruit, respectively, were detected in seven days. It is suggested that in peaches, even B in strong sinks like fruit is freely mobile in the phloem.

Acknowledgments

The work was supported by grants from the National Science Council of the Republic of China and Pomology Department of Cornell University.

References

Brown P H and Hu H 1996 Phloem mobility of boron is species dependent: Evidence for phloem mobility in sorbitol-rich species. Ann. Bot. 77, 497–505.

Bukovac M J and Wittwer S H 1957 Absorption and mobility of applied nutrients. Plant Physiol. 32, 428–435.

Chamel A, Andreani A-M and Eloy J-F 1981 Distribution of foliar-applied boron measured by spark source mass spectrometry and laser-probe mass spectrography. Plant Physiol. 67, 457–459.

Hanson, E J 1991 Movement of boron out of tree fruit leaves. Hort Science 26, 271–273.

212

Ho, L C 1988 Metabolism and compartmentation of imported sugars in sink organs in relation to sink strength. Ann. Rev. Plant Physiol. Mol. Biol. 39, 355–378.

Hu H, Brown P H and Labavitch J M 1996 Species variability in boron requirement is correlated with cell wall pectin. J. Exp. Bot. 47, 227–232.

Mengel K and Kirby E A 1982 Principles of Plant Nutrition. International Potash Institute, Switzerland. 654 p.

Shelp B J 1988 Boron mobility and nutrition in broccoli (*Brassica oleracea* var. italica). Ann. Bot. 61, 83–91.

Shu Z H, Oberly G H and Cary E E 1993 Time course study on the mobility and pattern of distribution of foliar-applied boron in peaches. J. Plant Nutr. 16, 1661–1673.

Tukey B 1936 Development of cherry and peach fruits as affected by destruction of the embryo. Bot. Gaz. 98, 1–24.

R.W. Bell and B. Rerkasem (eds.), Boron in Soils and Plants, 213–220.
© 1997 *Kluwer Academic Publishers.*

The effect of boron deficiency on development in determinate nodules: changes in cell wall pectin contents and nodule polypeptide expression

I. Bonilla[1], H. Perez[2], G. Cassab[2], M. Lara[2] & F. Sanchez[2]
[1] *Dpto. Biología, Facultad de Ciencias, Universidad Autónoma de Madrid, 28049 M. Spain*
[2]*Instituto de Biotecnología, Universidad Nacional Autónoma de México, Cuernavaca. México*

Key words: nitrogen fixation, pectin, polypeptide expresion, *Rhizobium*, symbiosis

Abstract

The effect of boron (B) deficiency on nodule development and nitrogen fixation in bean (*Phaseolus vulgaris*) was studied. Nodules from plants grown without B in the nutrient solution were smaller in size and had a weight lower than the controls. In addition, B starvation resulted in a severe decrease in acetylene reduction activity after two weeks of treatment. Light microscopy examination of B deficient nodules showed dramatic changes in cell shape, amyloplast number, and cell wall structure, mainly in those cells located in the middle of the central zone. The use of the tissue-printing technique indicated that pectins are decreased in B-deficient nodules. Finally, one and two dimensional analysis of the nodule polypeptides showed the premature expression of a 116 kDa polypeptide in B-deficient compared to control nodules. These data suggest that B is an obligatory requirement for normal determinate nodule development and functioning.

Introduction

In a previous report, we have established the relationship that exists between B availability and the N_2 fixation process in blue-green algae (Cyanobacteria) (Bonilla et al.,1990). In these microorganisms, B deficiency induced alterations in the heterocyst envelope which facilitated oxygen diffusion resulting in an inhibition of the nitrogenase activity (Garcia-Gonzalez et al., 1991). In addition, recent results have shown that B is also required for the legume-*Rhizobium* symbiotic process (Lukaszewski et al.,1992). Boron deficiency in pea (*Pisum sativum* L.) causes a decrease in the number of nodules and an alteration of indeterminate nodule development leading to an inhibition of nitrogenase activity. Electron micrographs of B-deficient nodules showed dramatic cell wall changes and alterations in both peribacteroid and infection thread membranes, suggesting a role of this microelement in the stability of these structures (Bolaños et al., 1994). Moreover, B plays an important role in mediating cell-surface interactions that lead to endocytosis of rhizobia by host cells

and hence to the correct establishment of the symbiosis between pea and *Rhizobium* (Bolaños et al., 1996).

Evidence is presented here that shows that B is not only a requirement for determinate nodule development and N_2 fixation in bean (*Phaseolus vulgaris*) but also that in B deficient nodules, the level of pectins decreases considerably, and hence the structure of cell walls is completely abnormal. Finally, one and two dimensional analysis of nodule polypeptides showed the premature and altered expression of a 116 kDa polypeptide in B-deficient compared to control nodules. These data suggest that B is an obligatory requirement for normal determinate nodule development and function.

Materials and methods

Bean (cv Negro Jamapa) seeds were surface sterilized, germinated on wet filter paper (Lara et al., 1984) and cultivated on vermiculite with Murashige medium (Murashige, 1974) without B. Plants were inoculated with *Rhizobium tropici* (strain CIAT 899), and grown

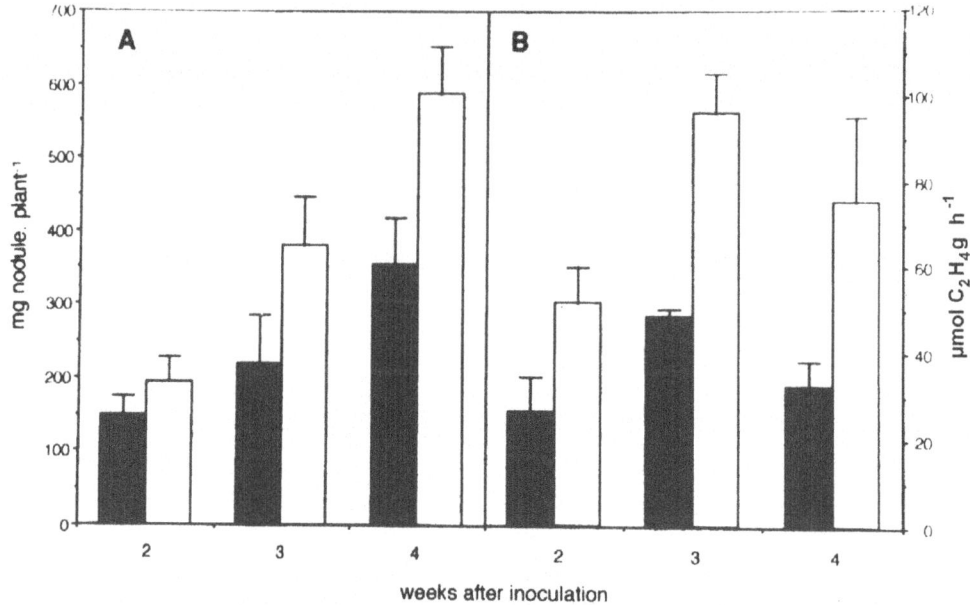

Figure 1. Effect of B deprivation (shaded bars - B deficient; open bars - control) on the weight (A) and nitrogenase activity (acetylene reduction) (B) of bean root nodules.

in a greenhouse with natural light at 25° C for 16 h per day with a 20° C night temperature. Relative humidity (RH) was kept between 60 and 70%.

All solutions were prepared and stored in polyethylene containers previously tested to prevent release of B under sterilizing conditions according to Mateo et al. (1986). Boron was removed from vermiculite by washing with 0.1 N HCl until no B was detected (detection limit was 1.8 μM). For control cultures, B (as H_3BO_3) was added to a final concentration of 9 μM.

Nitrogenase activity was measured by acetylene reduction as described by Dart et al. (1972). Ten plants for each determination were harvested weekly.

Nodule samples for light microscopy were fixed for 24 h at 40°C in 4% paraformaldehyde, 1% glutaraldehyde in phosphate buffer saline (PBS). Semi-thin sections (2 μm) were made with a glass knife in a LKB Bromma ultratome Nova and stained with toluidine blue for light microscopy observations.

Tissue-print western blot on nitrocellulose paper.

The nitrocellulose paper was soaked in 0.2 M CaCl$_2$ for 30 min and dried on paper towels. Freshly cut tissue was washed in distilled water for 30 s, dried on Kimwipes, and blotted onto the nitrocellulose paper for 30 s. The tissue print was treated for detection of

alkaline phosphatase-conjugated secondary antibody according to Cassab (1993). Two rat monoclonal antibodies, JIM5 and JIM7 raised against un-esterified and methyl-esterified pectins, were probed at a 1:250 dilution.

Analysis of DEAE-bound nodule soluble protein fractions

Two grams of nodules previously frozen in liquid nitrogen were pulverized in an electric grinder. Plant powder was mixed with 2.5 volumes of cold extraction buffer (30 mM Tris-HCl, 30% (w/v) sucrose, 2.5% (w/v) PVPP, pH 7.4). The slurry was stirred for 5 min and centrifuged (20 min at 26900 g, 4°C). All buffers were used ice-cold and every fractionation step was performed at 4°C to minimize proteolytic activity. The low ionic strength supernatant was mixed with one tenth volume of DEAE- Sephadel resin previously equilibrated with 30 mM Tris-HCl, 150 mM KCl, pH 7.3 according to Perez et al. (1994). The plant supernatant and the resin were mixed gently for 10 min, and then the resin was washed with 20 volumes of equilibrated buffer using a Buchner funnel. The damp resin was collected and incubated for ten minutes with 5 volumes of elution buffer (30 mM Tris-HCl, 500 mM KCl, pH 7.4). The slurry was filtered and the eluted proteins were precipitated with three volumes of

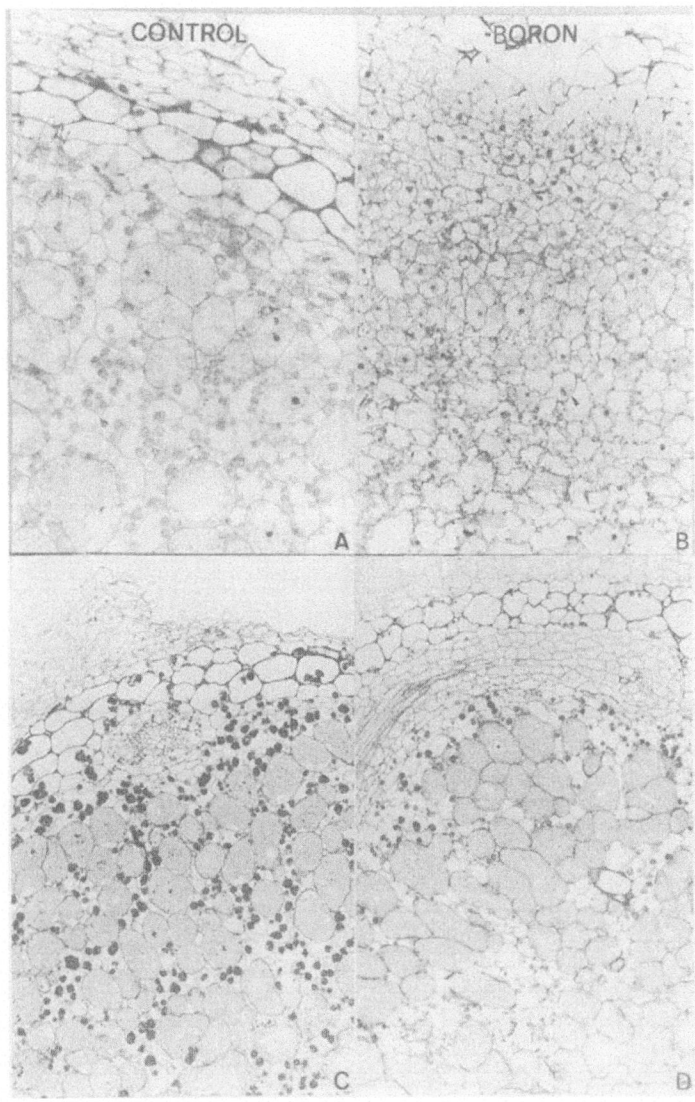

Figure 2. Light micrographs of bean nodules grown in the presence (A,C) and in the absence (B,D) of boron. A: In control nodules the cortex and infected zone are well differentiated after 2 weeks of development; ×200. B: In B-deficient nodules at 2 weeks of development, the cell walls of the infected and intersticial cells in the central zone are highly irregular in shape. There are also less amyloplasts present in the interstitial cells; ×40. C: Control nodule showing a high number of amyloplasts in interstitial cells after 3 weeks of development; ×630. D: B-deficient nodule showing a disorganized central zone with less amyloplasts in interstitial cells and larger infected cells after 3 weeks of development; ×630. All sections were stained with toluidine blue.

acetone (pre-chilled at −20°C). Acetone pellets were stored at −70°C until use.

One-dimensional SDS-PAGE 10% gels were prepared according to Laemmli (1970). Gels were loaded with 100 μg of protein. First dimension isoelectric focusing gels were run according to O'Farrel (1975) with the modifications for minigels recommended by Hoeffer Scientific Instruments. Gels were loaded with 150 μg of protein. The ampholine mixtures used were 75% (5–7 pH range) and 25% (3–10 pH range). One-dimensional gels were incubated for 15 min in O'Farrel solution '0' and processed in the second dimension according to Hoeffer Scientific Instruments. After electrophoresis, proteins were detected with Coomassie blue (Laemmli, 1970). Protein concentration was determined according to Bradford (1976).

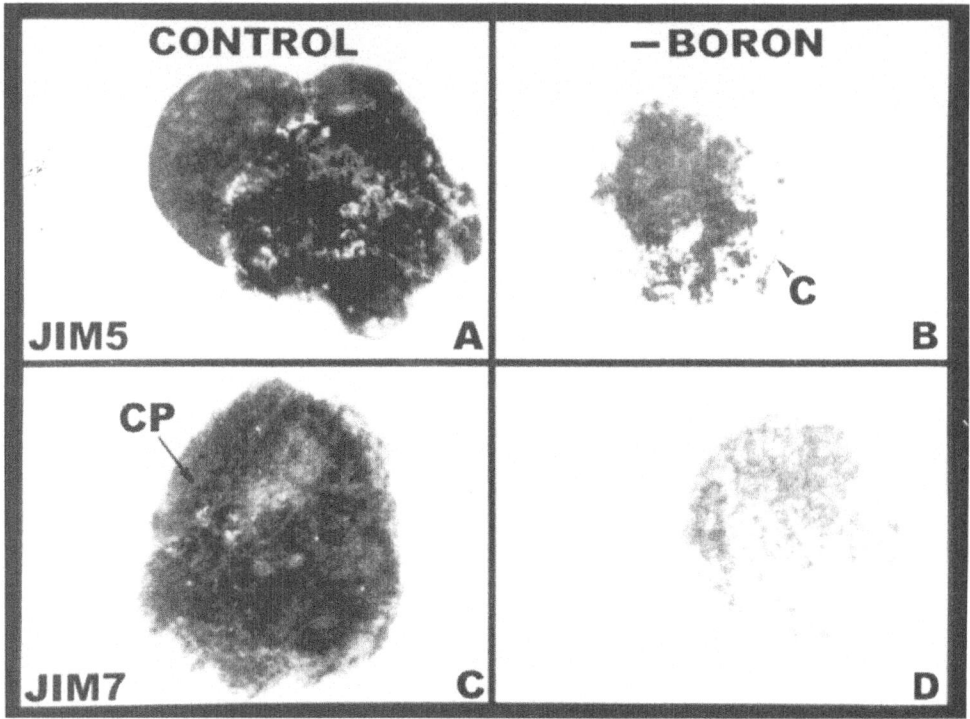

Figure 3. Effect of boron deficiency on the distribution of pectins in root nodule prints on nitrocellulose paper after 3 weeks of development. The tissue-prints were reacted with JIM5 (anti-un-esterified pectin) and JIM7 (anti-methyl-esterified pectin) monoclonal antibody and detected with alkaline phosphatase-conjugated anti-rat IgG antibodies. (A,C) Control nodules, (B,D) B-deficient nodules. In Figure 3B the feature labelled C is the cortex: in Figure 3C, CP identifies the central part of the nodule.

Table 1. Root weight (mg) in the presence or in absence of Boron.

Culture conditions	Time (d)			
	7	14	21	28
Control	739 ± 54	1150 ± 156	2228 ± 540	3753 ± 620
B deficient	506 ± 125	813 ± 208	1215 ± 358	1731 ± 250

Reproducibility

Data in the figures are mean values from four independent experiments.

Results

Plants grown under B deprivation showed a significant reduction in growth, mainly in root development compared to the control plants (Table 1). Both nodule number and weight (50% reduction after 2 weeks of treatment) were reduced in the B-deficient plants (Figure 1A). Boron starvation resulted in more than 50% inhibition of nitrogenase activity detected as acetylene reduction after 3 weeks of treatment and about 60% inhibition after 4 weeks (Figure 1B).

Light microscopy examination of B-deficient nodule thin sections (Figure 2B, D) showed dramatic changes and alterations in cell size and structure compared with control nodules at 2 (Figure 2A, B) and 3 weeks (Figure 2C, D) of treatment. In B-deficient nodules, after 2 weeks of treatment, a distal-proximal gradient (from the root axis) of cell size developed in the central tissue. The central tissue is surrounded by several layers of small cells, with heavily stained nuclei. Within this parenchymatous-like cell layer, the endodermal cells with a characteristic thickened and lignified secondary cell wall cannot be distinguished (Figure 2B) as compared to the control nodule (Figure 2A). In the proximal side of the central tissue,

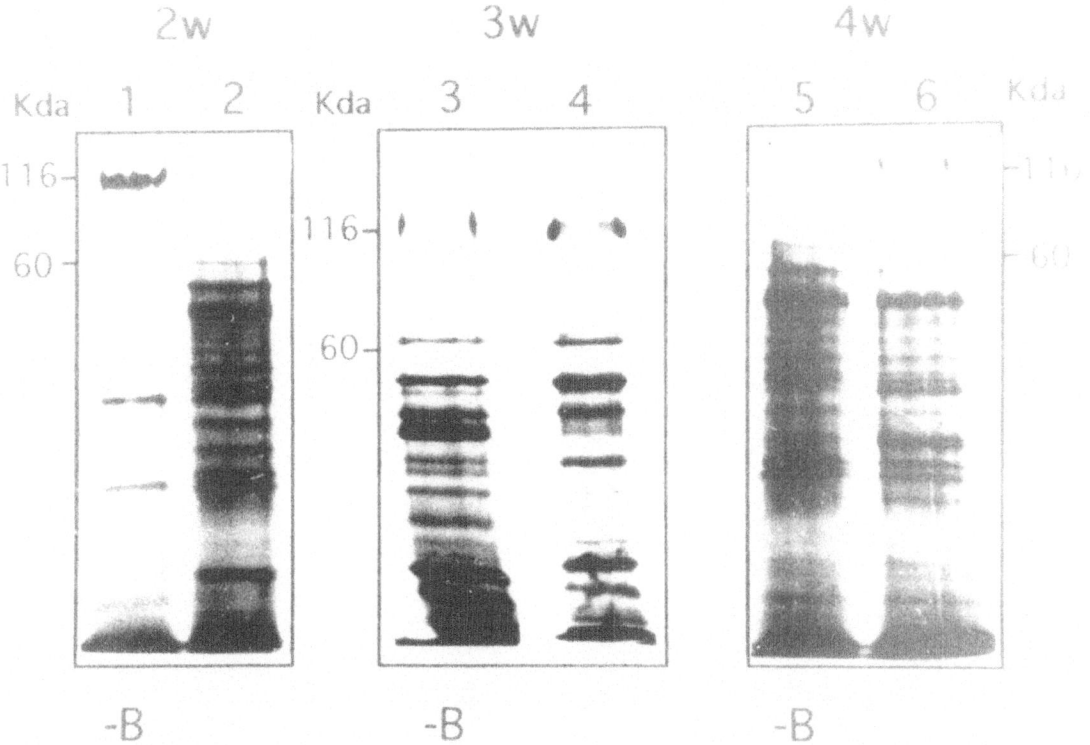

Figure 4. Polypeptides pattern of low-ionic-strength nodule extracts of both normal and B-deficient *Phaseolus vulgaris* L. Plant proteins were adsorbed and eluted from a DEAE resin as described in methods. Proteins were separated by SDS-PAGE on a 10% acrylamide gel and visualized by Coomassie blue staining. B-deficient nodule proteins: lanes 1.3.5; Control nodule proteins: lanes 2.4.6. Time of nodule collection is indicated at the top of each panel, 2 weeks (2 w), 3 weeks (3 w), 4 weeks (4 w).

uninfected cells have formed a boundary layer, where starch accumulation in the amyloplasts is clearly visible, and some infected cells with few bacteroids can also be observed (Figure 2B). Furthermore, cells in the central tissue have a smaller size and an irregular shape. A clear distinction between the infected and non-infected cells cannot be established because there are no cells with bacteroids or amyloplasts. In the distal side of the central tissue, clusters of small cytoplasmically-rich cells can be observed. These cells are apparently still mitotically active, their cell walls, in contrast to the central part, are not very deformed (Figure 2B). After 3 weeks, in B-deficient nodules, the cell walls of the infected cells of the central zone appear to have regions thicker than normal, others were thinner or even without apparent wall deposition (Figure 2D, arrow), and the uninfected cells seem to remain deformed and devoid of starch filled amyloplasts (Figure 2D) compared to control nodules (Figure 2C).

Because cell walls are frequently disturbed under B deficiency, we decide to investigate the presence of unesterified and methyl-esterified pectins in B-deficient root nodules. It can clearly be seen that the level of both pectin decreased considerably in B-deficient root nodules compared to control nodules (Figure 3).

One-dimensional protein gel electrophoresis showed striking differences between Boron-deficient and control nodules (Figure 4). These differences were more evident at 2 weeks post-inoculation. B-deficient nodule extracts showed an abundant 116 kDa protein not found in the control (Figure 4, lane 1). In contrast, the rest of proteins found in the control (Figure 4, lane 2), were absent or detected at decreased levels. At three-weeks, both B-deficient and control nodules showed qualitative and quantitative protein differences in the 60 kDa range and below (Figure 4, lanes 3, 4). Finally, after three to four weeks, the expression pattern of the 116 kDa protein was reversed in B-deficient nodules as evidenced by its lower amount (Figure 4, lane 3), and its complete absence in four week-old nodules (Figure 4, lane 5) compared to controls. On the contrary, the 116 kDa protein of control nodules appeared initially after three weeks (Figure 4, lane 4), and decreased after four weeks (Figure 4, lane 6).

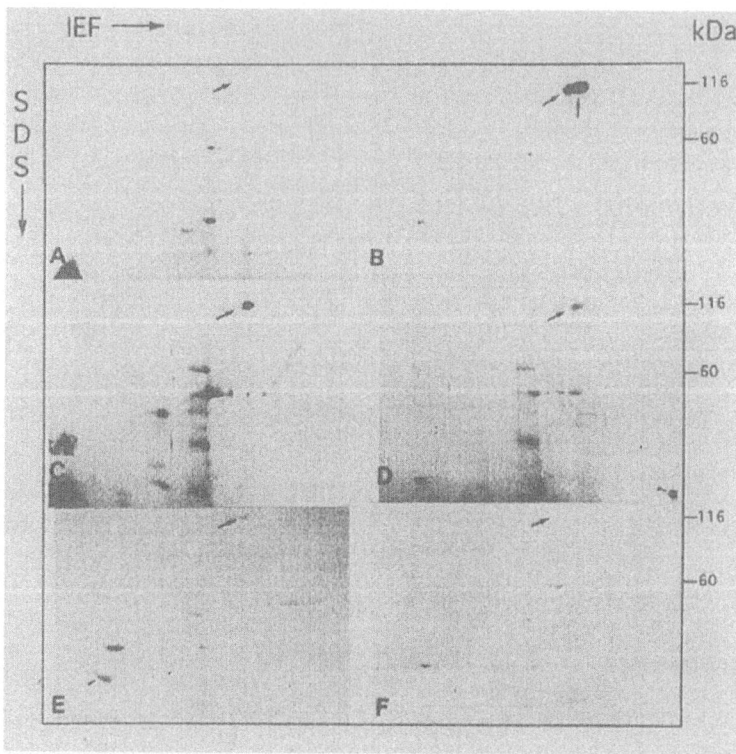

Figure 5. Two dimensional PAGE analysis of low-ionic-strength nodule extracts of both normal and B-deficient root nodules of *Phaseolus-vulgaris* L. Plant proteins eluted from DEAE with 500 mM KCl were concentrated with 4 volumes of ice cold acetone, and the pellets resuspended in O'Farrel buffer (see methods). The samples were separated by two-dimensional gels and stained with Coomassie blue. Panels A,C,E control nodule proteins; Panels B,D,F B-deficient nodule proteins. Panels A,B two week-old nodules. Panels C,D 3 weeks old nodules. Panels E,F 4 week old nodules. In all of cases the samples were loaded at the basic end. The pH ranges from 7.5 (left) to 4.5 (right) in the acidic gels.

Two-dimensional gel electrophoretic analysis of proteins from B-deficient and control nodules confirmed and extended the above results (Figure 5). In B-deficient nodules, the 116 kDa protein was highly evident at two weeks; moreover, there were two (possibly three) isoforms present at this stage (Figure 5, panel B, arrows). The most acidic spot(s) disappeared after three weeks (Figure 5, panel D), and after four weeks none of the 116 kDa proteins were detected (Figure 5, panel F). In control nodules, a single spot of the 116 kDa protein was initially observed after three weeks (Figure 5, panel C), this spot migrated at the same coordinates as the 116 kDa protein from three week-old B-deficient nodules (Figure 5, panel D). This spot was still observed, although in lower amounts, in four week-old nodules (Figure 5, panel E). Other evident differences in the protein pattern between control and B deficient nodules were also observed (Figure 5, panels D and E, arrowheads).

Discussion

In previous papers, we have shown that B is essential for indeterminate nodule development in pea (*Pisum sativum*) (Bolaños et al., 1994, 1996). Whereas indeterminate nodules are characterized by a persistent nodule meristem (Brewin, 1991), cell divisions in determinate nodules cease early during nodule development and the final spherical form of the nodules results from cell enlargement rather than cell division (Hirsch, 1992). Since B is mainly required in meristematic cells (Odhnoff, 1957) it might be expected that B is not specifically required in determinate nodules due to their lower rate of cell division later in development. However, from the data presented here, it is evident that B is also essential for determinate nodules.

Between 1–2 weeks after inoculation, in the central tissue of bean nodules grown under B-deprivation, a distal-proximal gradient (from the root axis) in cell size can be distinguished (Figure 2B). In these nodules, cells proximal to the meristematic region in the central nodule showed dramatic changes and alterations in cell size and morphology (Figure 2B). Also, infected cells of the central tissue of B-deficient nodules appear to have regions thicker, thinner or absent of cell wall deposition (Figure 2D).

We also observed that under B-deficient conditions the levels of both un-esterified and methyl-esterified pectin diminishes in root nodule prints (Figure 3). It is widely accepted that pectins play a key role in determining the plastic extensibility of plant cell walls (Varner and Lin, 1989). Evidence presented by Hu and Brown (1994) that B is physically located and structurally important in the cell wall is correlated with a rapid inhibition of cell expansion under B-deficiency.

In addition to previous results that demonstrated differential changes in cell wall structure occurring at different stages of nodule ontogeny (Higashi et al. 1986; Vasse et al., 1993;), the changes in cell wall structure and nitrogen fixing activity in nodules indicate a clear deleterious effect in nodule development and function by deficiency of B. Boron deficiency not only provides a novel and convenient approach to study nodule development and symbiotic nitrogen fixation, but also for more general and profound effects on plant cell wall integrity and metabolism.

On the other hand, B deficiency exerts an inhibition on IAA oxidase which results in an increase of auxin concentration (Bohnsack and Albert, 1977; Hirsch and Torrey, 1980). Auxins might be required for the very early events of nodule initiation acting probably synergistically with the nod factors (Kondorosi et al, 1993; Cooper and Long, 1994). Boron deficiency might provide an useful tool to study this open question.

Finally, the protein pattern in B-deficient nodules at 2 weeks postinoculation (Figure 4, lanes 1 and 2), resembles the protein expression pattern in stress responses found in other systems, *viz* an overall decrease in protein expression coupled with a very high expression of a few proteins (Linquist and Craig, 1988; Vierling, 1991). The modified expression of the 116 kDa protein, evidenced by a premature and abundant expression of distinct isoforms (Figure 5, panel B) could be the result of an extreme hormone imbalance, produced by the combined action of the B deficiency and the nodule initiation. These data support the hypothesis that B deficiency in the symbiotic interaction could be eliciting a stress response in the plant.

Acknowledgments

I. Bonilla was supported by a fellowship from Dirección General de Investigación Cient'ficay Tecnológica (Spain) and D.G.I.C.Y.T. n° PB95-0217-C02-01. This paper is dedicated to memory of Dr. Hector Perez for his friendship, enthusiasm and inspiration.

References

Bohnsack C W and Albert L S 1977 Early effectcs of boron deficiency on indolacetic acid oxidase levels of squash root tips. Plant Physiol. 59, 1047–1050.

Bolaños L, Brewin N J and Bonilla I 1996 Effects of boron on Rhizobium-Legume cell-surface interactions and nodule development. Plant Physiol. 110, 1249–1256.

Bolaños L, Esteban E, de Lorenzo C, Fernandez-Pascual M, de Felipe M R, Garate A and Bonilla I 1994 Essentiality of boron symbiotic dinitrogen fixation in pea (*Pisum sativum*)- *Rhizobium* nodules. Plant Physiol. 104, 85–90.

Bonilla I, Garcia-Gonzalez M and Mateo P 1990 Boron requirement in Cyanobacteria. Its possible role in the early evolution of photosyntetic organisms. Plant Physiol. 94, 1554–1560.

Bradford M M 1976 A rapid and sensitive method for the quantitation of microgram quantities of protein utilizating the principle of protein dye binding. Anal. Biochem. 72, 248–254.

Brewin N J 1991 Development of the legume root nodule. Ann. Rev. Cell Biol. 7, 191–226.

Cassab G I 1993 Localization of cell wall protein using tissue printing western blot techniques. Methods in Enzymol. 218, 682–688.

Cooper J B and Long S R 1994 Morphogenetic rescue of *Rhizobium meliloti* nodulation mutants by trans-zeatin secretion. Plant Cell 6, 215–225.

Dart P J, Day J M and Harris D 1972 Assay of nitrogenase activity by acetylene reduction. *In* Use of Isotopes for Study of Fertilizer Utilization by Legume Crops. FAO/IAEA Technical Report Series 149, 85–97.

Garcia-Gonzalez M, Mateo P and Bonilla I 1991 Boron requirement for envelope structure and function in Anabaena PCC 7119 heterocysts. J. Exp. Bot. 42, 925–929.

Higashi S, Kushiyama J and Abe M 1986 Electron microscopic observations of infection threads in driselase treated nodules of *Astragalus sinicus*. Can. J. Microbiol. 32, 947–952.

Hirsch A M 1992 Development biology of legume nodulation. New Phytol. 122, 211–237.

Hirsch A M and Torrrey J G 1980 Ultrastructural changes in sunflower root cells in relation to boron deficiency and added auxin. Can. J. Bot. 58, 856–866.

Hu H and Brown P H 1994 Localization of boron in the cell wall of squash and tobacco and its association with pectin. Evidence of a structural role of boron in the cell wall. Plant Physiol. 105, 681–689.

Kondorosi E, Hoffmann B, Endre G, Bogre L, Koncz C, Dutis D, Szecsi J, Kiss G, and Kondorosi A 1993 Involvement of hormones in nodule initiation: auxin sensitivity and hormone blance affect nodulation of *Medicago*. *In* New Horizons in Nitrogen Fixation. Eds R Palacios, J Mora, W Newton, pp 357. Kluwer Academic Publishers, the Netherlands.

Laemmli U K 1970 Cleavage of structural proteins during the assembly of head bacteriophage T4. Nature 277, 680–685.

220

Lara M, Porta H, Padilla J, Folch J and Sanchez F 1984 Heterogeneity of glutamine synthethase polypeptides in *Phaseolus vulgaris* L. Plant Physiol. 76, 1019–1023.

Lindquist, S and Craig, E A 1988 The heat shock proteins. Annu. Rev. Gen. 22, 631–677

Lukaszewski K M, Blevins D G and Randall D D 1992 Asparagine and boric acid cause allantoate accumulation in soybean leaves by inhibition of manganese-dependent allantoate amilohydrolase. Plant Physiol. 99, 1670–1676.

Mateo P, Bonilla I, Fernandez-Valiente E and Sanchez-Maeso E 1986 Essentiality of boron for dinitrogen fixation in *Anabaena sp.* PCC 7119. Plant Physiol. 81, 430–433.

Murashige T 1974 Plant propagation through tissue culture. Ann. Rev. Plant Physiol. 25, 135–166.

Odhnoff C 1957 Boron deficiency and growth. Physiol. Plant. 10, 984–1000.

O'Farrel P H 1975 High resolution two dimensional electrophoresis of proteins. J. Biol. Chem. 250, 4007–4021.

Perez H E, Sanchez N, Vidali L, Hernandez M, Lara M and Sanchez F 1994 Actin isoforms in non-infected roots and symbiotic root nodules of *Phaseolus vulgaris* L. Planta 293, 51–56. M.

Varner J E and Lin L-S 1989 Plant cell wall architecture. Cell 56, 231–239.

Vasse J, de Billy F and Truchet G 1993 Abortion of infection during the *Rhizobium meliloti*-alfalfa symbiotic interaction is accompanied by a hipersensitive reaction. The Plant Journal 43, 555–566.

Vierling E 1991 The roles of heat shock proteins in plants. Ann. Rev. Plant Physiol. Plant Mol. Biol. 42, 579–620.

R.W. Bell and B. Rerkasem (eds.), Boron in Soils and Plants, 221–227.
© 1997 *Kluwer Academic Publishers.*

Early effects of boron deficiency on physical cell wall parameters, hydraulic conductivity and plasmalemma-bound reductase activities in young *C. pepo* and *V. faba* roots

Peter Findeklee, Monika Wimmer & Heiner E Goldbach[1],*
Abteilung Agrarökologie, Universität Bayreuth, D-95448 Bayreuth, Germany
∗ *corresponding author*
[1] *Agrikulturchemisches Institut, Rheinische Friedrich-Wilhelms-Universität Bonn, Meckenheimer Allee 176, D53115 Bonn, Germany. E-mail: heiner.goldbach@t-online.de*

Key words: Boron, cell pressure probe, cell wall elasticity, *Cucurbita pepo*, hydraulic conductivity, reductase activity, *Vicia faba*

Abstract

Rapid effects of boron (B) starvation were investigated in squash (*Cucurbita pepo*) and bean (*Vicia faba*) roots. A transitory decrease of the cell wall elastic modulus (ε) and a concomitant increase of the cell hydraulic conductivity up to fourfold have been observed by measurement with the cell pressure probe. This is in line with a putative role of B in determinating the cell wall physical parameters by the formation of cross links in the pectate fraction.

A secondary rehardening of the cell wall was observed as well as a reduction of the hydraulic conductivity after 20 minutes. This implies that secondary effects become apparent soon after B deprivation. No loss of turgor occurred during the experimental period of up to 45 minutes. Taken together with the recovery of the hydraulic conductivity within 45 minutes, this clearly indicates that membrane desintegration and leakage are secondary or tertiary effects of B deficiency.

Plasmalemma-bound inducible reductase activity was reduced in *V. faba* root sections within 5 minutes after starting the deficiency treatment and showed some relation to the concentration of free boric acid in the external medium. When increasing the external concentration of boric acid, binding of Ca^{2+} to the negatively charged borate complexes was probably counteracted by a concomitant enhancement of proton release under our experimental conditions.

Introduction

The physiological role of boron (B) is not yet completely known and still a matter of controversial discussion. The reason for this lack of knowledge is due to the fact that:

i. there is no radioactive B isotope with a half-life time long enough to allow labelling of B binding compounds, and

ii. that B forms predominantly covalent bonds under physiological conditions, such that the reactions of boric acid or borate with its ligands are readily reversed.

Thus it is extremely difficult to extract B containing fractions from higher plants without grossly altering the conditions and thus shifting reaction equilibria as well as interfering with cell compartmentation. In freeze fractured cells, most of the B is located in cell walls, some in the plasmalemma, and no B could be detected in vacuoles (Martini and Thellier, 1993). In cell walls, B seems to be associated mostly with the pectic fraction (Hu and Brown, 1994; Kobayashi et al., 1996). The formation of borate complexes with vicdiols should lead to an increased number of negative charges in the apoplastic space. This might have implications for the binding of Ca^{2+} and cell wall physical properties as well, besides the physical effect caused by

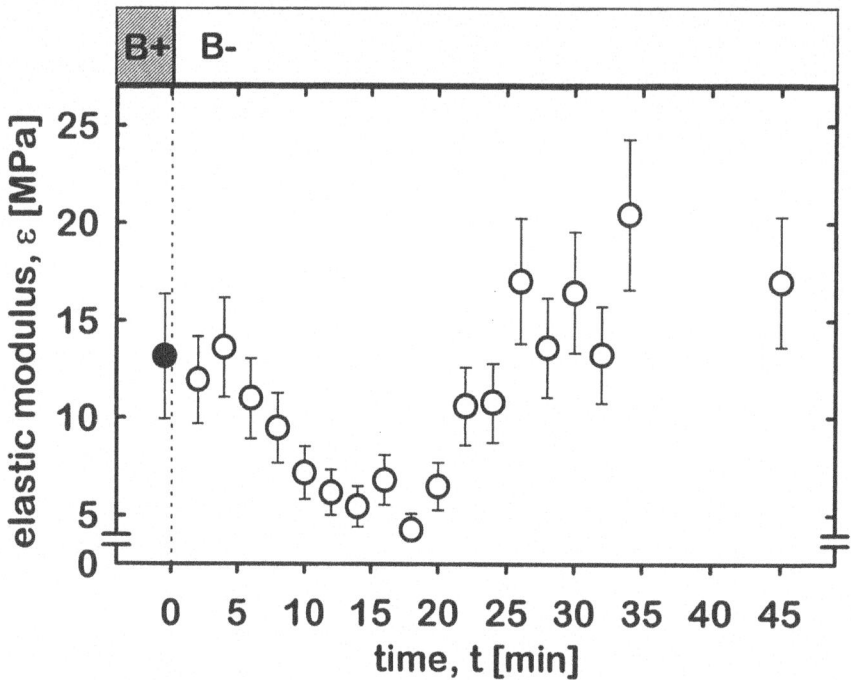

Figure 1. Change of volumetric elastic modulus (ε) of *Cucurbita pepo* cortical cells of the primary root after altering the B supply; plant age: 6 d. Data from one single experiment. Measurements were carried out in one single cell, standard deviation is calculated from experimental error and variation of cell volume.

cross-linking cell wall polymers. As pointed out recently (Goldbach, 1997) the free diffusible proportion of total B might be decisive for the onset of deficiency reactions additionally to an effect on the rheology of cell wall pectates (Kobayashi et al., 1996).

To determine which of boron's possible reactions are crucial to plant metabolism, rapid *in vivo* techniques are required to discriminate between primary and secondary reactions. Below, we will present further results on alterations which occur within minutes and thus should be closely linked to the primary target sites of B.

Material and methods

Seeds of squash, *Cucurbita pepo* L., F1 hybrid cv. 'Diamant', and bean, *Vicia faba* cv. 'Troy' were germinated in expanded clay granules (Lecaton, size 8–16 mm) until appearance of the first foliar leaf. They were irrigated with a 0.2 strength modified Hoagland solution as described earlier (Blaser-Grill et al., 1989).

Cell wall physical properties were determined by the use of the cell pressure probe according to Hüsken et al. (1978). The volumetric elastic modulus of the cell, (ε), is described by the equation (1): $\varepsilon = V\frac{\Delta P}{\Delta V}$, where V is the cell volume, ΔV is the volume pushed into or pulled out of the cell, and ΔP is the corresponding change in turgor. The glass microcapillary tip (diameter: 3 μm) was inserted into cortical cells of the second layer about 25 mm behind the root tip. Several relaxations of about 10 picolitres were carried out in both directions by shifting the cell sap/oil meniscus in the microcapillary to get the mean $\frac{\Delta P}{\Delta V}$. After changing to an almost B free solution, relaxations were again carried out at a high time resolution. The hydraulic conductivity of the cell (Lp) is described by the equation (2): $Lp = \frac{\Delta V \times \ln(2)}{\Delta P \times A \times T_e^w}$, where A is the cell surface area and ($T_{1/2}^w$ is the half-time of the relaxation, which is inversely related to the hydraulic conductivity. The mean A and V were obtained from cell dimensions measured with a stereo microscope. Measurements were terminated when turgor pressure decreased dramatically and unexpectedly indicating severe cell damage. For further details of the pressure probe see a recent review by Steudle (1993).

In a second experimental approach, apical sections of bean roots from plantlets (10 or 11 d after seed imbibition) were incubated in solutions containing: micronutrients, 0.5 mm ferricyanide and vary-

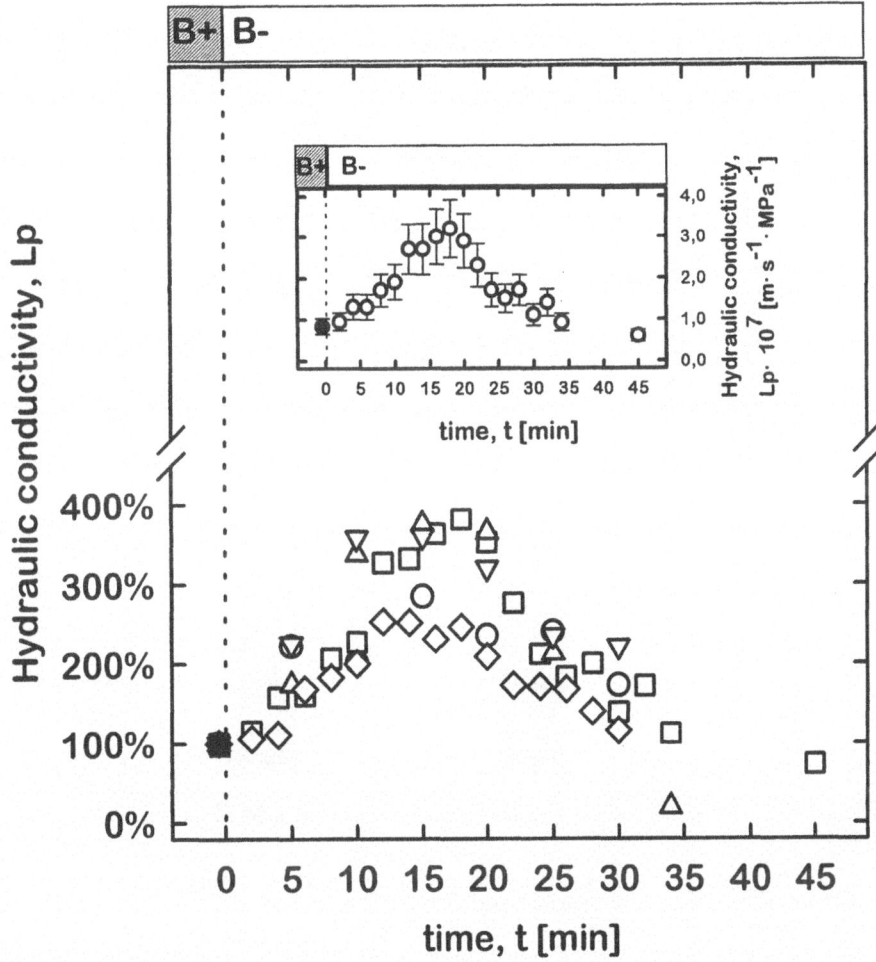

Figure 2. Relative changes of hydraulic conductivity (Lp) of *Cucurbita pepo* cortical cells of the primary root after altering the B supply; plant age: 4–6 d, data from five different experiments (the different symbols characterise the individual experiments). The series of measurements were carried out within the same cell. The insert shows one single experiment, standard deviation is calculated from experimental error and variation of cell surface area.

ing concentrations of boric acid at a ratio of about 300 mg root FW: 700 μL solution. The narrow ratio between solution and root leads to a rapid equilibration between uptake/exchange and external concentration. Thus, it should be possible to get estimates of the exchange rates for B and Ca^{2+}. Boron was determined by a miniaturised curcumin method (modified after Chevallerie-Haaf et al., 1986; publication in prep.) with a detection limit of 33.3 pmoles absolute (20 μL sample at a concentration of 1.66 μM). Ca^{2+} was measured by a slightly modified arsen-azo method as described by Ogawa et al. (1980). Ferricyanide reduction was determined photometrically by the change of absorbance at 420 nm. Statistical analysis was done where applicable by F-test and *t*-testing.

Results and discussion

Figure 1 shows the changes of elastic modulus (ε) of one individual experiment, i.e. from individual consecutive measurements *in the same cell*. Error bars indicate the experimental error. The elastic modulus was rapidly reduced following rinsing of roots with minus B solution and reached a minimum after 15 to 20 minutes. This has been confirmed recently in seven independent series of experiments (Findeklee and Goldbach, 1996). These results strongly support Loomis' hypothesis of B forming cross links between cell wall polymers by 2:1 (probably) apiose – borate ester bridges (Loomis and Durst, 1991; 1992) and recent findings of Matoh's group (Kobayashi et al.,

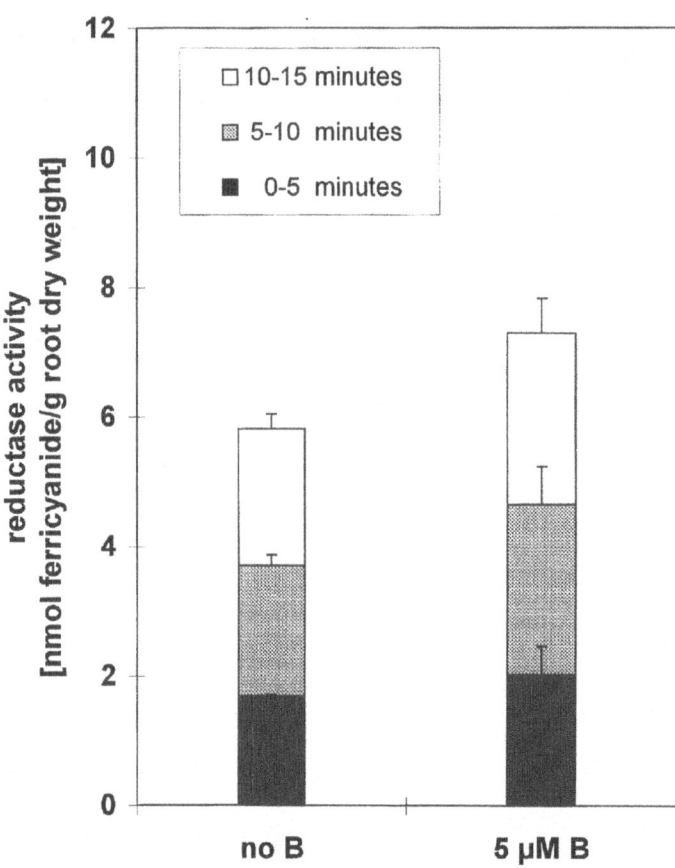

Figure 3. Ferricyanide reduction by *Vicia faba* root sections (10 d after seed imbibition) without addition of boron and at 5 μM boron at a ratio of 350 mg root FW to 700 μL solution. Presentation of a typical experiment, where the solution was replaced every 5 minutes up to 15 minutes; vertical bars indicate standard error (n=5); the experiment was repeated four times with similar results.

1996). It may be assumed that borate complexing to vic-diols of the rhamno-galacturonan II (RGII) fraction yields a gel-like structure (see Loomis and Durst, 1992) which reduces the mobility of embedded cellulose and hemicellulose fibres.

More detailed evaluation of the pressure relaxation curves show that the hydraulic conductivity of the cells increases about two to fourfold after removing B with a maximum between 10 and 20 minutes (Figure 2). The hydraulic conductivity behaves in an inverse manner to the transitory increase of ε as evaluated by five independent series of experiments. According to Henzler and Steudle (1995), water channels may (in part) be responsible for the hydraulic conductivity of the plasma membrane. The increased conductivity would also be in line with an enhanced release of K^+ (Cakmak et al., 1995; see also Goldbach, 1997) within a short period after removing B from the nutrient solution. It remains unclear whether this is a response to the altered physical behaviour of the cell wall and some kind of

'stretch activation' of water channels or whether water moves through K^+ channels which may open upon the small but repeatable hyperpolarisation found earlier in squash roots (Goldbach et al., 1991, for a review see Goldbach, 1997). Changes of the membrane potential followed about the same time scale as the alterations of cell wall physics and hydraulic conductivity of the cells.

The hydraulic conductivity decreased again 15 to 20 min. after starting the removal of B from the system (Figure 2) with a concomitant increase of ε (Figure 1). During the first hour of the deficiency treatment, we never observed a significant loss of cell turgor (Findeklee and Goldbach, 1996).

Taken together with the fact that the hydraulic conductivity was restored to or below the initial level within 30 minutes of the treatment (Figure 2), a loss of membrane integrity as an early deficiency response (see Cakmak et al., 1995) can safely be excluded. An enhanced membrane leakage due to free radical attack

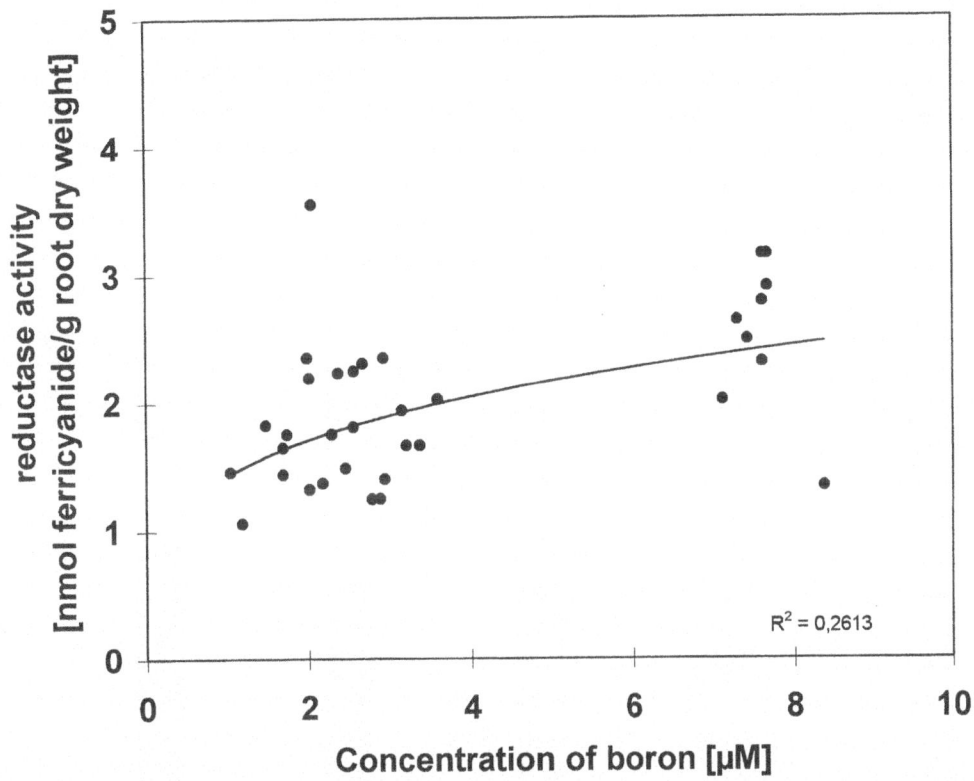

Figure 4. The figure shows the relation between external equilibrium concentrations of boron in a typical experiment and PL-bound reductase activity; variation between individual samples is due to variability between individual plantlets.

is thus likely a secondary or tertiary reaction to B deficiency and could never be found in our experiments during the first hour of treatment.

Besides the binding of boric acid/borate to the cell wall and the effect on its physical properties, B may rapidly influence membrane-bound reactions as well. These effects may be the consequence of an altered cell wall extensibility, which also might hold true for the hydraulic properties of the cell as shown above. But, as outlined in a recent review (Goldbach, 1997), this might also be a direct reaction to the concentration of free diffusible boric acid, which may (transitorily?) react to membrane-bound enzymes with appropriate vic-diol groups on either side of the membrane, as undissociated boric acid should easily cross the plasma membrane. Interestingly, Obermeyer et al. (1996) isolated a B-stimulated adenosine-triphosphatase (ATPase) fraction from ungerminated lilly pollen. It should also be taken into consideration that B may serve to facilitate the export of pectic cell wall precursors from the cytoplasm into the apoplast. The findings of Bonilla's group of an inhibited export of extensin precursors (Bonilla et al., 1994) are in line

with this assumption. It may thus be assumed that B serves multiple functions in plant metabolism.

We have observed earlier that the ferricyanide – reduction is rapidly inhibited under B deficiency (Goldbach et al., 1990), which was confirmed by Barr et al. (1993) as an inhibition of plasmalemma (PL)-bound NADH-oxidoreductase activity. Although it cannot be excluded that this is an early secondary event, it is easily determined and can be taken as an indicator for the initiation of deficiency reactions at the membrane level. We thus determined alterations of PL-bound reductase activity with about the same time resolution as the cell wall elasticity. For technical reasons, we had to use dissected bean roots. PL-bound reductase activity was reduced within the first 5 minute interval after transferring the apical root parts to a B 'free' nutrient solution ($<< 0.02$ μM B) (Figure 3). It can thus be stated that the PL-bound reductase reacts to a removal of B from the solution within the same time lapse as cell wall physical properties. In Figure 4, the ferricyanide reduction (i.e. PL-bound reductase activity) is plotted against the B equilibrium concentration resulting after each five minute incubation interval. Clearly there was

226

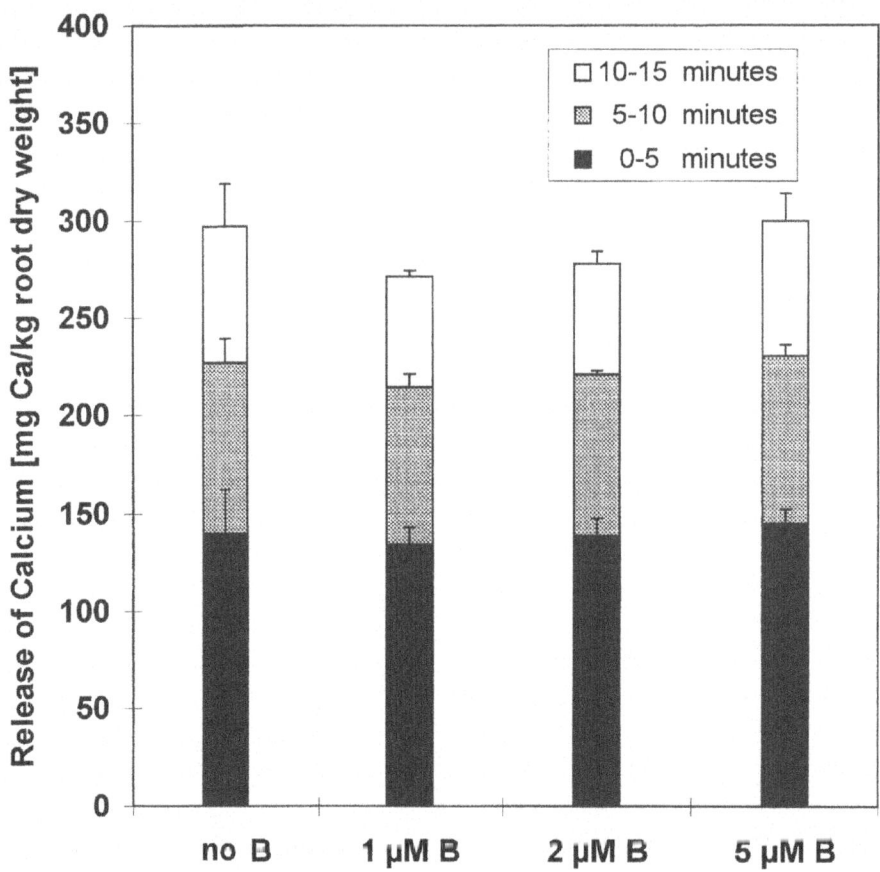

Figure 5. Release of Ca from *V. faba* root sections as related to boron supply: same experimental set up as in Figure 3; vertical bars indicate standard deviations for each five minute interval (n=5).

a correlation between the two parameters although the relationship may be an indirect one. Scattering of the points is caused by the variability between individual roots and experimental errors, which are due to the experimental set up with very small volumes. As we used unbuffered solutions in the past, an indirect pH-effect has also to be taken into consideration. Boric acid may thus react as any other weak permeating acid and stimulate proton pumping activities at the PL (see Goldbach, 1997).

This can be confirmed indirectly by a further experimental approach. We tested the influence of B supply on the apoplastic Ca^{2+} concentrations in the same experiments as in Figures 3 and 4. It can be deduced that a lower number of apoplastic borate complexes should result in a lower number of negative charges and thus some liberation of Ca^{2+} to the apoplastic solution. As can be seen in Figure 5, the increase of B supply leads to a lower equilibrium concentration of Ca^{2+} with 1 μM and 2 μM B as expected. A further

rise to 5 μM B, however, again released more Ca^{2+}. This can be interpreted by the enhanced net proton release (unpublished data) and thus an overlay of two counteracting effects.

It has to be summarized that PL-bound reductase activity is changed within the same time scale as the effects on cell wall elastic and plastic properties. Experiments with 24 h B starvation as used by Lukaszewski and Blevins (1996) or 7 d starvation as reported by Hu and Brown (1994) cannot separate between primary effects and secondary or tertiary alterations. As has been outlined, secondary reactions (rehardening of cell walls and reduction of hydraulic conductivity) can be noticed within 20 minutes or even less after starting the B deficiency treatment. Any experiment attempting to elucidate primary effects of B deficiency has thus to consider the obviously rapid effects it exerts on several physiological reactions (for a more thorough discussion, see Goldbach, 1997). Further experiments are under way to differentiate between primary and secondary reactions of cell walls and membranes to removal of external B supply.

Acknowledgements

The financial support by the grants No. Go 415/12-1 and 2 and Go 415/14-1 of the 'Deutsche Forschungsgemeinschaft' is gratefully acknowledged.

References

Barr R, Böttger M and Crane F L 1993 The effect of boron on plasma membrane electrone transport and associated secretion by cultured carrot cells. Biochem. and Molecular Biol. Int. 31, 31–39.

Blaser-Grill J, Knoppik D, Amberger A and Goldbach H E 1989 Influence of boron on the membrane potential in *Elodea densa* and *Helianthus annuus* roots and H^+ extrusion of suspension cultured *Daucus carota* cells. Plant Physiol. 90, 280–284.

Bonilla I, Cassab G I, Lopez L and Sanchez F 1994 Efecto de la deficiencia de boro en la estructura y funcionalidad de nodulos de judia (*Phaseolus vulgaris* L.). *In* Bases Fisiológicas, Bioquímicas y Moleculares de la Nutrición Mineral de las Plantas; V. Simposium Nacion 1 i Ibérico. Ed M C Alvarez, A Tinault et al., pp 104–109. Badajoz.

Cakmak I, Kurz H and Marschner H 1995 Short-term effects of boron, germanium and high light intensity on membrane permeability in boron deficient leaves of sunflower. Physiol. Plant. 95, 11–18.

Chevallerie-Haaf U, Meyer A and Henze M 1986 Photometrische Bestimmung von Bor im Grund- und Oberflächenwasser. Fresenius Z. Anal. Chem. 323, 266–270.

Findeklee P and Goldbach H E 1996 Rapid effects of boron deficiency on cell wall elasticity modulus in *Cucurbita pepo* roots. Bot. Acta 109, 463–465.

Goldbach H E 1997 A critical review on current hypotheses concerning the role of boron in higher plants: suggestions for further research and methodological requirements. J. Trace Microprobe Techniques 15. In press.

Goldbach H E, Hartmann D and Rötzer T 1990 Boron is required for the stimulation of the ferricyanide-induced proton release by auxins in suspension-cultured cells of *Daucus carota* and *Lycopersicon esculentum*. Physiol. Plant. 80, 114–118.

Goldbach H E, Blaser-Grill J, Lindemann N, Porzelt M, Hörrmann C, Lupp B and Gessner B 1991 Influence of boron on net proton release and its relation to other metabolic processes. Current Topics in Plant Biochem. and Physiol. 10, 195–220.

Henzler T and Steudle E 1995 Reversible closing of water channels in *Chara internodes* provides evidence for a composite transport model of the plasma membrane. J. Exp. Bot. 46, 199–209.

Hüsken D, Steudle E and Zimmermann U 1978 Pressure probe technique for measuring water relations of cells of higher plants. Plant Physiol. 61, 158–163.

Hu H and Brown P H 1994 Localization of boron in cell walls of squash and tobacco and its association with pectin: evidence for a structural role of boron in the cell wall. Plant Physiol. 105, 681–689.

Kobayashi M, Matoh T and Azuma J 1996 Two chains of rhamnogalacturonan II are cross-linked by borate-diol ester bonds in higher plant cell walls. Plant Physiol. 110, 1017–1020.

Loomis W D and Durst R W 1991 Boron and cell walls. Current Topics in Plant Biochem. and Physiol. 10, 149–178.

Loomis W D and Durst R W 1992 Chemistry and biology of boron. Biofactors 3, 229–239.

Lukaszewski K M and Blevins D G 1996 Root growth inhibition in boron-deficient or aluminum-stressed squash may be a result of impaired ascorbate metabolism. Plant Physiol. 112, 1135–1140.

Martini F and Thellier M 1993 Boron distribution in parenchyma cells of clover leaves. Plant Physiol Biochem 31, 777–786.

Obermeyer G, Kriechbaumer R, Strasser D, Maschessnig A and Bentrup F W 1996 Boric acid stimulates the plasma membrane H^+-ATPase of ungerminated lily pollen grains. Physiol. Plant. 98, 281–290.

Ogawa Y, Harafuji H and Kurebayashi N 1980 Comparison of the characteristics of four metallochromic dyes as potential calcium indicators for biological experiments. J. Biochem. 87, 1293–1303.

Steudle E 1993 Pressure probe techniques: basic principles and application to studies of water and solute relations at the cell, tissue and organ level. *In* Water Deficits: Plant Responses from Cell to Community. Ed J A C Smith and H Griffith. pp 5–36 Bios Scientific Publishers, Oxford.

R.W. Bell and B. Rerkasem (eds.), Boron in Soils and Plants, 229–235.
© 1997 *Kluwer Academic Publishers.*

A fluorescein boronic acid conjugate as a marker for borate binding sites in the apoplast of growing roots of *Zea mays* L. and *Helianthus annuus* L.

K-H Glüsenkamp[1], H Kosegarten[2], K Mengel[3], F Grolig[3], A Esch[2] & H E Goldbach[4]

[1] *ImmunoTec (Immunologisch-Technische Produkt- und Beratungs-GmbH) Elbertstraße 10 D 45768 Marl, Germany*
[2] *Institut für Pflanzenernährung, Justus-Liebig-Universität Giessen Südanlage 6, D 35390 Giessen, Germany*
[3] *Institut für Allgemeine Botanik und Pflanzenphysiologie, Justus-Liebig-Universität Giessen, Senckenbergstraße 17, D 35390 Giessen, Germany*
[4] *Agrikulturchemisches Institut, Rheinische Friedrich-Wilhelms-Universität Bonn, Meckenheimer Allee 176, D 53115 Bonn, Germany. E-mail: heiner.goldbach@t-online.de*

Key words: Boron, borate esters, borate-binding, cell wall, fluorescein boronic acid conjugate, fluorescence

Abstract

A fluorescein boronic acid conjugate has been used as a pH-sensitive dye in the root apoplast and because of its likely links to boric acid-binding sites. The marker (Glüsenkamp et al.; German Patent Application) is non-toxic to cell growth up to concentrations of 100 μM and stains almost exclusively the apoplastic space of sunflower and maize roots. The meristematic zone behind the calyptra is hardly stained, which is in line with the assumption of a lack of demand for boron (B) during cell divison. By contrast, the expansion zone shows a more intense staining in a part of the longitudinal walls which is indicative of an intensive secretion of B-binding cell wall material into this part. A 'banding' of the stain in sunflower roots hints to a pulsating extension growth or at least intermittent secretion of B binding moieties. Root hairs seem to be more intensively stained than other cells of this region, especially at their tips. Experiments are under way to further characterize the bonds and to apply other boronic acid-coupled dyes for the histological identification of boric acid-binding ligands.

Introduction

Fluorescein isothiocyanate (FITC) is a fluorescent dye whose fluorescence ratio is pH dependent and, when coupled to dextrane, can be used to monitor pH-changes in the apoplast of leaves (Hoffmann and Kosegarten, 1995). Its disadvantage for use in the root apoplast is the lack of retention of the dye in this compartment. Expensive confocal microscopy equipment would be required for pH measurements (Taylor et al. 1996). We thus developed a fluorescein boronic acid conjugate (Glüsenkamp et al., 1997) which is retained in the apoplastic compartment. The surplus of the dye can be removed by simple rinsing. This allows highly sensitive pH-measurements in the range between pH 4.5 and pH 7.0.

The dye shows a binding behaviour similar to boric acid, as boronic acid is likely to bind to *vic,cis*-diols in the same manner as boric acid does. The distance between the vicinal oxygen atoms should favour the formation of one-sided esters (see review by Loomis and Durst, 1992). The FITC-boronic acid adduct should thus be useful as a marker for boric acid-binding moieties in intact tissues and extracted ligands.

We have recently found evidence for a control of cell wall physical properties by boric acid (Findeklee and Goldbach, 1996), and it would be extremely useful to identify the respective binding sites by a specific dye. In the present communication, first photographs are presented which show the applicability of the boronic acid-coupled fluorescein in identifying boric acid-binding sites in young growing root tissues.

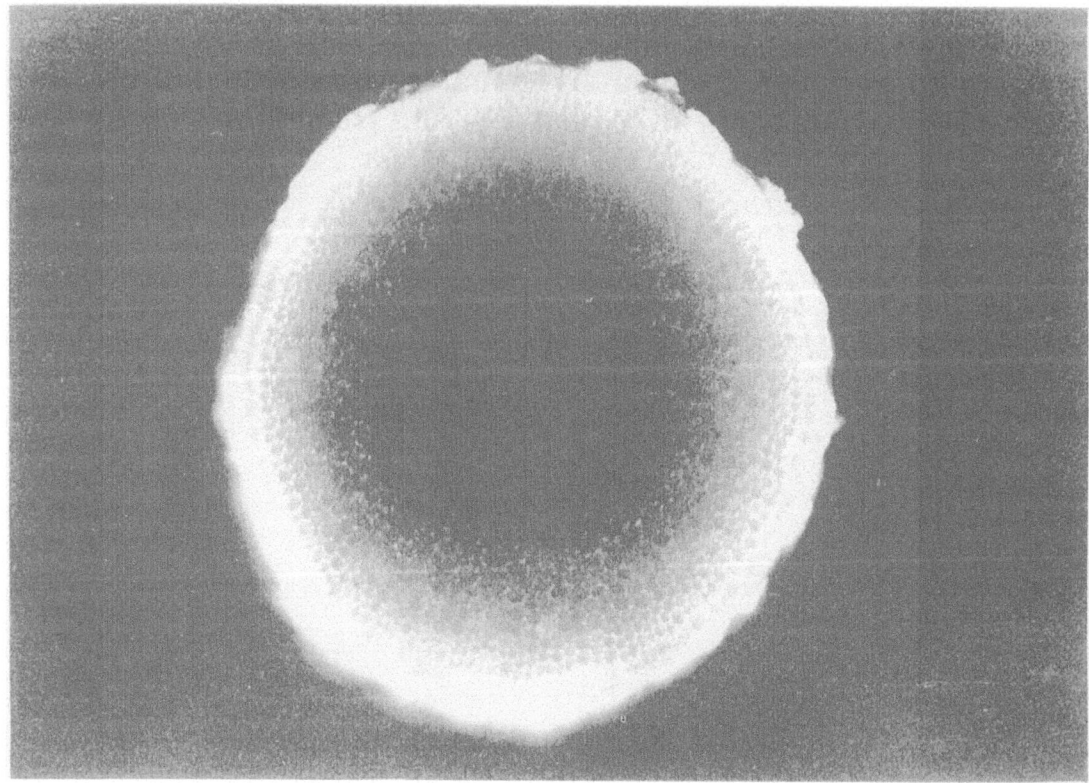

Figure 1. Boronate-coupled FITC - fluorescence in a young root of *Helianthus annuus* L. (cv. Erika), after 12 h *in-vivo* incubation with the dye and rinsing with fresh nutrient solution, cross section at the root apex.

Materials and methods

Young roots of *Helianthus annuus* (cv. Erika) and *Zea mays* (cv. Helix) were grown in a modified Hoagland solution as described elsewhere (Kosegarten and Englisch, 1994) and supplemented with 100 μM of the boronic acid-coupled fluorescein for 12 h in a complete nutrient solution (but without added boric acid). After incubation, the surplus dye was removed by rinsing with complete fresh unlabelled nutrient solution. Cuttings were made by means of a razorblade and slices were transferred to glass mounts. The root segments were inspected by epifluorescence microscopy using filterbloc Ploemopak™ L3 (excitation BP 450–490 nm; dichroic FT 510 nm; emission BP 525/20), XBO 75-W/2 Xenon lamp and NPL Fluotar 6.3/0.2. Fluorescence micrographs were taken with a camera (MPS-46; Leica Heerbrugg AG, Heerbrugg, Switzerland) using film (TMAX 400-Eastman Kodak Ca., Rochester, N.Y.) at 1600 ASA.

Results and discussion

On the basis of its chemical behaviour, the boronic acid-coupled fluorescein can be used as a marker for boric acid binding sites at tissue and probably cell level. The marker is non-toxic to cell growth up to concentrations of 100 μM and stains almost exclusively the apoplastic space (Figure 1), which is in agreement with the observation of a high proportion of B - binding moieties in this compartment [esp. rhamnogalacturonan II (RGII): Kobayashi et al., 1996]. Rhamnogalacturonan II is probably a fraction which is closely related to newly secreted cell wall precursors and later on transformed with cell aging (Matoh et al., 1996).

It might be argued that boronic acid conjugates will not be able to form di-ester bonds such as those formed by boric acid (see Loomis and Durst 1991, 1992). Loss of RGII cross linking (Kobayashi et al., 1996) is likely the reason for the changes observed in cell wall physical properties (Findeklee and Goldbach 1996). As only one pair of binding sites is available in the boronic acid-coupled dye, it is likely that it will not be able to link two RGII chains together as boric

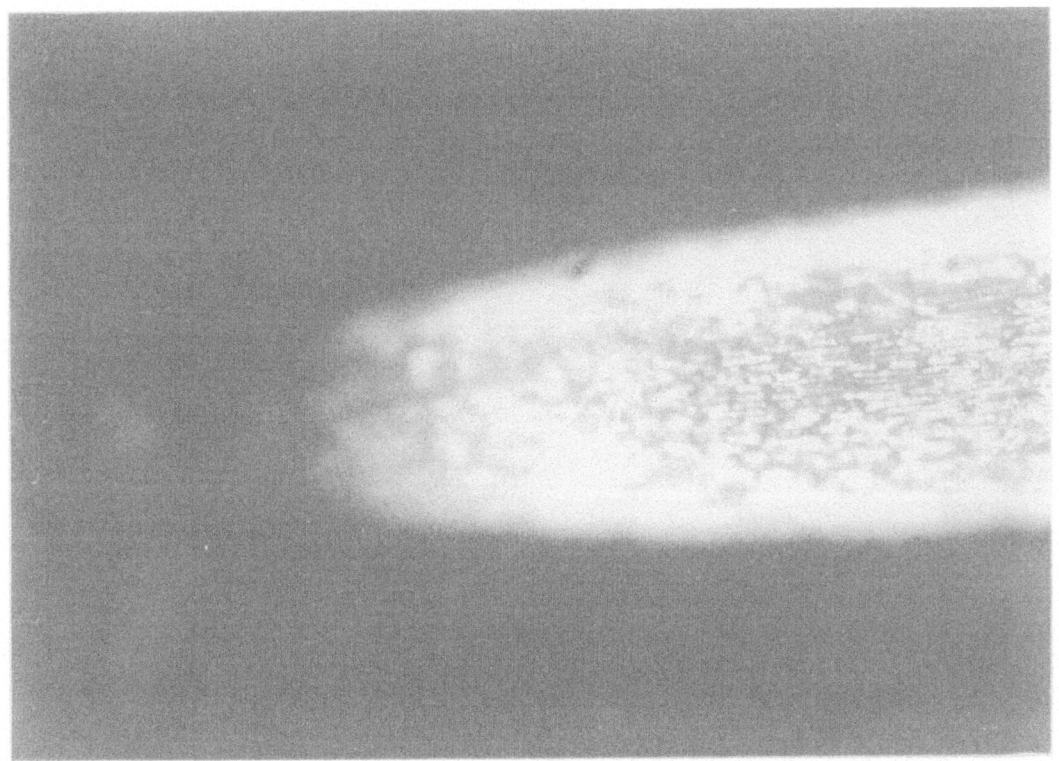

Figure 2a.

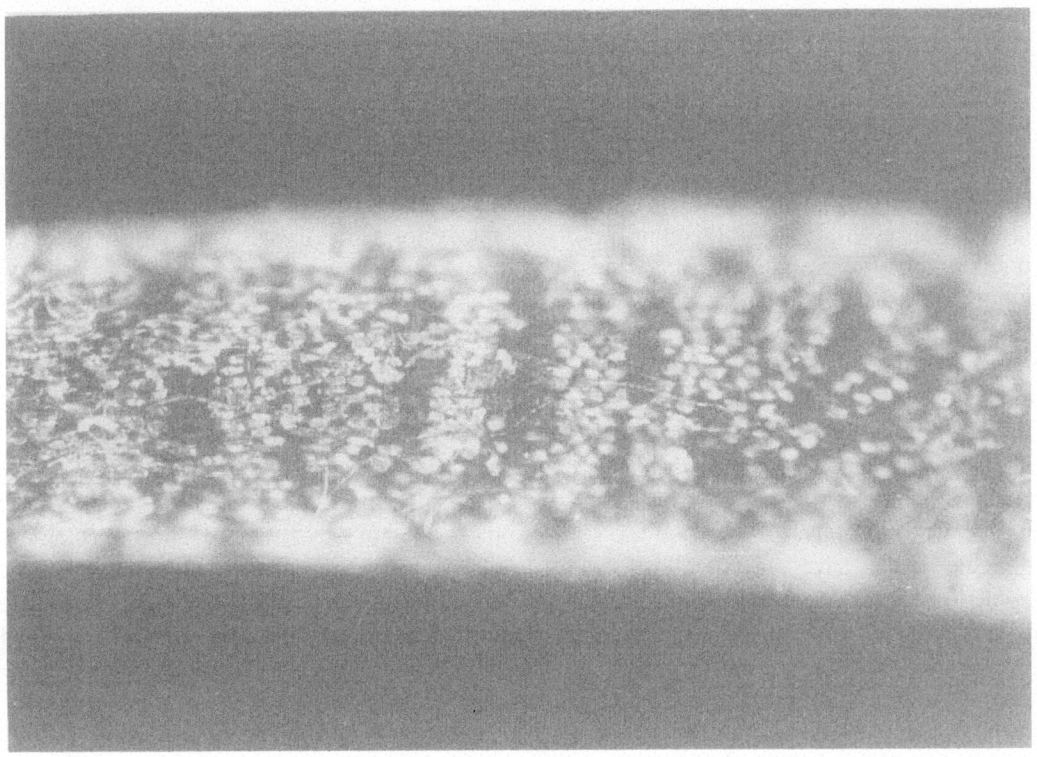

Figure 2b.

232

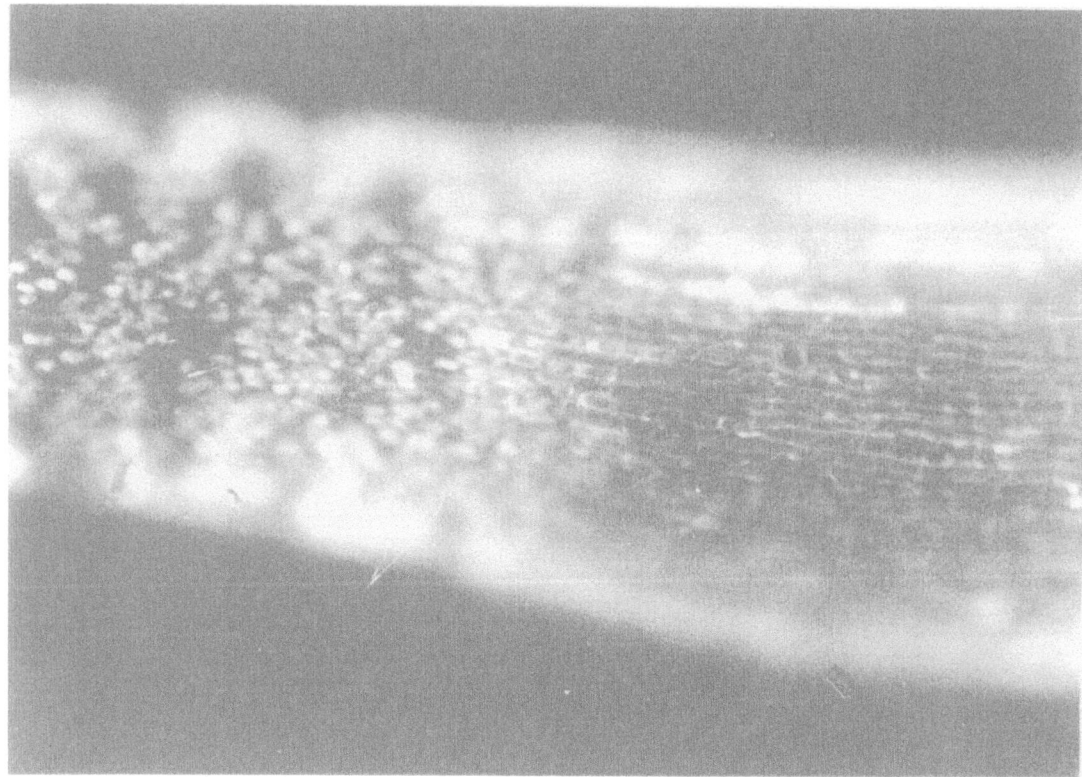

Figure 2c.

acid does. The distance of the oxygen atoms in the dye should favour the formation of one sided esters with vic-diols. These might not be able to fulfil exactly the same function as borate esters. The binding sites, however, will be the same as for boric acid. Apparently, the dye efficiently competes with boric acid, as it was coupled to the cell wall even after precultivating in boric acid and could not be removed by rinsing with unlabelled solution. The reason for this might be a great surplus of B-binding sites or that the binding of the dye might be even favoured due to steric reasons or other molecular interactions. Interestingly, Torssell (1956) described a growth promoting effect of arylboric acids on wheat roots and Avena coleoptiles and attributed it to the formation of one sided esters instead of two-sided borate esters. Unfortunately, however, he used graminaceous species which have vegetative growth with a very low demand for B. More detailed analysis of the binding behaviour of boronated dyes is thus required.

The meristematic zone behind the calyptra hardly stains (Figure 2a), which would support the earlier indications of Loomis and Durst (1991) of an uninhibited cell divison in the absence of B and their notion that only expanding tissues have a high demand for B. It is obvious that the cells of the expansion zone are heavily stained (Figure 2b and 2c), which shows the presence of boric acid-binding moieties. The cell walls of these cells, however, are not uniformly stained which would be in line with the preferential secretion of newly formed cell wall material (including pectic substances) to the expanding part of the cell walls. It also shows some kind of 'banding', so one can assume that not all cells of the extension zone grow simultaneously at the same pace and that some kind of pulsating growth is taking place, probably similar to growing pollen tubes of several species (Geitmann et al., 1996).

As can be seen in Figure 2c and 2d, the cell walls bind less dye after finishing their main extension growth, but they are more uniformly stained. Figure 2c shows clearly the different dye binding pattern of the elongation zone and the root part where the root hairs begin to grow. More dye seems to be bound by root hairs than by the rest of the epidermal cells (Figure 2d, Figure 3). Transverse sections revealed that less fluorescein boronic acid conjugate is fixed to the meristematic cells than to more basal ones (Figure 1 and 3).

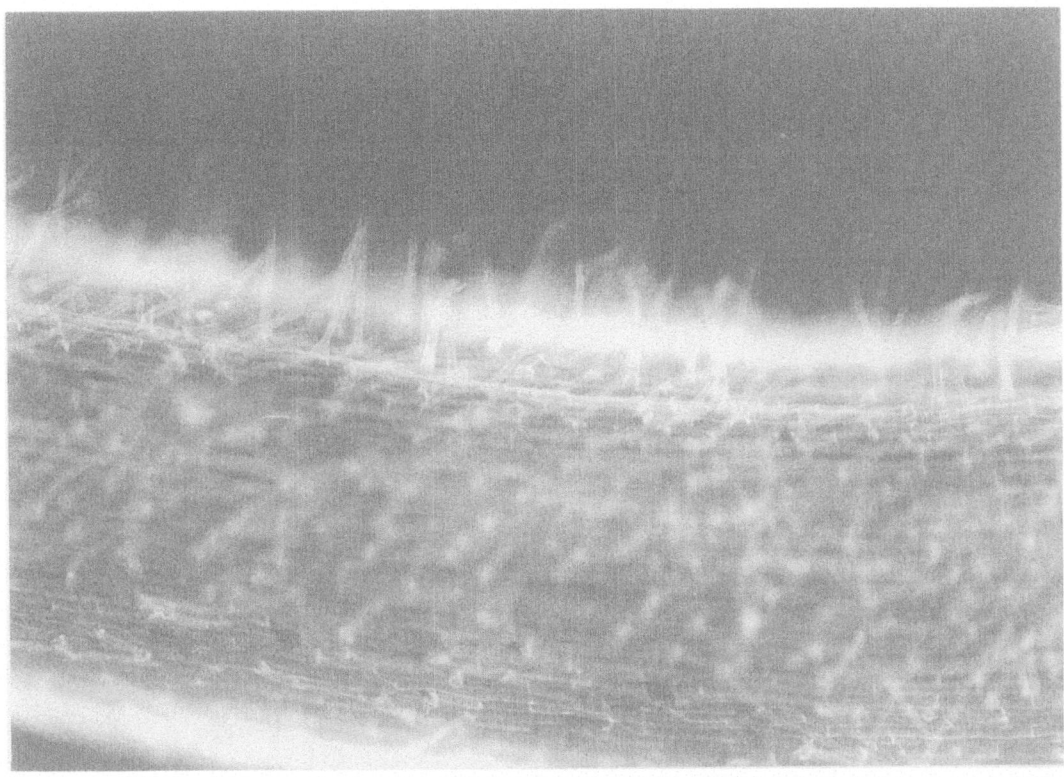

Figure 2d. Boronate-coupled FITC - fluorescence in one young root of *Helianthus annuus* L. (cv. Erika), after 12 h *in-vivo* incubation with the dye and rinsing with fresh nutritent solution. (a) Apical root part showing almost no binding of the dye. (b) Root part basally adjacent to Figure 2a, zone of expansion growth; note bright fluorescence in a part of the cells; the staining seems to be restricted to a part of the expanding cells; a 'banding' can be noticed when looking at the outline of the roots. (c) Root part adjacent to Figure 2b, transition zone between expansion and root hair zones: roots in the root hair zone are less intensively but more uniformly stained than in the expansion zone. (d) Root part adjacent to Figure 2c; root hairs seem to be more intensively stained than the cell walls of the epidermal and cortical cells.

Root hairs stain heavily at their tips (Figure 3), which is indicative of an enhanced secretion of boric acid-binding fractions into this area. It hints to the preferential tip growth of root hairs such as found in pollen tubes (Geitmann et al., 1996). This behaviour was found in maize (Figure 3) as well as in sunflower roots. It is less clear in Figure 2d, but was seen in other transverse cuttings of sunflower roots (not shown).

So far, the dye was used only for *in vivo* staining, and it became clear that it will be bound with sufficient strength to the boric acid - binding ligands. Strong retention at the outer parts of the root (rhizodermis and cortical tissue) may be due to a great surplus of boron-binding ligands there which would limit the staining of the interior part of the root (Figure 3).

We will thus initiate experiments where the dye is used for staining fixed tissue or cryosections to identify whether the preferential staining at the outer parts is either due to a different distribution of boric acid-binding fraction(s) such as RG II or to a limited penetration of the dye into the interior part of the roots. Due to its hydrophilic character, the dye will probably not enter the central cylinder.

It might be argued that the fluorescence of the boronic acid coupled dye is pH-dependent and thus different staining intensities would be a consequence of differences in pH. This is, however, unlikely to be the cause for the staining patterns shown in Figures 2 and 3, as the observed fluorescence intensities do not correlate with the known pH-profile along the root. In maize, a pH of 4.9 was found in the expansion zone (Taylor et al. 1996) and of 5.3 in the root hair zone (Kosegarten and Esch, personal communication). The root apex, which shows generally higher pH-values (Mühling and Wimmer, personal communication), was hardly stained, whereas the fluorescence increased in the expansion zone and in the root hair zone which are generally more acidic.

Work is in progress to couple different ion-specific dyes to boronic acid according to our patented pro-

234

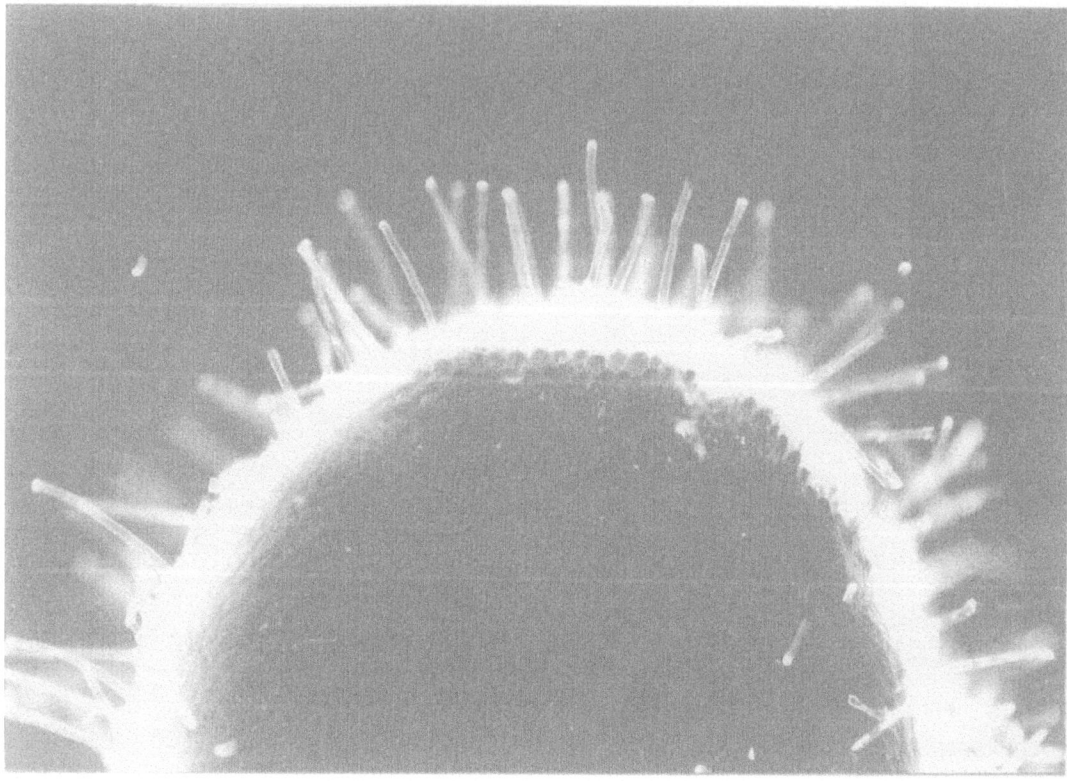

Figure 3. Cross section of a *Zea mays* L. (cv. Helix) root in the root hair zone after 12 h treatment with a boronate - coupled FITC; fluorescence is clearly more intensive at the tips of the root hairs

cedure in order to monitor ion concentrations in the apoplast of young living roots. Such dyes would be extremely valuable tools for the identification of even minor compounds which have *vic,cis*-diol groups and are thus able to bind boric acid or borate, respectively. The method may be applied for histological staining as well as for the identification of boric acid-binding ligands in extracted compounds, e.g. on gels, on TLC plates as well as in column chromatography. The dye was developed quite recently, thus binding and transport behaviour of boronic acid adducts are yet to be characterised in detail.

Acknowledgements

Part of the present work was supported by grants of the Deutsche Forschungsgemeinschaft (DFG) to *K.M.* [Me 209/46-1] and *H.G.* [Go 415/14-1] in the context of the special program 'The apoplast: site of reaction and transport', which is gratefully acknowledged. The authors are indebted as well to Monika Wimmer and Peter Findeklee for valuable help in the preparation of the manuscript.

References

Findeklee P and Goldbach H E 1996 Rapid effects of boron deficiency on cell wall elasticity modulus in *Cucurbita pepo* roots. Bot. Acta 109, 463–465.

Geitmann A, Li Y Q and Cresti M. 1996. The role of the cytoskeleton and dictyosome activity in the pulsatory growth of *Nicotiana tabacum* and *Petunia hybrida* pollen tubes. Bot. Acta 109, 102–109.

Glüsenkamp K H, Kosegarten H and Steinweg D 1997 Verfahren zur schonenden Markierung von Geweben mit analytischen Sonden und Affinitätsliganden. German Patent Application.

Hoffmann B and Kosegarten H 1995 FITC-dextran for measuring apoplast pH and apoplastic pH gradients between various cell types in sunflower leaves. Physiol. Plant. 95, 327–335.

Kobayashi M, Matoh T and Azuma J 1996 Two chains of rhamnogalacturonan II are cross-linked by borate-diol ester bonds in higher plant cell walls. Plant Physiol. 110, 1017–1020.

Kosegarten H and Englisch G 1994 Effect of various nitrogen forms on the pH in leaf apoplast and on iron chlorosis of *Glycine max* L. Z. Pflanzenern. Bodenkunde 157, 401–405.

Loomis W D and Durst R W 1991 Boron and cell walls. Current Topics in Plant Biochem. Physiol.10, 149–178.

Loomis W D and Durst R W 1992 Chemistry and biology of boron. Biofactors 3, 229–239.

Matoh T, Kawaguchi S and Kobayashi M 1996 Ubiquity of a borate-rhamnogalacturonan II complex in the cell walls of higher plants. Plant Cell Physiol. 37, 636–640.

Taylor D P, Slattery J and Leopold A C 1996 Apoplastic pH in corn root gravitropism: a laser scanning confocal microscopy measurement. Physiol. Plant. 97, 35–38.

Torssel K 1956 Chemistry of arylboric acids. VI. Effects of arylboric acids on wheat roots and the role of boron in plants. Physiol. Plant. 6, 652–664.

R.W. Bell and B. Rerkasem (eds.), Boron in Soils and Plants, 237–241.
© 1997 *Kluwer Academic Publishers.*

In vitro reconstitution of the boron-polysaccharide complex purified from cultured tobacco BY-2 cells

Masaru Kobayashi & Toru Matoh
Laboratory of Plant Nutrition, Department of Agricultural Chemistry, Faculty of Agriculture, Kyoto University, Kyoto, 606-01, Japan

Key words: Boron, cultured cells, pectic polysaccharide, rhamnogalacturonan II, tobacco BY-2

Abstract

A boron-polysaccharide complex isolated from cell walls of cultured tobacco (*Nicotiana tabacum*) BY-2 is a dimeric rhamnogalacturonan II (RG-II) cross-linked by borate-diol esters, the same as in the radish (*Raphanus sativus*) root cell wall. This complex can be reconstituted spontaneously *in vitro* simply by mixing monomeric RG-II and boric acid at pH 4. Germanic acid substitutes for boric acid in the reconstitution to some extent.

Introduction

Although boron (B) is an essential microelement for higher plants, its primary function is not known yet (Loomis and Durst, 1992). We have proposed that determination of the location of B in the cell seems to be a prerequisite for identifying the functions of B. On the other hand, it had been claimed that the sites where boric acid is localized may be merely a consequence of the diester bonding of boric acid with cis-diol groups of sugars and phenolics. Therefore, B might bind to cell wall polysaccharides at random. However, a particular B-polysaccharide complex (BPC) was isolated and purified from radish root cell walls (Matoh et al., 1993a). Subsequent reports revealed that the complex is composed of two chains of monomeric rhamnogalacturonan II (RG-II), which are cross-linked by boric acid with borate-diester bonding (Kobayashi et al., 1995, 1996). The RG- II was first isolated from sycamore (*Acer pseudoplatanus*) cell walls by Darvill et al. (1978), without knowing its physiological role. Following our report, Ishii and Matsunaga (1996) and O'Neill et al. (1996) confirmed our finding in sugar beet (*Beta vulgaris*), sycamore and pea (*Pisum sativum*) cell walls, respectively. A wider survey for the polysaccharide-associated B in cell walls of higher plants strongly suggested that RG-II is the exclusive carrier for B in cell walls (Matoh et al., 1996). This report describes reconstitution of the complex *in vitro* and discussed the mode of binding *in vivo*.

Materials and methods

Cell culture and preparation of cell walls

Cultured tobacco BY-2 cells (*Nicotiana tabacum* cv. BY-2) were subcultured as described previously (Matoh et al., 1992). The concentration of B in the medium was 1 mg B L^{-1} as boric acid. One-week old cells were harvested by filtration under suction and washed three times with 20 volumes (v/w) of 0.5 mM $CaCl_2$.

The cells were suspended in four volumes (v/w) of 99.5% ethanol and filtered. The filtered cake was washed successively by suspending in 10 volumes (v/w) of 99.5% ethanol twice 1 h each, in 10 volumes of a mixture of chloroform and methanol (2:1, v/v) once overnight, and finally in 10 volumes of acetone once for 1 h. The residue was allowed to air-dry in a hood and used as a cell wall preparation.

Purification of a boron-polysaccharide complex (BPC) from tobacco cell walls

A batch of 2 g of the cell wall preparation was used for each purification. The B-polysaccharide complex was solubilized and purified with column chromatography using DEAE-Sepharose and Superdex 75, as described for radish roots (Kobayashi et al., 1996). Fractions rich in B were pooled, dialyzed against water and lyophilized.

Preparation of the monomeric RG-II and reconstitution of the BPC

Boron was removed from the BPC by incubating in 0.1 M HCl as reported for the radish root BPC (Kobayashi et al., 1996).

Monomeric RG-II was dissolved in 100 mM sodium acetate buffer (pH 4.0) at concentrations specified in the appropriate figure legend and the same concentration of boric or germanic acid solutions were mixed. The mol wt of RG-II was taken as 5,000 (Kobayashi et al., 1996). A 5-μL aliquot of the mixture was applied to size-exclusion chromatography (YMC-pack Diol-120, 8 × 300 mm) with a refractory index detector (TOSOH RI-8020). The column was equilibrated and eluted with 50 mM sodium acetate buffer (pH 5.2) containing 0.2 M NaCl at a flow rate of 1 mL min^{-1}. The proportions of the RG-II converted to the dimer with B were determined by their respective peak areas.

Germanic acid was prepared by alkaline fusion of germanium dioxide with sodium carbonate followed by passage through a column of Amberlite IR-120 (H^{+}) to remove sodium ions.

Assay methods

Boron was determined by the 2,4-dinitro-1,8-naphthalenediol method (Matoh et al., 1993b). ^{11}B-NMR spectra were recorded with a Bruker ARX-500 spectrometer operated at 160.1 MHz at 27°C and peaks were assigned according to Oi et al. (1992). Glycosyl-residue composition was analyzed as described previously (Kobayashi et al., 1996)

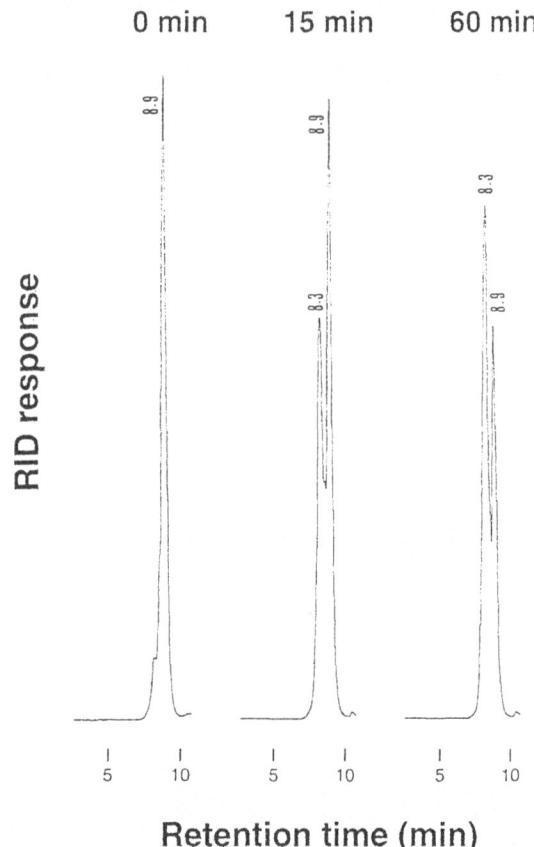

Figure 1. Time course of the reconstitution of the B-dimeric RG-II. The monomeric RG-II and B were incubated in 40 μL of 50 mM sodium acetate buffer (pH 4.0) at a final concentration of 5 mM each. A 5-μL aliquot was subjected to size-exclusion chromatography at the time indicated.

Results

Occurrence of B in a borate-rhamnogalacturonan II complex

More than 80% of the tobacco cell wall B was recovered in a single peak in DEAE-Sepharose chromatography. Glycosyl-residue and glycosyl-linkage composition analyses of the tobacco BPC revealed the presence of such diagnostic monosaccharides for RG-II as 2-O-methylfucose, 2-O- methylxylose, apiose, aceric acid and 2-keto-3-deoxysugars and such linked residues as 3,4-linked fucosyl, 3-linked rhamnosyl and 2,3,4-linked rhamnosyl (data not presented). These results substantiate that the sugar moiety of the tobacco BPC is RG-II.

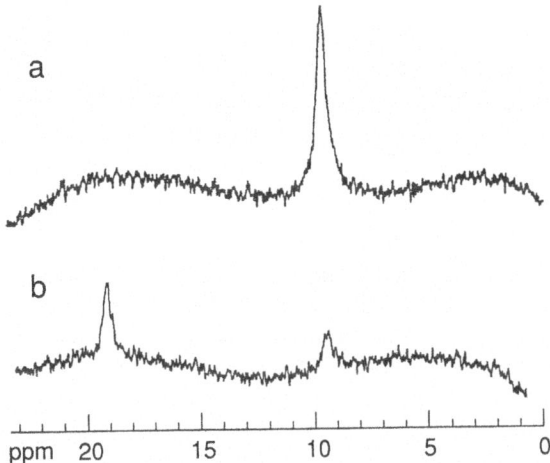

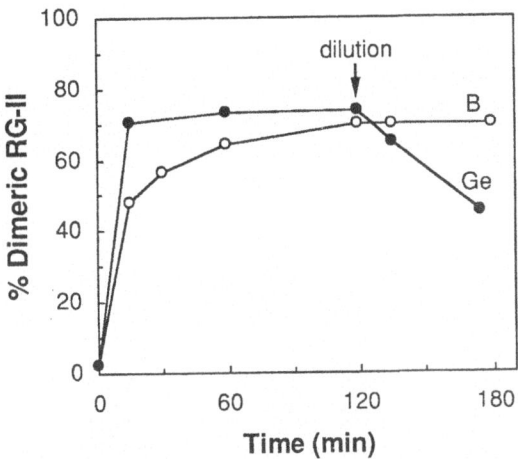

Figure 2. ^{11}B-NMR spectra of (a) the tobacco boron-polysaccharide complex (BPC) (5 mM) and (b) the reconstituted B dimeric RG-II. Tobacco monomeric RG-II and boric acid each at a final concentration of 20 mM were mixed and incubated for 3 h at a room temperature. The chemical shift was expressed relative to external boron trifluoride etherate.

Figure 4. Formation of the dimeric RG-II with boric and germanic acid. The monomeric RG-II and boric or germanic acid were incubated in 50 mM sodium acetate buffer (pH 4.0) at a final concentration of 5 mM each. At 120 min (indicated by arrow), the mixture was diluted 5-fold with 50 mM sodium acetate buffer (pH 4.0).

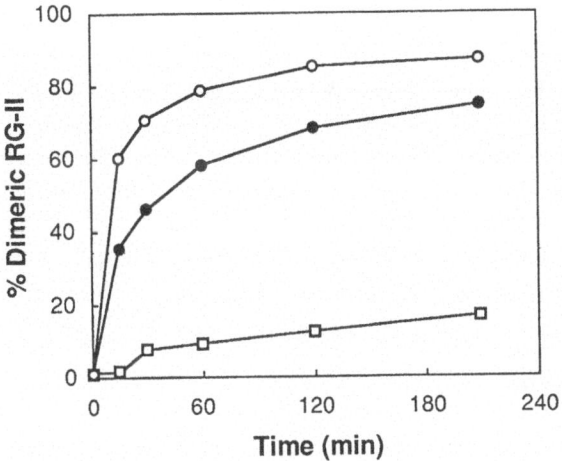

Figure 3. The rate of the formation of B-dimeric RG-II complex. Equimolar concentrations of the monomeric RG-II and boric acid were incubated in 50 mM sodium acetate buffer (pH 4.0) at a final concentration of 2 mM (open square), 10 mM (closed circle) or 20 mM (open circle). At time intervals a 5-μL aliquot was subjected to size exclusion chromatography equipped with refractory index detector. The rate of the dimeric RG-II formation was expressed by the change in the mol percent of RG-II converted into the dimer.

Reconstitution of the borate-rhamnogalacturonan II complex in vitro

Monomeric RG-II eluted from the YMC Diol-120 column at a retention time of 8.9 min (Figure 1). Incubation of the monomeric RG-II at a final concentration of

5 mM with 5 mM boric acid at pH 4.0 yielded polysaccharides eluted at 8.3 min, which is the position of the native BPC (Figure 1). A ^{11}B-NMR spectrum of the mixture gave a signal at 9.6 ppm (Figure 2), which corresponds to the signal of the native BPC (Matoh et al., 1993a). Precise structure of the reconstituted BPC is not known, however, the increase in apparent mol wt, the shift of the ^{11}B -NMR signal to the higher field, and the incorporation of B into the reconstituted BPC at the same ratio as the native BPC (data not presented) led us to conclude that the BPC was reconstituted spontaneously *in vitro*. Figure 3 shows a time-dependent increase in mol% of RG-II converted into the dimer, which may represent the rate of the reconstitution. More immediate reconstitution was observed when higher concentrations of RG-II and boric acid were present. The optimum pH for the reconstitution was around 4 and the rate decreased when the pH was below 3 and above 5 (data not presented).

When germanic acid was mixed with the monomeric RG-II in place of boric acid, a new peak in the size-exclusion chromatography appeared at the same position of the BPC (data not shown), suggesting that the dimeric RG-II was formed with germanic acid. The rate of the dimer formation was faster with germanic acid than boric acid, while it dissociated immediately upon dilution (Figure 4). The prompt dissociation suggests that the monomeric RG-II and the dimeric one are under equilibrium and when the concentration of the monomer is reduced by dilution, the dimer with

240

germanic acid dissociates immediately to the monomer form.

The borate-RG-II complex was reconstituted *in vitro* simply by mixing the monomeric RG-II with boric acid. However, the optimum pH and required concentrations of the monomeric RG-II and B seem to be unusual. The maximum formation of the dimeric B-RG-II complex occurred around pH 4 (in this report) and between pH 3 and 3.4 (O'Neill et al., 1996). O'Neill et al. (1996) suggested that the apiosyl residues of RG-II were the binding sites for B. Apiose has been proposed to be the best candidate for fixing B among sugars, however, even the B-apiose bonding is more stable under neutral to alkaline pH (Loomis and Durst, 1992). This apparent discrepancy may be explained by the anionic character of RG-II. As the backbone of RG-II is a polygalacturonan, the negative charge of the fragments increases as pH rises and may repel the two fragments from each other to hinder the dimer formation. Monovalent cations (Na^+, K^+ and $NH4^+$) stimulated the reconstitution to some extent (data not presented). This supports the assumption, because the cations may mask the negative charge of the fragments. The BPC content was 0.6% (w/w) of tobacco cell walls (data not presented), and assuming that one g of the dried cell walls occupies a volume of 10 mL of fresh cell walls, the concentration of RG-II in fresh cell walls is calculated to be 0.1 mM. However, as presented in Figure 3, such a low concentration of RG-II did not bring about spontaneous dimer formation, although in intact cell walls nearly all the B and RG-II occurred as the complex.

Discussion

Skok (1957) found that germanic acid can substitute for B to some extent in the growth of B-deprived sunflower plants. Loomis and Durst (1992) reported that in cultured carrot cells the minimum requirement by the cells for Ge was 10 times higher than B. The authors speculated that it was because germanic acid coordinated to the binding site less tightly than boric acid did. They confirmed the effect of germanic acid on B deficiency by establishing cell lines which could grow using germanic acid without boric acid. The ability of germanic acid to form a dimeric RG-II complex *in vitro* may explain the substitution effect of Ge when B is deprived from culture media. This, in turn, suggests that the formation of the cross-linkage of pectic chains is the essential function of B.

A B-RG-II complex was reconstituted *in vitro*, however, the conditions may not reflect the mode of the B-RG-II binding *in vivo* exactly. It is possible that some other mechanism(s), such as proteinous factors, other inorganic elements than B, and extent of the methylesterification of galacturonic acid residues around the RG-II region, may affect readiness of the complex formation *in vivo*. O'Neill et al (1996) found that supplements of Sr, Pb and Ba ions at a final concentration of 0.5 mM to the reconstitution medium stimulated the dimer formation and suggested that elements having an ionic radius larger than 1.1 Å were stimulative to the dimer formation. This suggests that steric factors are important for the dimer formation. Research should be developed to establish the conditions for the reconstitution of the borate-dimeric RG-II complex *in vitro*. These information will be of value to elucidate the physiological function of the B-RG-II complex in cell metabolism.

Acknowledgements

We express our sincere thanks to Dr. M. O'Neill, CCRC, The University of Georgia, for providing a pre-print of his work and for discussion. We thank Professors J. Azuma and J. Sekiya, Kyoto University, for their advice and discussion. A part of this work was supported by a Grant-in-Aid (No. 08660073) from the Ministry of Education, Science and Culture, Japan to T. M.

References

Darvill A G, McNeil M and Albersheim P 1978 Structure of plant cell walls. VIII. A new pectic polysaccharide. Plant Physiol. 62, 418–422.

Ishii T and Matsunaga T 1996 Isolation and characterization of a boron- rhamnogalacturonan-II complex from cell walls of sugar beet pulp. Carbohydr. Res. 284, 1–9.

Kobayashi M, Matoh T and Azuma J 1995 Structure and glycosyl composition of the boron-polysaccharide complex of radish roots. Plant Cell Physiol. 36 (supplement), s 139.

Kobayashi M, Matoh T and Azuma J 1996 Two chains of rhamnogalacturonan II are cross-linked by borate-diol ester bonds in higher plant cell walls. Plant Physiol. 110, 1017-1020.

Loomis W D and Durst R W 1992 Chemistry and biology of boron. BioFactors 3, 229–239.

Matoh T, Ishigaki K, Mizutani M, Matsunaga W and Takabe K 1992 Boron nutrition of cultured tobacco BY-2 cells. I. Requirement for and intracellular localization of boron and selection of cells that tolerate low levels of boron. Plant Cell Physiol. 33, 1135–1141.

Matoh T, Ishigaki K, Ohno K and Azuma J 1993a Isolation and characterization of a boron-polysaccharide complex from radish roots. Plant Cell Physiol. 34, 639–642.

Matoh T, Kawaguchi S and Kobayashi M 1996 Ubiquity of a borate-rhamnogalacturonan II complex in the cell walls of higher plants. Plant Cell Physiol. 37, 636–640.

Matoh T, Matsunaga W, Mizutani M and Ohno K 1993b A sensitive and microscale method for boron determination of plant materials. Jpn. J. Soil Sci. Plant Nutr. 64, 71–73.

Oi T, Tanaka T and Kakihara H 1992 Esterification of boric acid with 1,2-propanediol, 3-amino-1,2-propanediol, and (±)-3-dimethylamino-1,2-propanediol as studied by ^{11}B NMR spectroscopy. Bull. Chem. Soc. Jpn. 65, 1903–1909.

O'Neill M A, Warrenfeltz D, Kates K, Pellerin P, Doco T, Darvill A G and Albersheim P 1996 Rhamnogalacturonan-II, a pectic polysaccharide in the walls of growing plants cells, forms a dimer that is covalently cross-linked by a borate ester. In vitro conditions for the formation and hydrolysis of the dimer. J. Biol. Chem. 271, 22923–22930.

Skok J 1957 The substitution of complexing substances for boron in plant growth. Plant Physiol. 32, 308–312.

R.W. Bell and B. Rerkasem (eds.), Boron in Soils and Plants, 243–249.
© 1997 *Kluwer Academic Publishers.*

Immunocytochemistry of the borate-rhamnogalacturonan II complex in cell walls of radish roots

Miki Takasaki, Sayaka Kawaguchi, Masaru Kobayashi, Keiji Takabe[1] & Toru Matoh
Laboratory of Plant Nutrition, Department of Agricultural Chemistry, and [1]Laboratory of Wood Structure, Department of Wood Science and Technology, Faculty of Agriculture, Kyoto University, Kyoto, 606-01, Japan

Key words: antibody, boron, boron-rhamnogalacturonan II complex, cell wall, radish roots

Abstract

A polyclonal antibody toward the borate-rhamnogalacturonan-II (RG-II) complex was raised in rabbits. The antibody recognizes RG-II exclusively in cell walls. Immunocytochemical studies reveal that the epitope is ubiquitous in cell walls of all the cells in radish (*Rhaphanus sativus*) roots. The label was denser proximal to the plasma membrane and was not detected in the middle lamella. These results suggests that the borate bridge between RG-II chains is important for the network formation of newly produced pectic polysaccharides, and that the bridge may not be involved in cell to cell adhesion.

Introduction

Although boron (B) is an essential element for higher plants, its primary function is not known yet (Loomis and Durst, 1992). We have proposed that determination of the location of B in a cell is a prerequisite if we are to identify the functions of B. In accordance with this proposal, our laboratory isolated a particular B polysaccharide complex from radish root cell walls (Matoh et al., 1993) and revealed subsequently that the sugar component of the complex was rhamnogalacturonan II (RG-II) (Kobayashi et al., 1995) and that the complex was comprised of two chains of monomeric RG-II with two mol of borate (Kobayashi et al., 1996). Rhamnogalacturonan II was first isolated from the cell wall of cultured sycamore cells as a fragment of pectic polysaccharide which is resistant to endopolygalacturonase hydrolysis (Darvill et al., 1978). Our results suggest that pectic polysaccharide chains are covalently cross-linked together at the RG-II region through borate-diester bonding. Recently, Ishii and Matsunaga (1996), O'Neill et al. (1996) and Pellerin (1996) confirmed our finding in sugar beet pulps, sycamore and pea cell walls, and red wine, respectively. The B-RG-II complex occurs in at least 24 species of mono- and dicotyledonous higher plants, and in these tissues

contents of B and RG-II are closely related suggesting that RG-II exclusively provides the binding site for B in higher plant cell walls (Matoh et al., 1996). In order to identify the function of B in cell walls, information on the localization of the complex is useful. Therefore, an antibody toward the B-RG-II complex was raised and immunocytochemical analysis was performed.

Materials and methods

Plant material

Seeds of radish (*Rhaphanus sativus* cv. Aokubi, obtained from Takii Seed Co, Kyoto 600-91, Japan) were grown hydroponically.

Preparation and blot assay of antibodies

Antigen, the B-RG-II complex, was purified from radish root cell walls following the method described previously (Kobayashi et al., 1996). Antibody was raised in rabbits following the method of Moore et al. (1986). The antisera were partially purified by precipitation with ammonium sulfate and Protein-A (Pharmacia) column chromatography. The antisera was kept

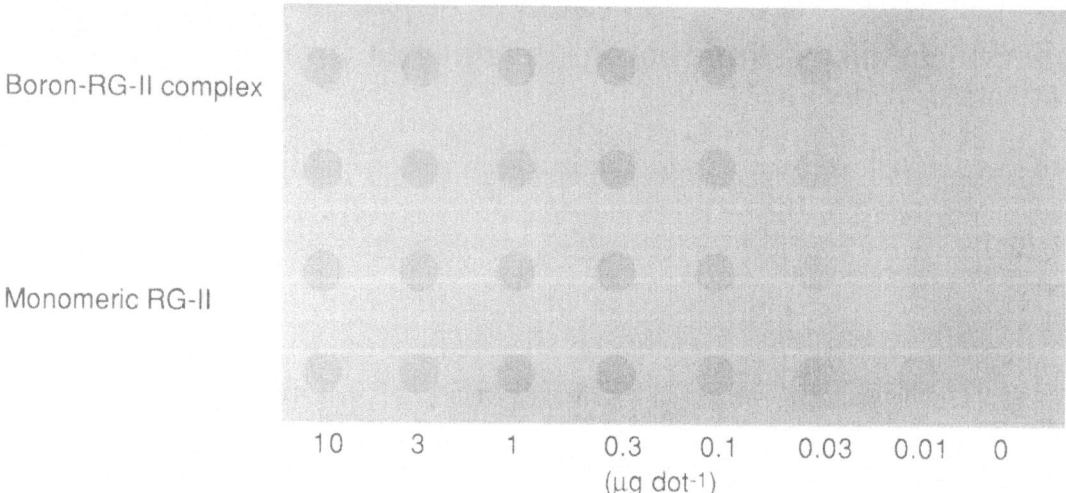

Figure 1. Dot blot showing the reactivity of antisera toward the B-RG-II complex and the monomeric RG-II. The B-RG-II complex (upper two lines) and the monomeric RG-II (lower two lines). The first and second lines of each sample are replicates.

in 10 mM Na phosphate buffer (pH 7.2) containing 0.8% NaCl (henceforth referred to as PBS) with NaN$_3$ (0.1%, w/v).

A titre of the antibody was determined using a dot blot assay. Nylon membrane (Amersham Hybond N$^+$) and a blocking agent Block Ace (Dai Nihon Seiyaku, Osaka 565, Japan) were used throughout. Antibody binding was detected by goat anti-rabbit IgG sera conjugated with horseradish peroxidase (Biomakor) and the diaminobensidine assay (Liners et al., 1989).

Specificity of the antibody

Monomeric RG-II was prepared as described previously (Kobayashi et al., 1996). Pectic polysaccharides were extracted from radish root cell walls following the method of Jarvis (1982) with cyclohexanediamine tetraaceteic acid (CDTA). The CDTA-extracted and alkaline CDTA-extracted pectin were precipitated by adding 4 volumes of ethanol, and collected by centrifugation (10,000 g for 5 min).

Radish root cell walls (0.3 g) were hydrolyzed with 0.5% (w/v) Pectinase SS (Kyowa Chemical Products Co Ltd) in 30 mL of 20 mM Na acetate buffer (pH 4.0) for 48 h, and the digest was applied to a column of DEAE-Sepharose (1.6 × 30 cm, Cl$^-$ form). The column had been equilibrated with 20 mM Tris-HCl buffer (pH 8.0) and eluted with a linear gradient of NaCl (0 to 0.5 M) made in the column buffer (1.5 L). The activity of the antibody toward each peak was assayed using the dot blot assay.

Microscopy

Roots of 8-day old radish seedlings were cut into small pieces and the segments were fixed in 4% paraform aldehyde and 0.1% glutaraldehyde in 100 mM Na phosphate buffer (pH 7.4) overnight at 4°C, then washed with the buffer (four times, 15 min each). The tissues were passed through an ethanol series and embedded in LR White resin (London Resin Co Ltd, London) and polymerized in gelatin capsules overnight at 60° C.

For light microscopy, thin sections (1 μm) were cut with a microtome (ULTRACUT, Reichert-Jung, Vienna, Austria) and mounted on glass slides. Ultrathin sections (60–90 nm) were used for electron microscopy. The section was immunolabeled with the antisera. The sections were labelled with goat anti-rabbit IgG conjugated to 15 nm colloidal gold (Amersham), diluted 25-fold with PBSG, for 30 min. For light microscopy, the signal was enhanced with silver precipitation (IntenSE m kit, Amersham) following the manufacture's instruction. Preimmune sera was used for a control.

Results and discussion

Specificity of the antibody

Figure 1 shows dot blot analyses of diluted antisera with the radish BPC (the dimeric RG II cross-linked with B) and the monomeric RG-II of radish. The sera

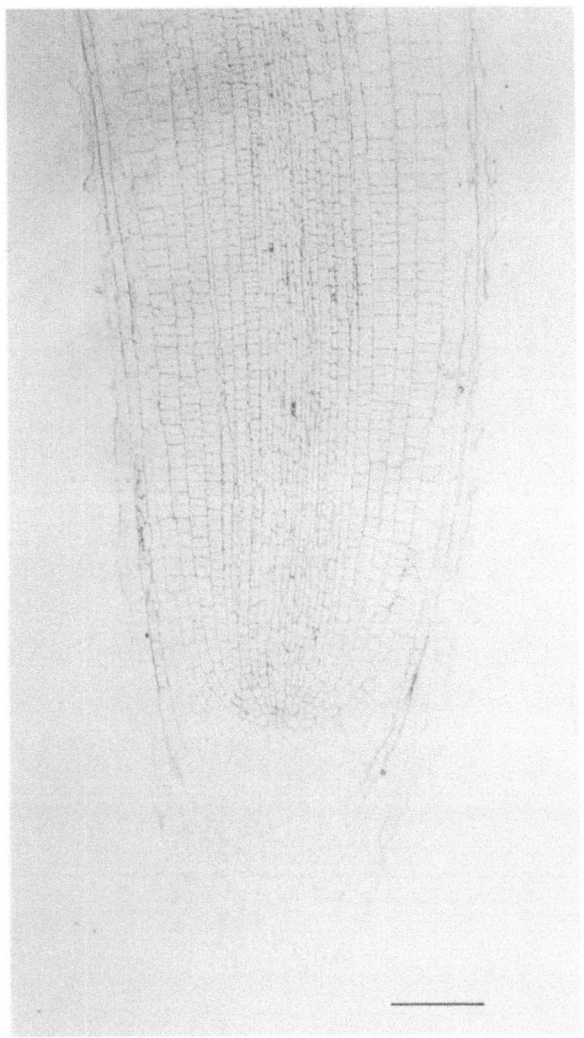

Figure 2. Silver precipitates due to immunogold labeling in radish root apices with an antibody toward the boron-RG-II complex in a longitudinal section. Measurement bar is 100 μm.

recognized both dimeric and monomeric RG-II, indicating that the antisera recognizes the sugar component of the complex. The signal was enhanced according to the increase in the antigen concentration. Preimmune sera and antisera previously immunoprecipitated with the BPC did not react with the antigen (data not presented).

Polysaccharide fragments obtained from chromatogram on a DEAE-Sepharose column of the hydrolysate of radish root cell walls by Pectinase SS, including those not retained by the column, were all negative toward the antisera, except for the BPC (data not presented). The BPC fraction was then acid-hydrolyzed to prepare the monomeric RG-II, and rechromatographed on the same column. Monomer-

ic RG-II eluted earlier than the BPC, and the antisera reacted only with it, and did not react with polysaccharides which eluted at the position of the BPC. Thus, this result indicates that the BPC preparation is not contaminated with other fragments which react with the antibody.

The CDTA-extracted pectin reacted weakly, while alkaline CDTA-extracted pectin reacted strongly toward the antisera. As 2-keto-3-deoxysugars were detected in both pectic fractions (data not presented), they may contain the RG-II region (Stevenson et al. 1988). These signals were not detected when the antisera had been immunoprecipitated with BPC (data not presented). From these examinations, it is likely that the antibody recognizes RG-II exclusively.

Immunocytochemistry

Figure 2 shows localization of the BPC epitope in radish seedling roots in a longitudinal section. The antibody bound to cell walls of all cells in all tissues of the roots, although the columellar root cap and epidermal cells were less intensely labeled than the endodermal cells. In these assays, the shape of cells, i.e., cell walls, are visualized with silver particles, and when preimmune sera was used instead, cell walls were not visible at all (data not presented). If B fulfills its essential function through connecting pectic polysaccharide chains at the RG-II region, it is likely that the distribution is ubiquitous. The omnipresence of the epitope contrasts to the RG-I epitope, which was not detected in the stele tissues of the younger part of *Arabidopsis thaliana* roots (Freshour et al., 1996). This result suggests that the RG-I and RG-II region may not share the same pectic polysaccharide chain, at least when cells are immature.

The RG-II epitope was not detected in the middle lamella nor in the intercellular spaces in radish roots (Figure 3A). The immunogold seems to be denser in the cell walls facing air space than in those facing each other (Figure 3A versus B). This may explain the difference in the delivery frequency of RG-II into cell walls. Many authors detected the epitope of RG-I (Moore et al., 1986; Moore and Staehelin, 1988) and of de-esterified pectin (Knox et al.,1990; Liners and Van Cutsem, 1992) in the middle lamella. The absence of the RG-II epitope in the middle lamella suggests that B may not work to cement cells. Immunogold particles primarily labeled the inner half of the cell walls (Figure 3A, B). When cell wall was thin, the gold particle was found very close to the plasma membrane (Fig-

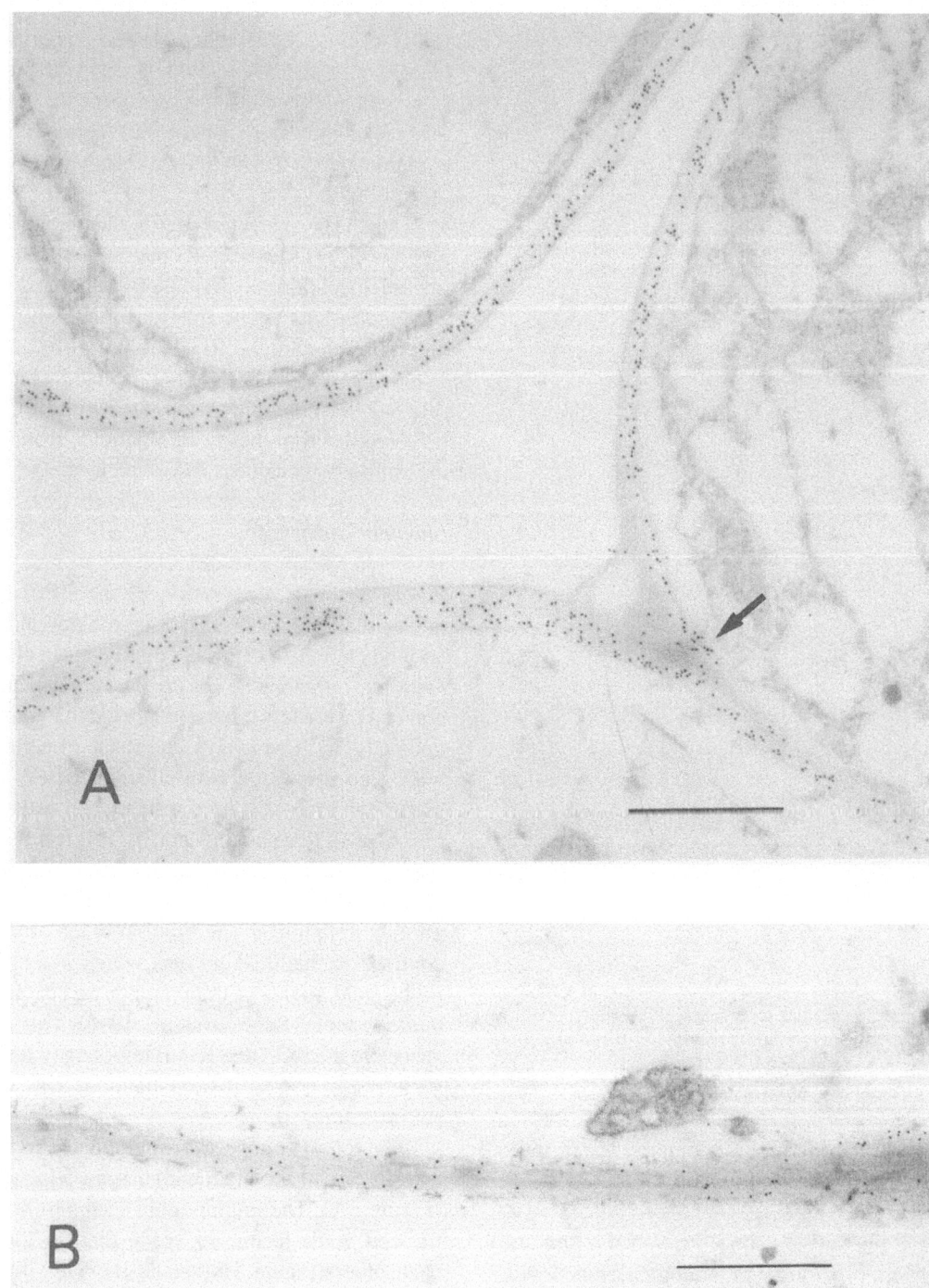

Figure 3a,b. Immunogold labeling with an antibody toward the B-RG-II complex in radish root apices in the cortex region (A, B), and in the stele region (C). D, Immunogold labeling with preimmune antisera in the cortex region. Measurement bar is 1 μm.

ure 3C). Williams et al. (1996) indicate that the RG-II epitope is located at plasma membrane. In mature cells, the RG-I antibody labeled the quarter of the cell wall closest to the plasma membrane (Freshour et al., 1996). Knox et al. (1990) noted that their monoclonal antibody which recognizes the deesterified epitope of pectin was located on the inner surface of the primary cell walls adjacent to the plasma membrane in the root apex of carrot. Similar distribution of RG-II, RG-I, and deesterified epitope of pectic polysaccharides in

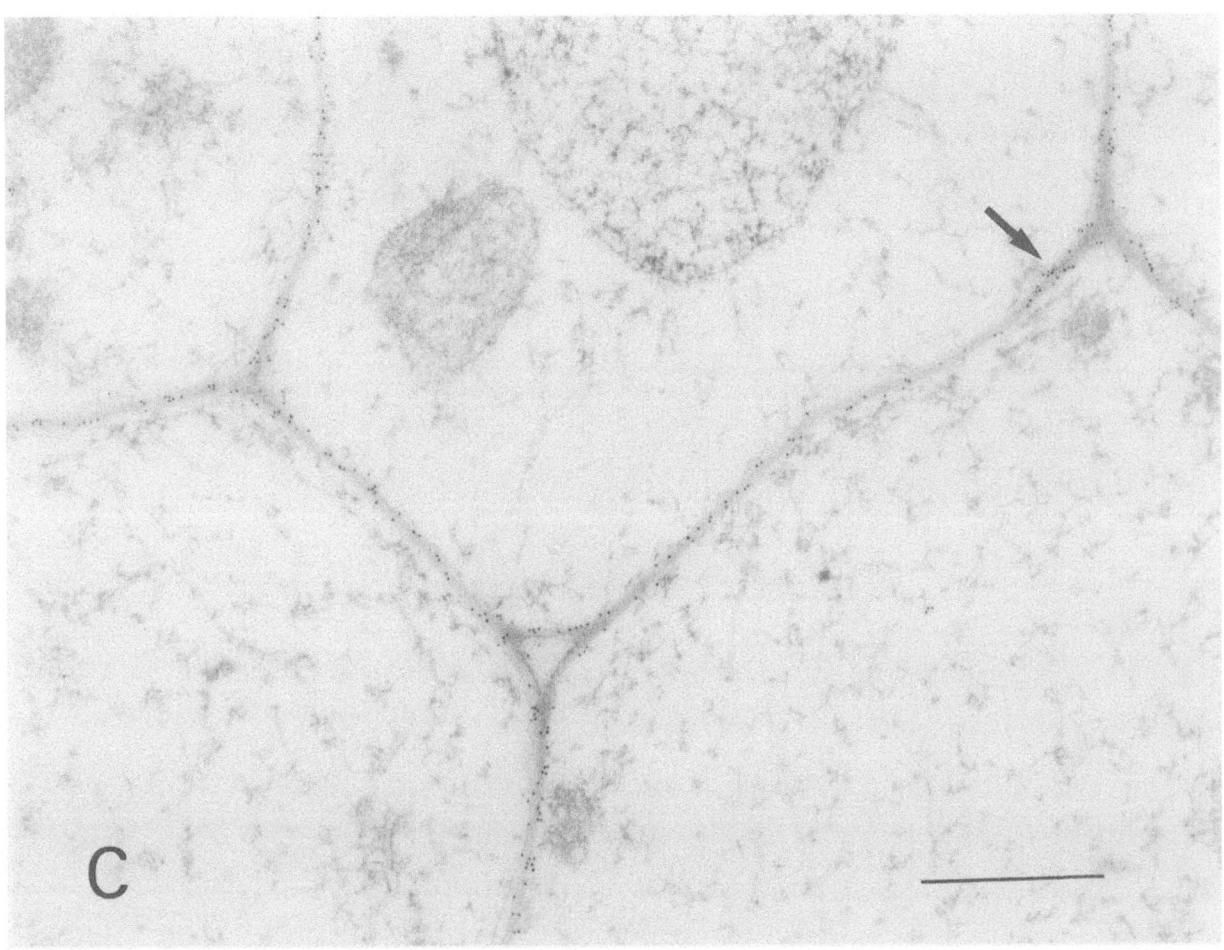

Figure 3c.

the primary cell walls suggests that cross-linking with Ca ions and boric acid is important for the network formation of newly produced pectic polysaccharides. The gradient in distribution of the RG-II across the cell wall suggests that during the process of maturation of a cell wall, the epitope of RG-II disappears. Kobayashi et al. (1997) detected monomeric RG-II in the extracellular polysaccharides of cultured tobacco BY-2 cells at the middle log phase when the cells grew vigorously. Together, these results suggest that some pectic polysaccharides may be decomposed during cell wall maturation and the RG-II fragments may be released into cell walls and into surrounding media, perhaps after removal of B from the complex.

Boron deficiency brings about structural abnormalities, especially in cell walls. Spurr (1957) demonstrated that B apparently affects the rate and process of carbohydrate condensation into wall material. The anatomical changes in cell walls convinced him that B

is a morphogenetic agent affecting the development of specific form of the cell walls of higher plants. In our research, slightly denser label (arrows in Figure 3A, C) was found in the region of cell corners, which reminds us that some of the collenchyma cells in B deficient celery plants fail to develop typical 'corner' thickening (Spurr, 1957). Thus, involvement of B in cell wall intactness is likely. Boric acid works to form a network of pectic polysaccharides in cell walls, however, direct evidence for the function of B is still lacking, since the significance of the integrity of cell walls to cell viability is little understood. Effects of inorganic elements, such as B and Ca, on the intactness of higher plant cell walls should be examined further.

248

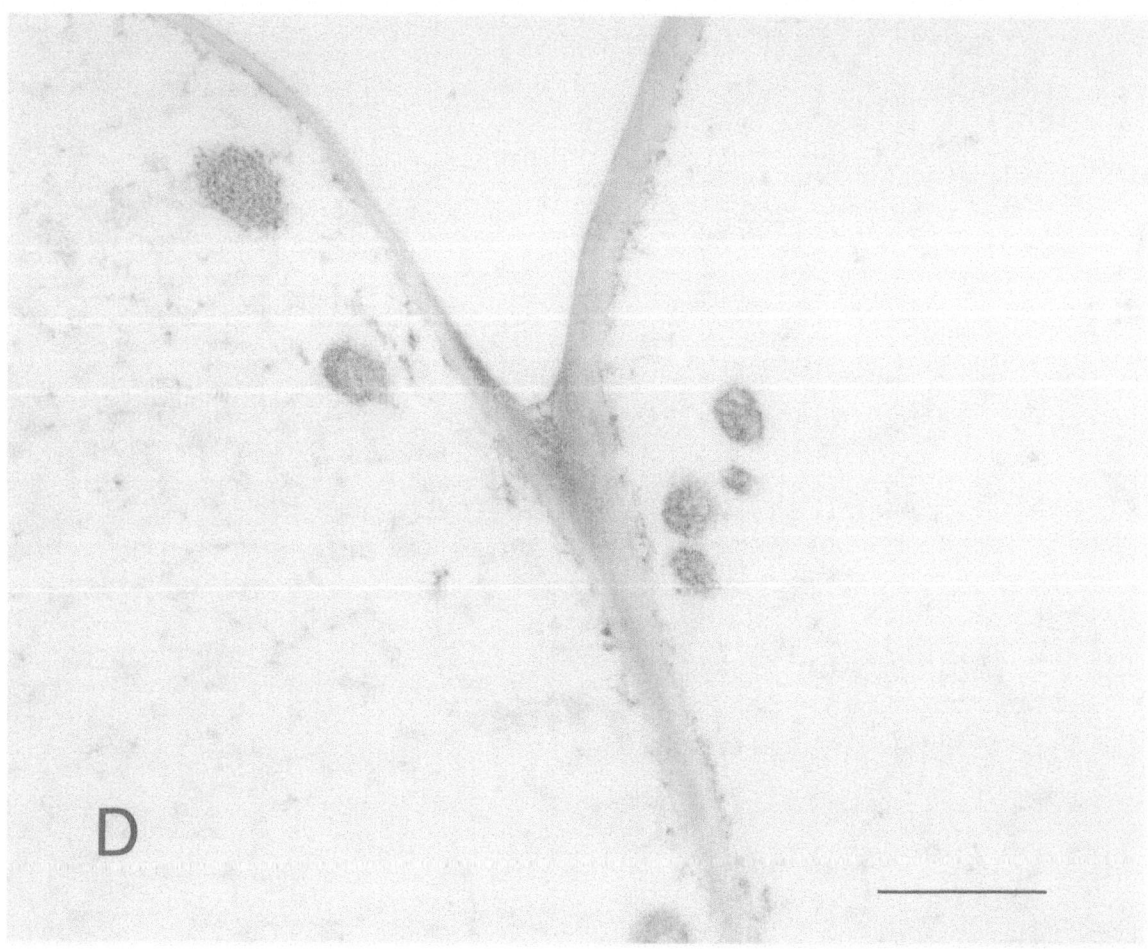

Figure 3d.

References

Darvill A G, McNeil M and Albersheim P 1978 Structure of plant cell walls VIII. A new pectic polysaccharide. Plant Physiol. 62, 418–422.

Freshour G, Clay R P, Fuller M S, Albersheim P, Darvill A G and Hahn M G 1996 Developmental and tissue-specific structural alterations of the cell-wall polysaccharides of *Arabidopsis thaliana* roots. Plant Physiol. 110, 1413–1429.

Ishii T and Matsunaga T 1996 Isolation and characterization of a boron rhamnogalacturonan-II complex from cell walls of sugar beet pulp. Carbohydr. Res. 284, 1–9.

Jarvis M C 1982 The proportion of calcium-bound pectin in plant cell walls. Planta 154, 344–346.

Knox J P, Linstead P J, King J, Cooper C and Roberts K 1990 Pectin esterification is spatially regulated both with in cell walls and between developing tissues of root apices. Planta 181, 512–521.

Kobayashi M, Matoh T and Azuma J 1995 Structure and glycosyl composition of the boron-polysaccharide complex of radish roots. Plant Cell Physiol. 36S, 139.

Kobayashi M, Matoh T and Azuma J 1996 Two chains of rhamnogalacturonan II are cross-linked by borate-diol ester bonds in higher plant cell walls. Plant Physiol. 110, 1017–1020.

Kobayashi M, Ohno K and Matoh T 1997 Boron nutrition of cultured tobacco BY-2 cells. II. Characterization of the boron-polysaccharide complex. Plant Cell Physiol. 38, 676–683.

Liners F and Van Cutsem P 1992 Distribution of pectic polysaccharides throughout walls of suspension-cultured carrot cells. An immunocytochemical study. Protoplasma 170, 10–21.

Liners F, Letesson J-J, Didembourg C and Van Cutsem P 1989 Monoclonal antibodies against pectin, Recognition of a conformation induced by calcium. Plant Physiol. 91, 1419–1424.

Loomis W D and Durst R W 1992 Chemistry and biology of boron. BioFactors 3, 229–239.

Matoh T, Kawaguchi S and Kobayashi M 1996 Ubiquity of a borate rhamnogalacturonan II complex in the cell walls of higher plants. Plant Cell Physiol. 37, 636–640.

Matoh T, Ishigaki K, Ohno K and Azuma J 1993 Isolation and characterization of a boron-polysaccharide complex from radish roots. Plant Cell Physiol. 34, 639–642.

Moore P J and Staehelin L A 1988 Immunogold localization of the cell-wall matrix polysaccharides rhamnogalacturonan I and xyloglucan during cell expansion and cytokinesis in *Trifolium pratense* L.; implication for secretory pathways. Planta 174, 433–445.

Moore P J, Darvill A G, Albersheim P and Staehelin L A 1986 Immunogold localization of xyloglucan and rhamnogalacturonan

I in the cell walls of suspension-cultured sycamore cells. Plant Physiol. 82, 787–794.

O'Neill M A, Warrenfeltz D, Kates K, Pellerin P, Doco T, Darvill A G and Albersheim P 1996 Rhamnogalacturonan-II, a pectic polysaccharide in the walls of growing plant cells, forms a dimer that is covalently cross-linked by a borate ester. J. Biol. Chem. 271, 22923–22930.

Pellerin P, Doco T, Vidal S, Williams P, Brillouet J-M and O'Neill M A 1996 Structural characterization of red wine rhamnogalacturonan II. Carbohydr. Res. 290, 183–197.

Spurr A R 1957 The effect of boron on cell-wall structure in celery. Am. J. Bot. 44, 637–650.

Stevenson T T, Darvill A G and Albersheim P 1988 Structural features of the plant cell wall polysaccharide rhamnogalacturonan-II. Carbohydr. Res. 182, 207–226.

Williams M N V, Freshour G, Darvill A G, Albersheim P, and Hahn M G 1996 An antibody Fab selected from a recombinant phage display library detects deesterified pectic polysaccharide rhamnogalacturonan II in plant cells. Plant Cell 8, 673–685.

R.W. Bell and B. Rerkasem (eds.), Boron in Soils and Plants, 251–??.
© 1997 *Kluwer Academic Publishers.*

Distribution of water soluble boron in soils of different agro-climatic regions of Karnataka state, India

M.S. Badrinath, Mir Khamer Ali, H.M. Chidanandappa, K. Sudhir & R. Siddaramappa
Department of Soil Science and Agricultural Chemistry, University of Agricultural Sciences, GKVK, Bangalore 560 065, India.

Key words: boron, hot water soluble boron, soil types

Abstract

In Karnataka state in India, agriculture, horticulture and forestry are practiced on the following soil orders: Alfisols, Vertisols, Aridisols, Inceptisols, Ultisols, Entisols and Oxisols. The soil temperature regime is isohyperthermic in all the six regions where diversified farming is practiced. Through an intensive soil survey, the entire state has been surveyed, soil profiles examined, and the hot water soluble boron (B) evaluated for all the regions. The distribution of hot water soluble B was higher in all the zones at surface layers except in northern dry zone, central dry zone, southern transition zone and northern transition zone where it was a lower. The higher levels of B in the sub-surface layers was observed in the sub surface horizons in the dry lands of the arid climatic zone. Considering all soils surveyed, there was no relationship between hot water soluble B and rainfall, organic carbon, pH, clay or sand.

Introduction

Boron (B) is required in relatively small quantities for crops but is nevertheless often deficient for crop production. However, excess B is toxic to plant growth. Boron content is therefore one of the factors considered in determining water quality for irrigation. Agriculture, horticulture and forestry in Karnataka is practiced on a range of soils including Alfisols, Ultisols, Entisols, Inceptisols, Vertisols, Aridisols and Oxisols. Not much work has been done on B in soils of the state and hence an attempt was made here to evaluate systematically the distribution of hot water soluble B in different depths of 10 pedons representing the main soil orders under the six agro-climatic regions of Karnataka.

Material and methods

Ten soil profiles covering all the 10 agro-climatic zones of Karnataka were selected for the study. The location of the sampling sites and geological descriptions (Table 1) are shown. The soils collected according to horizons were processed as per the standard procedure.

The water hot soluble B was estimated by a colorimetric method. Twenty grams of air dried soil was placed in a conical flask and 40 mL of distilled water was added and refluxed by boiling for 5 minutes. The soil-water suspension was centrifuged to get a clear solution. The B concentration of the solution was determined by the curcumin method as described by Jackson (1973).

Results and discussion

Region 1

This region includes the north eastern transition and north eastern dry zones comprising Bidar, Gulburga and parts of Raichur districts. Hot water soluble B is higher (1.92 mg kg^{-1}) in the surface A1 horizon and decreased with depth in Bt3 horizon (0.53 mg kg^{-1}). Organic matter showed a similar trend. The soil texture ranged from sandy clay loam to sandy clay as the soil depth increased. Clay content in different depths might alter the B in soils of these two zones.

Table 1. Geographical description of soil sampling sites in different agro-climatic zones of Karnataka, India.

Site	Zone[1]	Location	Altitude (m.a.s.l.)	Longitude (E)	Latitude (N)	Rainfall (mm)	Geology	Soil taxonomy
1	NET	Bidar	800–900	76° 30′	11° 00′	860	Laterite	Rhodic Paleustalf
2	NED	Bhimarayanagudi	300–450	76° 30′	15° 45′	767	Basalt	Typic Haplustert
3	ND	Gangavathi	450–800	74° 30′	14° 30′	585	Granite	Typic Haplustert
		Bijapur	450–800	74° 30′	14° 30′	585	Basalt	Typic Haplustert
		Siriguppa	450–800	74° 30′	14° 30′	585	Magmatic	Typic Haplustert
4	CD	Hiriyur	800–900	75° 30′	12° 50′	611	Granite	Chromic Haplustert
5	ED	Bangalore	800–900	76° 35′	12° 10′	776	Laterite	Kandic Paleustalf
6	SD	Mandya	800–900	76° 05′	11° 30′	734	Granite gneiss	Typic Rhodustalf
7	ST	Shimoga	800–900	75° 25′	11° 45′	869	Granite	Rhodic Paleustalf
8	NT	Dharwad	800–900	74° 10′	14° 30′	780	Schistose	Typic Haplustert
9	H	Mudigere	800–900	74° 15′	12° 00′	2209	Laterite	Paleustalf
		Chethalli	800–900	74° 15′	12° 00′	2209	Granite gneiss	Typic Rhodustalf
		Sakleshpur	800–900	74° 15′	12° 00′	2209	Granite gneiss	Typic Rhodustalf
10	C	Brahmavara	<300	74° 05′	12° 30′	3893	Laterite	Typic Kandiustalf

[1] NET - North Eastern Transition, NED - North Eastern Dry, ND - Northern Dry, CD - Central Dry, ED - Eastern Dry, SD - Southern Dry, ST - Southern Transition, NT - Northern Transition, H - Hilly, C - Coastal.

Region 2

This region is the northern dry zone of the state comprising Bijapur, Bellary and parts of Raichur, Dharwad and Belgaum districts. Hot water soluble B was lowest at 0.46 mg kg^{-1} in the surface Ap horizon and in the A$_4$ layer. The higher levels of B observed in A$_2$ and A$_3$ horizons at 1.71 and 0.95 mg kg^{-1} may be due to the lack of leaching in the sub soils of these dry lands which because of high clay content and alkaline pH are expected to retard the vertical movement of B.

Region 3

This region includes the district of Chitradurga, Tumkur, Mandya, Bangalore, Kolar and parts of Hassan, Chikamaglur and Mysore in the zones of central, Eastern and Southern parts of the Karnataka state. The distribution of hot water soluble B in the different horizons of soils of the central zone was observed to be 1.68 mg kg^{-1} in the surface (Ap) layer increasing to 2.05 mg kg^{-1} in the A$_2$ horizon of the pedon. The soil is deep and texturally clay loam. Increasing acidity is noticed as the soil depth increases while it also registers higher levels of B in this zone. In the eastern dry zone, the soil depth is deep and texturally sandy clay loam to clay. Soil pH ranges from 5.41 to 6.21 as the depth increases. The distribution of B is higher in the Ap horizon and lower in the Bt$_3$ horizon of the pedon. The distribution pattern seen in the southern dry zone

pedon was higher in the surface soil and lower in the last horizon at 140 cm depth.

Region 4

This region includes southern and northern transition zones of Karnataka with parts of the districts of Shimoga, Chikamaglur, Hassan, Mysore in the southern transition zone and parts of Dharwad and Belgaum in northern transition zone. The B distribution is lower in surface soil of these zones compared to the lower depths of these pedons. Soil pH increased as the depth increases in southern transition zone whereas it is inconsistent in northern transition zone. Texturally soils are sandy clay loam and sandy loam in northern and southern transition zone, respectively.

Region 5

This region comprises Coorg and parts of Hassan, Chikamaglur, Shimoga, Uttara Kannada and Dharwad districts. This is the hilly zone of Karnataka and the crops grown are padi rice (*Oryza sativa*), plantation and horticulture crops. Soils are lighter in texture at the surface and contain more clay as the depth increases. Organic matter is high in surface soils (Ap) with 28 g organic carbon kg^{-1} soil and it decrease with depth. A similar trend was noticed with respect to increase in soil pH status. Boron was high in the Ap horizon compared to the lower horizon Bt$_2$. Texture, pH and

Table 2. Water soluble boron, pH, organic carbon and soil texture in 10 soil profiles from different agro-climatic zones of Karnataka.

Zone	Soil depth (cm)	Horizon	Soil pH 1:2.5 soil: water	Organic carbon ($g\,kg^{-1}$)	Sand ($g\,kg^{-1}$)	Clay ($g\,kg^{-1}$)	Hot water soluble Boron ($mg\,kg^{-1}$)
North	0–15	A_1	8.07	65.0	660	270	1.92
Eastern	15–30	Bt_1	8.00	42.0	710	260	1.29
Transition	30–57	Bt_2	8.08	40.0	660	250	0.88
	57–140	Bt_3	7.95	20.0	480	430	0.53
North	0–13	A_1	8.50	66.0	620	240	1.48
Eastern	13–38	A_2	8.70	43.0	610	240	3.03
Dry	38–65	A_3	8.76	40.0	540	280	5.45
	65–140	Ac	8.80	32.0	140	620	0.90
Northern	0–26	Ap	9.17	66.0	360	330	0.46
Dry	26–67	A_2	9.37	49.0	330	370	1.71
	67–119	A_3	8.48	32.0	180	540	0.95
	119–140	A_4	8.65	17.0	240	470	0.61
Central	0–10	Ap	9.54	78.0	350	360	1.68
Dry	10–140	A_2	8.93	17.0	320	450	2.05
Eastern	0–11	Ap	5.41	64.0	730	220	1.50
Dry	11–35	Bt_1	5.31	46.0	670	300	0.91
	35–67	Bt_2	5.62	35.0	560	350	1.13
	67–140	Bt_3	6.21	32.0	390	520	0.87
Southern	0–9	Ap	6.72	72.0	720	240	6.60
Dry	9–13	A_2	6.57	84.0	710	240	1.00
	13–68	Bw_1	6.75	32.0	650	280	1.10
	68–97	Bw_2	7.01	26.0	730	250	1.10
	97–140	Bw_3	7.40	20.0	640	290	1.02
Southern	0–17	Ap	5.88	20.0	760	180	1.06
Transition	17–24	A_2	6.32	17.0	890	150	0.73
	24–40	A_3	5.05	17.0	730	180	0.83
	24–90	Ac_1	5.43	14.0	780	170	0.78
	90–140	Ac_2	6.56	9.0	730	200	1.15
Northern	0–13	Ap	6.72	64.0	160	570	0.35
Transition	13–52	A_2	7.28	61.0	120	600	0.31
	52–140	A_3	7.85	29.0	190	660	0.56
Hilly	0–26	Ap	5.83	28.0	700	100	1.05
	26–34	A_2	6.15	93.0	630	120	0.90
	34–96	Bt_1	5.20	70.0	730	110	0.64
	96–140	Bt_2	4.91	46.0	460	460	0.85
Coastal	0–13	Ap	5.27	177.0	490	390	0.90
	13–140	Bt	6.24	58.0	530	350	0.74

organic matter contribute to the differences observed in B status under soils of hilly zone.

Region 6

This forms the coastal zone of Karnataka where monoculture padi rice is being practiced in all the seasons. This region includes parts of Uttara Kannada and the entire district of Dakshina Kannada. The pedon has two horizons, Ap and Bt. Soil pH increases whereas organic matter decreases as the soil depth increases. Sandy clay is the soil texture of costal zone soils. The hot water soluble B is 0.90 mg kg^{-1} in Ap horizon and decreases as the depth increases in the pedon. Padi rice, pulses, oilseeds, plantation and horticulture crops are grown in the region.

254

Table 3. Correlation between hot water soluble boron, soil properties and rainfall.

	pH	Org C.	Sand	Clay	Boron	Rainfall
pH	1.0000					
Org.C.	−0.2060	1.000				
Sand	−0.5026	0.0651	1.0000			
Clay	0.2687	−1.896	−0.7947	1.0000		
Boron	0.2233	0.0203	0.1844	−0.1830	1.0000	
Rainfall	−0.4746	0.5384	0.1122	−0.1090	−0.1655	1.0000

Conclusions

There was no relationship between hot water soluble B in the soils of Karnataka and key properties (pH, organic carbon, sand and clay content) or rainfall, (Table 3). However, soil hot water soluble B levels provide a basis for understanding the distribution of B in soil parent material (Table 2). This will help us to predict the areas of B sufficiency and deficiency. Based on the generated information on the distribution of boron in all the six regions (includes ten zones) strategies can be worked out to overcome risk of toxicity or deficiency of B in soils of Karnataka.

Acknowledgements

The senior author is grateful to the Australian Agency for International Development for financial support under its International Seminar Support Scheme.

References

Jackson ML 1973 Soil Chemical Analysis. Prentice Hall of India, New Delhi. 498 p.

R.W. Bell and B. Rerkasem (eds.), Boron in Soils and Plants, 255–259.
© 1997 *Kluwer Academic Publishers.*

Characterization of boron adsorption in soils of Tamil Nadu (S. India)

R. Krishnasamy[1], J. Jaisankar[1] & M. Suresh[2]
[1] *Department of Soil Science and Agricultural Chemistry*
[2] *Department of Statistics, Tamil Nadu Agricultural University, Coimbatore-641 003, India.*

Key words: Adsorption, calcium carbonate, isotherms, kinetic equations, organic matter, reaction time

Abstract

Both boron (B) toxicity and deficiency have been recorded on soils in the state of Tamil Nadu. In this study, we investigated the B sorption behaviour of 11 agricultural soils of Tamil Nadu as influenced by solution concentration, reaction time and removal of organic matter and $CaCO_3$. Adsorption of B was almost complete after 24 h in all the soils. Boron adsorption in these soils was best described by parabolic diffusion and Elovichian kinetics. Though adsorption of B conformed to both the Langmuir and the Freundlich adsorption isotherms, all the data fitted well with the Freundlich adsorption isotherm. The B adsorbability (K values) as measured by the Freundlich adsorption isotherm varied from 0.18 to 1.96 mg B kg^{-1}. The K values of Freundlich adsorption isotherm of soils after removal of organic matter as well as $CaCO_3$ were lower than those of untreated soils. The strongly significant relationships between clay, organic matter and CEC and the Freundlich parameter K indicates that these soil properties control B adsorption in the soils.

Introduction

Boron (B) deficiency is of great concern in areas receiving heavy rainfall, while its toxicity may be a problem in arid regions. As plants respond primarily to the B activity in soil solution, understanding the mechanism of B adsorption on soil materials is vital. Boron applied to soils is adsorbed to a variable extent and an equilibrium exists between B in the solid and liquid phases (Hingston, 1964). The adsorption of B is one of the most important factors determining the release and fixation of applied B and thus deciding the efficiency of B fertilization. Aside from the concentration of B in soil solution, B adsorption has been reported to be influenced by soil pH (Gupta, 1968), sesquioxide (Hatcher et al., 1967), clay content (Hatcher and Bower, 1958; Biggar and Fireman, 1960) and organic matter (Elrashidi and O'Connor, 1982; Yermiyahu et al., 1995).

Although organic matter is an active component in soils, only a few studies of B- organic matter interactions have been conducted (Yermiyahu et al., 1995). Boron adsorption processes on soil organic matter are not yet fully understood. Despite the fact that many of the cultivated soils contain 10 to 80 g organic matter kg^{-1}, this low content may have a significant effect on B distribution between the solid and liquid phases in soils. Goldberg and Forster (1991) reported that calcite plays important role in B adsorption by calcareous soils. However, the importance of $CaCO_3$, in B adsorption is still uncertain and must be established.

Hence the present investigation was carried out to study the adsorption of B in soils of Tamil Nadu, with the following objectives: to characterise the adsorption behaviour of B in soils before and after removal of organic matter and $CaCO_3$ and; to investigate the relationship between the soil properties and the various parameters associated with B adsorption in soil.

Materials and methods

Soils

Eleven surface soil samples (0–25 cm), representing four soil orders viz., Vertisol, Inceptisol, Alfisol and Entisol, comprising the major soil series of Tamil Nadu (S. India) were collected, air dried and passed through a 2.0 mm sieve. The homogenised soil samples were then analysed for selected physical and chemical properties

Table 1. Physical and chemical characteristics of 11 soils from Tamil Nadu, India. See materials and methods for a description of soil analysis methods. Treated soils had organic carbon and $CaCO_3$ substantially removed.

Soil No.	Soil location	Classification	pH (1:2.5 soil:water)	Untreated		Treated						
				Free CaCO$_3$ (g kg^{-1})	Org.C (g kg^{-1})	Free CaCO$_3$ (g kg^{-1})	Org.C (g kg^{-1})	Al$_2$O$_3$ (g kg^{-1})	Fe$_2$O$_3$ (g kg^{-1})	CEC (cmol (p$^+$) kg^{-1})	Sand (g kg^{-1})	Clay (g kg^{-1})
1.	Kangeyampalayam	Ustifluvents	7.80	16.8	5.5	3.2	0.9	40.8	49.2	23.8	671	197
2.	Kallipalayam	Paleustalfs	8.20	69.9	3.9	17.6	0.8	48.9	86.1	28.4	730	220
3.	Krishnagiri	Haplustalfs	8.00	8.4	1.8	0.4	0.5	37.9	32.1	12.8	771	132
4.	Kurumbalur	Haplustalfs	7.80	50.4	5.3	10.0	1.1	57.2	42.8	40.1	510	360
5.	Palladam	Ustochrepts	7.70	24.0	1.8	2.8	0.4	68.8	51.2	18.4	750	160
6.	Pattukottai	Haplustalfs	6.50	4.8	5.3	1.2	0.3	138.0	113.0	35.5	709	219
7.	Pilamedu	Chromusterts	8.05	66.8	6.0	6.4	1.1	81.5	43.5	44.0	450	440
8.	Pongalivalasu	Ustochrepts	7.60	8.0	5.1	2.0	0.3	25.8	59.2	14.4	680	200
9.	Pootuthakku	Ustochrepts	7.50	12.0	5.5	1.6	0.5	78.3	41.7	34.5	720	220
10.	Somayanur	Haplustalfs	7.80	19.2	6.4	4.4	1.3	24.0	96.0	21.1	730	210
11.	Thiruvanamalai	Ustochrepts	7.80	39.6	8.3	12.4	1.8	86.8	38.2	39.8	600	300

(Table 1) by the methods described by Piper (1966) and Jackson (1973).

Rate of reaction

Batch studies were used to investigate the kinetics of the adsorption process. Calcium chloride (0.01 M) solution containing 40 mg B L^{-1} (as boric acid) was added to each 10 g of soil in 50 mL centrifuge tubes in triplicate. The soil suspensions were then equilibrated at $27\pm1°C$ for 1, 2, 4, 8, 16, 32 and 64 h in a reciprocating shaker. At the end of the reaction time, the suspension was centrifuged at 15,000 rpm for 20 min, filtered and B was determined in the filtrate by the carmine method (Hatcher and Wilcox, 1950). The amount of B adsorbed was calculated from the difference between the concentrations of the initial solution and the equilibration solution.

Modelling on reaction kinetics

The kinetic models used to describe B adsorption from the eight soils during 64 h were tested for goodness of fit by least-square regression analysis. They are:

i) Zero order: $qt = q_o + K_o t$ where K_o is the zero order rate constant [mg B kg^{-1} s^{-1}]

ii) First Order: $\ln qt = \ln q_o + K_1 t$ where K_1 is the first order rate constant (s^{-1})

iii) Second order: $1/qt = 1/q_o - K_2 t$ where K_2 is the second order rate constant [(mg B kg^{-1}) s^{-1}]

iv) Third order: $1/qt^2 = 1/q_o^2 - Kt$ where K_3 is the third order rate constant [(mg B kg^{-1}) s^{-2}]

v) Parabolic diffusion: $q = \alpha + Kd \sqrt{t}$ where Kd is the diffusion rate constant [(mg B kg^{-1})$^{0.5}$]

vi) Elovich: $qt = q_o + (1/\beta) \ln (\alpha\beta) + (1/\beta) \ln t$, where α is the initial B adsorption rate [(mg B kg^{-1}h)] and; β B desorption constant [(mg B kg^{-1}) $^{-1}$]

In all the equations, q_0 and q_1 are the amounts of B adsorbed (mg B kg^{-1}) at time zero and t, respectively. To determine the equation that best described the adsorption of B in soils, a standard error of estimate was calculated for each equation. A relatively high value of the coefficient of determination (R^2) and low standard error were used as criteria for the best fit (Chien and Clayton, 1980). The standard error was calculated as follows:

$$SE = [\sum (q - q')^2 / (N - 2)]^{0.5}$$

where q and q' are the measured and calculated amounts of B in soil respectively, at time t, and N is the number of measurements (Steel and Torrie, 1960).

Role of organic matter and $CaCO_3$ in B adsorption

This was studied by investigating B adsorption in treated and untreated soil samples. Soil organic matter was removed following the procedure of Marzadori et al. (1991) and the content of organic carbon in soils before and after oxidation was determined by wet digestion.

CaCO$_3$ was removed following the procedure of Goldberg and Forster (1991).

Adsorption studies

Adsorption experiments were conducted with original and treated soils (organic matter oxidized and CaCO$_3$ removed soils) using the method described by Elrashidi and O'Connor (1982). A 10 mL solution containing varying levels of B (0 to 120 mg kg^{-1}) as boric acid in 0.01 M NaCl for CaCO$_3$ free soil and 0.01 M CaCl$_2$ for the rest of the two adsorbents was added to triplicate 50 mL centrifuge tubes containing 10 g of each soil. The contents were equilibrated using a reciprocatory shaker at 27±1°C for 24 h. At the end of the equilibrium period, the tubes were centrifuged, and the concentration of B in the equilibrium solution was determined. Boron adsorbed by soil was calculated as the difference between the initial B and equilibrium B concentration.

Boron adsorption was modelled with both the Langmuir and Freundlich adsorption isotherms. Simple correlation coefficients were made to determine whether any relationship existed between various soil properties and B adsorption parameters.

Results and discussion

Soil properties

The study included soils representing major soil series of Tamil Nadu, varying widely in texture (loamy sand to clay), free CaCO$_3$ (8.0 to 69.9 g kg^{-1}), organic C (1.8 to 8.3 g kg^{-1}), Fe$_2$O$_3$ (32.1 to 113 g kg^{-1}), Al$_2$O$_3$ (24 to 138 g kg^{-1}) and cation exchange capacity (CEC) (12.8 to 44 cmol (p$^+$) kg^{-1}). Thus the soils possessed widely varying physico-chemical properties apart from pH which provided sufficient scope for the study of B adsorption.

Kinetics of B adsorption

For all soils, the B adsorption was characterised by an initial fast reaction followed by a slow process. Adsorption of B was almost complete after 24 h in all these soils though complete equilibrium was not attained until 32 h had elapsed (Table 2). The concentration of boron in the equilibrium solution after 64 h ranged from 15.5 to 34.0 mg L^{-1}. Bingham et al. (1971) demonstrated that equilibrium was only attained under conditions of prolonged reaction periods. Mezuman

Table 2. Decline in solution boron (B) concentration (mg B L^{-1}) due to adsorption in soils of Tamil Nadu. Batch sorption studies conducted with an initial B concentration of 40 mg B L^{-1}.

Soil No.	Reaction time (h)						
	1	2	4	8	16	32	64
2	34.2	31.7	28.7	28.0	26.2	25.7	25.7
4	27.8	26.0	24.4	23.7	22.2	21.2	21.4
5	39.0	36.5	36.5	35.5	35.0	34.0	34.0
7	27.5	20.0	18.0	16.5	15.5	16.0	15.5
8	38.0	36.0	34.0	33.0	30.5	30.0	30.0
9	32.2	28.9	27.9	25.6	24.0	24.2	24.0
10	34.5	32.4	29.2	28.4	27.0	26.9	27.0
11	30.9	28.2	26.7	24.4	23.4	23.9	23.4

and Keren (1981) used 24 h whereas Elrashidi and O'Connor (1982) adopted 12 h for equilibrium time for their experiments in B adsorption. The boron equilibrium concentration was negatively correlated with clay content (r = $-$0.928**). This is also consistent with the previously published literature (Elrashidi and O'Connor, 1982) which showed that fine textured soils usually had higher B adsorption capacities than the coarse textured ones. The role of these soil parameters is further discussed under adsorption of B.

The amount of B adsorbed by eight soils was poorly described by the zero, first, second and third order reaction equations. The parabolic diffusion equation described B adsorption for the 11 soils only over a limited part of the isotherm (1 to 8 h), showing discontinuity beyond 8 h reaction time. This suggested that two different mechanisms are controlling the rate of B adsorption by these soils (Taylor et al., 1995). The relatively higher R^2 (r = 0.900**) together with the relatively low SE (0.95) over the entire time range indicated that adsorption of B by these soils was best described by Elovichian kinetics.

Adsorption of B

L type B adsorption isotherm curves were obtained for the soils in the present study. Though the B adsorption data conformed to both the Langmuir and the Freundlich equations, the fit was excellent with the latter.

Although Freundlich equation is empirical, this could be used to explain the adsorption of compounds on soils and it is best suited for a heterogenous system like the soil-nutrient water system. While the Freundlich constant, K, denoted the quantity of the B adsorbed when its concentration in the equilibri-

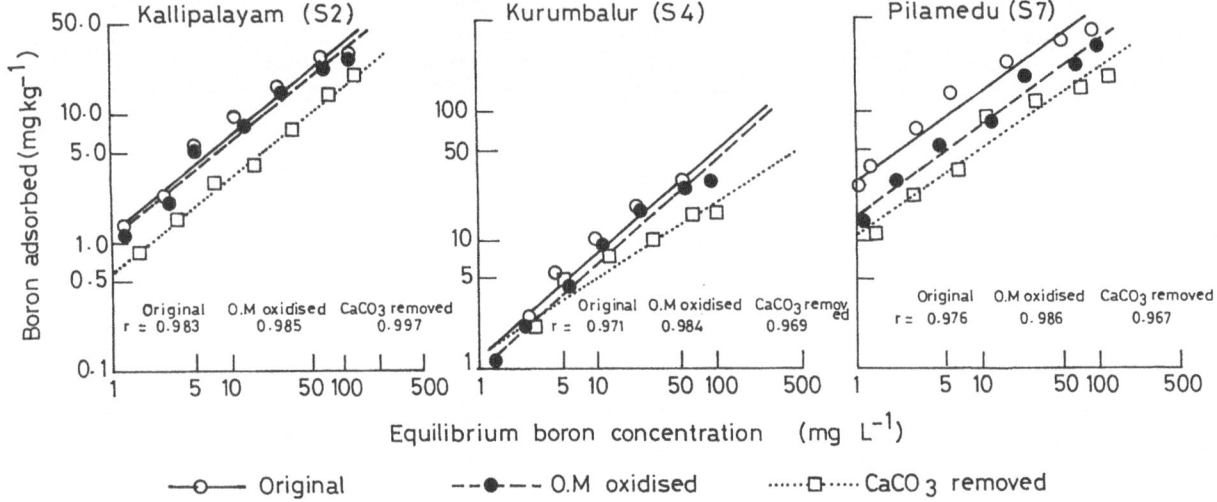

Figure 1. Freundlich isotherms for B adsorption in two alfisols (S2 and S4) and a vertisol (S7) of Tamil Nadu, India.

Table 3. Effect of oxidising soils to remove organic carbon and treatment to remove $CaCO_3$ on Freundlich adsorption isotherm constants (K,n) in soils of Tamil Nadu, India. r denotes the goodness of the fit of the Freundlich model to B sorption data.

Soil No.	Original soil			OM oxidised soil			$CaCO_3$ removed soil		
	K (mg kg^{-1})	n	r	K (mg kg^{-1})	n	r	K (mg kg^{-1})	n	r
1.	1.04	1.16	0.992	0.61	1.39	0.994	0.43	1.24	0.997
2.	1.20	1.33	0.983	1.13	1.37	0.985	0.55	1.32	0.996
3.	0.18	1.38	0.997	0.14	1.28	0.994	0.16	1.36	0.999
4.	1.28	1.26	0.971	1.10	1.27	0.984	1.22	1.62	0.969
5.	0.38	1.59	0.991	0.38	1.66	0.987	0.30	1.71	0.990
6.	1.07	1.30	0.993	0.46	1.24	0.996	1.02	1.55	0.983
7.	1.96	2.30	0.976	1.64	1.47	0.986	1.18	1.56	0.967
8.	1.19	1.31	0.922	0.56	1.64	0.963	0.66	1.69	0.967
9.	1.30	1.32	0.983	0.55	1.29	0.997	1.14	1.80	0.978
10.	1.09	1.33	0.996	1.06	1.39	0.983	0.75	1.22	0.965
11.	1.43	1.14	0.993	0.35	1.09	0.993	0.70	1.31	0.998

um solution was at unity, the value of constant 'n' served as a measure of the energy of adsorption (Bailey and White, 1970). The quantities of B adsorbed when plotted as a function of equilibrium concentration on a log-log scale resulted in a straight line (Figure 1). Boron adsorption was successfully described with the Freundlich model. Similar success with the Freundlich model has been reported by many workers (Elrashidi and O'Connor, 1982; Nicholaichuk et al., 1988). Freundlich 'K' values for the adsorption reaction of eleven soils ranged from 0.18 to 1.96 mg B kg^{-1} (Table 3). The values of K obtained are in consonance

with the results obtained by Elrashidi and O'Connor (1982) and Nicholaichuk et al. (1988). This may be attributed to the high CEC, clay and sesquioxide content of all the soils studied which play a key role in B adsorption (Goldberg and Glaubig, 1985). Among the soils studied, soils 3 and 5 recorded low K values as they were sandy in texture and low in CEC. Maximum adsorption capacity appeared to increase with the fineness of the soil texture (Biggar and Fireman, 1960). The clay and organic carbon content in the soil were significantly related with the K values of Freundlich equation (r = 0.839** and r = 0.787**, respectively).

Similar results were expressed by Elrashidi and O'Connor (1982). Several investigators (e.g. Yermiyahu et al., 1995) have indicated that organic matter is one of the active factors in B adsorption.

In soils treated to remove organic matter, the K values of Freundlich adsorption isotherms were generally lower than that of original soils. This showed that organic adsorptive surfaces had greater affinity to retain B than mineral surfaces. Though there was a reduction in the adsorptive capacity (K values) due to oxidation of organic matter, the trend of K values followed the same pattern as that of original soils. The formation of complexes between dihydroxy organic compounds and B can presumably explain the observed behaviour, showing that organic carbon plays an important role in B adsorption. The results were also consistent with those of Elrashidi and O'Connor (1982).

Removal of $CaCO_3$, led to the marked reduction in adsorption capacity (K values) for all the soils. Similar observations were recorded by Goldberg and Forster (1991). Although B adsorption decreased with $CaCO_3$ removal, no significant correlation was observed between $CaCO_3$ and K values of Freundlich sorption isotherm in the present study. This was probably a result of limited $CaCO_3$ value range within the soils studied.

Conclusion

The B adsorption data clearly shows that soils not only vary in their capacity to retain B but also in the energy with which they adsorb it. These two factors affect the ability of a soil to release B and to maintain supply in the soil solution. The study indicates that the equilibrium period of 24 h was sufficient to study the B adsorption characteristics of soils. The kinetics of B adsorption by these soils was best described by the Elovichian model. Though adsorption of B conformed to both the Langmuir and Freundlich adsorption isotherms, the fit was excellent with the latter. Clay, organic matter content and CEC of the soils were significantly correlated with K values of Freundlich equation. Removal of either organic matter or $CaCO_3$ reduced the adsorption capacity of the soils of Tamil Nadu.

Acknowledgements

The authors wish to thank Dr Ravendra Naidu, CSIRO, Land and Water, Adelaide, Australia for his constructive and valuable comments, Dr C.R.Ranganathan and Dr Honorora J. Francis of the Tamil Nadu Agricultural University, Coimbatore for their help, and the Australian Agency for International Development financial assistance, under its International Seminar Support Scheme.

References

Bailey G L and White J L 1970 Factors influencing the adsorption, desorption and movement of pesticides in soil. Residue Review 32, 29–92.

Biggar J N and Fireman M 1960 Boron adsorption and release by soils. Soil Sci. Soc. Am. Proc. 24, 115–120.

Bingham F T, Page A L, Coleman N T and Flach K 1971 Boron adsorption characteristics of selected amorphous soils from Mexico and Hawaii. Soil Sci. Soc. Am. Proc. 35, 546–550.

Chien S H and Clayton W R 1980 Application of Elovich equation to the kinetics of phosphate release and sorption in soils. Soil Sci. Soc. Am. J. 44, 265–268.

Elrashidi M A and O'Connor G A 1982 Boron sorption and desorption in soils. Soil Sci. Soc. Am. J. 46, 27–31.

Goldberg S and Forster H S 1991 Boron sorption on calcareous soils and reference calcites. Soil Sci. 152, 304–310.

Goldberg S and Glaubig R A 1985 Boron sorption on aluminium and iron oxide minerals. Soil Sci. Soc. Am. J. 49, 1374–1379.

Gupta W C 1968 Relationship of total and hot water soluble boron and fixation of added boron to properties of Podzol soils. Soil Sci. Soc. Am. Proc. 32, 45–48.

Hatcher J T and Bower C A 1958 Equilibria and dynamics of boron adsorption by soils. Soil Sci. 85, 319–323.

Hatcher J T, Bower C A and Clark M 1967 Adsorption of boron by soils as influenced by hydroxy aluminium and surface area. Soil Sci. 104, 422–426.

Hatcher J T and Wilcox L V 1950 Colorimetric determination of boron using carmine. Anal. Chem. 22, 567–569.

Hingston J J 1964 Reaction between boron and clays. Aust. J. Soil Res. 2, 83–95.

Jackson M L 1973 Soil Chemical Analysis. Prentice Hall of India, New Delhi. 498 p.

Marzadori C, Antisari V L, Ciavatta C and Sequi P 1991 Soil organic influence on adsorption and desorption of boron. Soil Sci. Soc. Am. J. 55, 1582–1585.

Mezuman U and Keren R 1981 Boron adsorption by soil using a phenomenological adsorption equation. Soil Sci. Soc. Am. J. 45, 722–726.

Nicholaichuk W, Leyson A J, Jame Y W and Campbell C A 1988 Boron and salinity survey of irrigation projects and the boron adsorption characteristics of some Saskatchewan soils. Can. J. Soil Sci. 68, 77–90.

Piper C S 1966 Soil Plant Analysis. Hans Publishers, Bombay.

Steel R D G and Torrie J H 1960 Principles and Procedures of Statistics. McGraw-Hill Co., New York. 481 p.

Taylor R W, Hassan K, Mehadi A A and Shuford J W 1995 Kinetics of zinc sorption by soils. Commun. Soil Sci. Plant Anal. 26, 1761–1771.

Yermiyahu U, Keren R and Chen Y 1995 Boron sorption by soil in the presence of composted organic matter. Soil Sci. Soc. Am. J. 59, 405–409.

R.W. Bell and B. Rerkasem (eds.), Boron in Soils and Plants, 261–264.
© *1997 Kluwer Academic Publishers.*

Relationship between water table depth, soil boron and sunflower genotypic response on a saline soil

S.G. Patil, S. Setkar & M. Hebbara
Saline Water Project, Agricultural Research Station, Gangavati- 583 227, Karnataka, India.

Key words: boron, genotypic response, leaching behaviour, plant B, water table

Abstract

Studies conducted on the effect of water table (WT) depth on boron (B) distribution, plant B and B leaching from soil revealed that both hot-water soluble B (HWSB) and cold water soluble B (CWSB) increased with soil depth and proximity to WT. Hot-water soluble B as an index of plant available B remained high (> 1.5 mg kg^{-1}) even with a deeper WT and increased further under shallow (<65 cm) WT. Increased levels of HWSB and CWSB in the soil were positively correlated with increased electrical conductivity and increased plant B and negatively with seed yield of sunflower (*Helianthus annuus*) cultivars. Leaching of B was less efficient in deeper than shallow WT and required twice the amount of water required for leaching of salt.

Introduction

Excess B is a potentially serious constraint to crop growth in areas of low rainfall and high evaporation (Evans and Sparks, 1973). According to Ryan et al. (1977), plants respond mainly to cold water soluble B and total B has no significance. Plant response to B varies and both deficiency and excess B can cause yield reduction in crops. Boron concentration of 20 to 100 mg kg^{-1} dry wt in leaves is adequate for plant growth but sensitive crops may exhibit toxic symptoms at well below 200 mg kg^{-1} dry wt (Gupta, 1984). The toxic level of B is alleviated by leaching but required three times more water than the leaching of salts (Bingham et al., 1972). However, no reports are available on water table (WT) effects on B distribution, crop response and leaching and this investigation concentrates on these aspects.

Materials and methods

A field experiment to evaluate response of sunflower (*Helianthus annuus*) cultivars to B as influenced by WT depth was conducted at the Agricultural Research Station, Gangavati during 1995-96. The experimental

area located on slightly sloping land (0.5%) has a natural gradient in WT and salinity. On this uni-directional slope, profile samples representing WT depths of 65, 80, 90 and 110 cm were collected and initial soil properties were determined (Jackson 1973). Cold water soluble (CWS) B was extracted under 1:2.5 (soil:water) conditions while HWSB and total B were extracted with hot water (1:2) and 6N HCl, respectively. All B determinations were made colorimetrically by carmine (Hatcher and Wilcox, 1950).

Achenes of 10 sunflower cultivars varying in maturity from early to late were sown along the slope to observe variations in grain yield in relation to WT.

In situ leaching of salts and B was carried out in fields with WT at 120 and 70 cm. Micro-plots of 1 × 1 m with a buffer of 3 × 3 m were prepared and initial soil properties were determined. Effective depths of irrigation water (diw) of 5 to 45 cm in shallow (70 cm) and 5 to 72 cm in deeper (120 cm) WT were imposed in 5 cm increments. After each increment, soil samples were collected for salinity and B. The water used in leaching had the following properties: electrical conductivity (ECe), 0.20 dS m^{-1}; pH, 8.2; negligible B; Ca^{2+}>Na$^+$>Mg^{2+} and HCO$_3^-$>Cl$^-$>SO$_4^{2-}$.

Table 1. Mean physico-chemical properties of the experimental site and ground water composition.

Parameter	Water table depth (cm)			
	110	90	80	65
Soil properties[A]				
pH (1:2.5 extract)	8.52	8.47	8.34	8.30
ECe (dS m^{-1})	8.24	6.05	8.45	13.20
Textural class		Clay loam		
HWSB (mg kg^{-1})[B]	2.2	2.1	2.4	2.7
Total B (mg kg^{-1})	20.2	22.3	23.8	23.7
CaCO$_3$ (g kg^{-1})	16.9	15.2	16.0	15.0
Organic carbon (g kg^{-1})	4.3	4.9	4.3	4.0
CEC (cmol(p$^+$)kg^{-1})		60.7		
Ground water composition				
pH	7.90	7.55	7.45	7.70
EC (dS m^{-1})	1.75	3.70	8.00	15.7
Boron (mg L^{-1})	0.50	0.80	1.50	3.30
SAR (mmol/L)$^{0.5}$	8.50	11.0	20.8	30.4

[A] All the values represent means for the profile (0–90 cm)
[B] HWSB, Hot-water soluble B.

Table 2. Distribution of salts and boron (B) fractions as affected by depth of the water table.

Soil depth (cm)	Depth of water table below the ground surface (cm)											
	110			90			80			65		
	ECe	HWSB	CWSB	ECe	HWSB	CWSB	ECe	HWSB	CWSB	ECe	HWSB	CWSB
0–15	2.3	1.51	0.54	2.51	1.70	0.63	6.40	2.01	0.96	22.1	2.17	1.90
15–30	3.5	1.80	0.59	4.14	1.90	0.79	6.40	2.08	1.00	9.30	2.54	1.01
30–45	6.1	2.07	0.85	5.32	2.00	0.83	6.90	2.15	1.09	10.6	2.73	1.12
45–60	9.3	2.40	0.95	6.65	2.20	0.98	8.50	2.57	1.17	13.3	3.00	1.27
60–75	15.2	2.95	1.14	9.31	2.50	1.07	10.9	2.80	1.26	12.8	2.88	1.30
75–90	14.1	2.54	1.27	9.04	2.45	1.11	12.8	2.94	1.32	13.0	2.88	1.30

HWSB - Hot water soluble (mg B kg^{-1} soil)
CWSB - Cold water soluble (2:5 soil:water extract, mg B kg^{-1} soil)
ECe - Electrical conductivity of saturation extract (dS m^{-1}).

Table 3. Relationship between electrical conductivity (ECe)(dSm^{-1}), hot water soluble B (HWSB)(mg B kg^{-1}), shoot B (mg B kg^{-1} dry wt) and yield (g 6 m^{-1}) of sunflower cultivars grown at four depths to the water table.

Genotype groups	Water table depth (cm)															
	110				90				80				65			
	ECe	HWSB	Shoot B	Yield	ECe	HWSB	Shoot B	Yield	ECe	HWSB	Shoot B	Yield	ECe	HWSB	Shoot B	Yield
Early[A]	1.9	1.78	32	71	4.3	1.98	41	62	7.3	2.18	47	62	18.2	2.82	52	25
Mid-late[B]	1.45	1.76	37	296	3.5	2.04	42	130	7.7	2.27	44	95	17.9	2.79	54	27
Late[C]	1.40	1.86	37	236	4.4	2.04	42	131	7.1	2.31	47	45	23.3	2.65	55	25

[A] Early cultivars: PKVSH-27, PKVSf-9; Co-1, Morden
[B] Mid-late cultivars: Co-2, EC-68414, PAC36, ICI-308
[C] Late cultivars: PAC-8699, SUF-7.

Table 4. Leaching efficiency (%)[1] of salts and boron as affected by depth to water table and irrigation water.

Diw[2] (cm)	70				120			
	Soil depth (cm)				Soil depth (cm)			
	0–15	0–30	0–60	0–90	0–15	0–30	0–60	0–90
Salts								
5	9	4	15	19	−15	−14	−1	−1
10	21	16	5	12	−36	−30	3	3
20	49	38	34	28	29	13	7	4
30	55	42	37	33	39	21	10	7
40	58	52	49	42	41	35	14	8
50	71	57	40	36	42	37	17	11
60	–	–	–	–	42	39	20	15
70	–	–	–	–	49	47	23	16
80	–	–	–	–	62	45	24	19
Mean	44	35	31	30	28	21	13	9
Boron								
5	−4	3	4	3	−3	−3	−1	1
20	15	9	−2	−7	14	9	5	1
30	28	22	6	1	27	23	9	5
40	41	35	25	6	34	34	13	9
50	53	49	32	11	21	30	15	10
60	–	–	–	–	49	39	17	13
70	–	–	–	–	28	25	18	14
80	–	–	–	–	21	18	19	17
Mean	24	20	11	3	20	18	11	8

[1] Leaching efficiency (%) = $(1 - ECef/ECei) \times 100$, where: $ECei$ = initial soil salinity (dS m^{-1}); $ECef$ = soil salinity after leaching
[2] Diw - Depths of irrigation water applied.

Results

Boron distribution in soil

Irrespective of WT, there was a linear relationship between B-fractions and soil salinity (Table 2) and CWSB correlated more closely with soil salinity ($r=0.94$**) than HWSB ($r=0.72$**). Hot-water soluble B and CWSB increased with depth (Table 2) irrespective of WT. Similarly $CaCO_3$ increased and organic carbon decreased. The increasing $CaCO_3$ however, had no relation with HWSB. An inverse relationship ($r= -0.76$ **) was observed between organic carbon and HWSB but not with CWSB.

Genotypic response to B and soil salinity

Early maturing cultivars in general, had lower yield potential and increase in soil salinity and plant B did not affect yield until WT was shallow (65 cm). A mean plant B of 32 mg kg^{-1} dry wt observed in deeper WT was increased to 52 mg kg^{-1} dry wt under shallow WT (Table 3). Mid-late cultivars generally accumulated more B than early cultivars, but the effect of higher HWSB and increased salinity on yield levels was far greater than the effect on B accumulation in the plant. Late types accumulated more B than early but similar concentrations to the mid-late cultivars.

Leaching of salts and boron

Salt leaching was observed to be higher in surface soil layers as well as under shallow WT (Table 4). A depth of 50 cm irrigation water leached 71 per cent of salts in the surface layer (0–15 cm) but leaching was reduced to 42 per cent in the deeper layers with WT at 120 cm.

Boron was less leachable (53%) from the soil surface than salts (71%) in shallow WT. In contrast, leach-

264

ing efficiency of both B and salts remained similar in deeper WT but more B was removed from surface in shallow WT (Table 4).

Discussion

Both HWSB and CWSB remained fairly high (>1.7 mg kg^{-1}) even with deeper WT. The rise in WT increased soil salinity and HWSB, increasing plant B and depressing yield. In early sunflower cultivars, reduction in yield was more gradual up to 80 cm WT with a serious decline with a WT closer to the soil surface. In mid-late types, increase in HWSB due to rising WT encouraged higher B uptake. A rise of WT within 80 cm increased plant B from 44 to 53 mg kg^{-1} dry wt but remained well below the suggested critical level, 200 mg kg^{-1} dry wt (Gupta, 1984).

Leaching of B was less effective than leaching of salts in soils especially with WT of 70 cm. However, B leaching became more efficient in shallow WT but only 24 per cent of B was flushed out as against 44 per cent of salts. Bingham et al. (1972) reported 3 times more water was required to achieve the same relative B removal as for salt. Alternate methods like lowering WT, breeding B and salt-tolerant species should be considered for managing B affected saline soils.

References

Bingham F T, Marsh A W, Branson R and Mahler Ferry G 1972 Reclamation of salt-affected high B soils in Western Karn County. Hilgardia 41, 195–24.
Bowen J E 1977 The fine art of using enough but not too much boron. Crops Soils Mag. Aug–Sept. 1977.
Evans C M and Sparks D L 1983 On the chemistry and minerology of boron in pure and mixed systems: A review. Commun. Soil Sci. Plant Anal. 14, 827–846.
Gupta U C 1984 Boron nutrion of alfalfa, red clover and timothy grown on podzol soils of Eastern Canada. Soil Sci. 137, 16–22.
Hatcher J T and Wilcox L V 1950 Colorimetric determination of boron using carmine. Anal. Chem. 22, 567–569.
Jackson M L 1973 Soil Chemical Analysis. Prentice Hall of India, New Delhi. 498 p.
Ryan J, Miyamoto S and Stroehlein J L 1977 Relation of solute and sorbed boron to the boron hazard in irrigation water. Plant and Soil 47, 253–256.

R.W. Bell and B. Rerkasem (eds.), Boron in Soils and Plants, 265–270.
© 1997 *Kluwer Academic Publishers.*

Soil contamination with boron due to irrigation with treated municipal waste water

C. D. Tsadilas

National Agricultural Research Foundation – Institute of Soil Classification and Mapping, Theofrastos 1 Street, 413 35 Larissa, Greece

Key words: boron toxicity, cotton, maize, irrigation, waste water

Abstract

In two separate field experiments lasting two years, the influence of irrigation of cotton (*Gossypium hirsutum* L.) and maize (*Zea mays* L.) with treated municipal wastewater on the soil boron (B) contamination and B uptake by the crops was studied. For both crops the experiments were completely randomized blocks with five treatments each replicated four times. The treatments were: control, involving irrigation with natural water but no mineral fertilisation (C); irrigation with natural water plus complete mineral fertilisation (MF); irrigation with wastewater but no mineral fertilisation (W); irrigation with wastewater plus partial mineral fertilisation (WSF) and; irrigation with wastewater plus complete mineral fertilization (WTF). Irrigation was applied through a drip system. The results showed that wastewater had a high B concentration (above 1 mg L^{-1}) and increased B concentration in both crop leaves and soil. Hot water soluble boron (Hws-B) values increased from below 1 mg kg^{-1} in the control treatment up to above 2.0 mg kg^{-1} in some treatments including irrigation with wastewater. Leaf B concentration was increased in cotton from 67 mg kg^{-1} in the control up to 81 mg kg^{-1} dry wt in treatments irrigated with wastewater while B concentration in corn leaves was increased from about 9 to 21 mg kg^{-1}. These B concentrations were not toxic to cotton and maize since the yield of both crops was increased significantly in comparison to the control treatment.

Introduction

Wastewater reuse is nowadays receiving more attention as a reliable source of water. The reuse of wastewater is mainly for agricultural and landscape irrigation, industrial recycling, groundwater recharge, recreational uses, nonpotable urban uses and in some limited cases to the potable reuse (Tsobanoglous and Burton, 1991). For agricultural reuse, wastewater can be considered as both as irrigation water and as a partial substitute for inorganic fertilisers (Pescot, 1992). However, wastewater in most cases contains many substances that, in excess, can be toxic to the plants. Therefore its reuse requires very careful handling. The nutrients contained in wastewater usually include nitrogen (N), phosphorus (P), and a lesser extent potassium (K), zinc, boron (B) and sulfur. Unregulated wastewater reuse can cause problems of toxicities for plant growth.

Boron is an essential element for plant growth for which the concentration range in soil and irrigation water corresponding to deficiency and toxicity is relatively narrow (Gupta et al., 1985). On the other hand, soil B decontamination is very difficult and in any case it may be economically prohibitory (Leyshon and Jame, 1993). So the use of B containing materials must be managed carefully in order to avoid harmful effects on plants and soils. In the literature many studies have referred to the use of wastewater in agriculture and its impacts on soil properties such as salinity, phosphorus, potassium and heavy metal concentration (Pescot, 1992). However, information on the impacts of the use of wastewater on soil contamination due to B is very limited. So the purpose of this study was to investigate the influence of irrigation of agricultural crops with treated municipal wastewater on soil contamination of B and its uptake by the crops.

Table 1. Variation of boron concentration in wastewater.

Date	B concentration, mg L^{-1}
31-5-96	1.11
18-6-96	1.29
1-7-96	1.25
15-7-96	1.33
2-8-96	1.22
19-8-96	1.01
3-9-96	1.12
18-9-96	1.01
25-9-96	0.14
15-10-96	1.59
30-10-96	1.39

Materials and methods

Wastewater used in this study comes from the waste treatment plant of Larissa, Central Greece. This plant is scheduled to serve a population of about 200,000 people and produces 15000 m^3 wastewater day^{-1}. Two experiments were conducted during the years 1995 and 1996 on a Typic Xerochrept located near the wastewater treatment plant. The first experiment included cotton (*Gossypium hirsutum*, var. Corina) and the second one maize (*Zea mays*, hybrid doge). The experimental designs were for both crops, completely randomised blocks with the following treatments: control, irrigation with natural water but no mineral fertilisation (C); irrigation with natural water plus complete mineral fertilisation (MF); irrigation with wastewater but no mineral fertilisation (W); irrigation with wastewater plus partial mineral fertilisation (WSF) and; irrigation with wastewater plus complete mineral fertilization (WTF) according to levels suggested for these crops by the local authorities. This consists of 160 kg N ha^{-1} and 35 kg P ha^{-1} for cotton and 350 kg N ha^{-1} and 53 kg P ha^{-1} for maize. All treatments were replicated four times.

The experimental plots (4 × 4 m) were irrigated with a drip system. Irrigation system consisted of plastic pipes which were located at a distance 2 m one from the other. The distance between the drippers on the pipes was 1 m. At the end of each irrigation which lasted for about 24 h, wet soil areas from the drippers within pipe lines overlapped. This was not true between pipe lines. So soil wetting was not even. For this reason, soil samples were taken from sites located in the wet areas at equal distances from the drippers.

During the growing season of each year about 650 mm of natural or wastewater were applied through irrigation to cotton and 920 mm to corn plus about 115 mm water through rainfall.

Four soil samplings were carried out at one month intervals during the growing season starting from June in the second year. The samples were composite from five places in the plots and were taken from a depth 0 to 30 cm. At the bloom stage (mid-July for cotton and at the late July for maize), leaf samples were taken from both crops. Soil samples were air dried, crushed, sieved (2 mm) and analysed for hot water soluble boron (HwsB) (Bingham, 1982). Leaf samples were dried at 70°C, ground (40 mesh), placed in a high form porcelain crucible and heated at 500°C for 6 h (Jones et al., 1991). Cotton leaf samples were separated into stalks and laminas which were analyzed separately. The ash was diluted with 0.1 M HCl. Wastewater samples were taken twice a month during the second year. Boron in soil and leaf extracts and in wastewater was determined by the azomethine-H procedure (Bingham, 1982). In the samples of wastewater, electrical conductivity, concentration of ammonium and nitrate nitrogen, P and K and heavy metals (Cd, Pb, Ni, Zn, Cu, and Mn) was determined. Ammonium was determined by the indophenol blue method, nitrate by the copperized cadmium reduction method (Page et al., 1982) and heavy metals by atomic absorption spectrometry.

Results and discussion

The soil on which the experiments were conducted, was a Typic Xerochrept with pH 8.4, 355 g clay kg^{-1}, 13 g organic matter kg^{-1}, about 30 g free carbonates kg^{-1}, 25 mg extractable P kg^{-1} and 165 cmol exchangeable K kg^{-1}. The soil was very well drained. Electrical conductivity of wastewater ranged from 70 to 95 mS m^{-1}, ammonium nitrogen between 0 to 3 mg L^{-1}, nitrate nitrogen between 6 to 8 mg L^{-1}, P from 4 to 9 mg L^{-1} and K from 20 to 35 mg L^{-1}. Heavy metal concentration was very low and in many cases below the detection limit.

Boron concentration in wastewater (Table 1) ranged from 1.01 to 1.33 mg L^{-1} during the experimentation period in which the crops were irrigated (June to middle of September) and between 0.14 to 1.59 at other times. This high B concentration in municipal wastewaters is attributed mainly to the household detergents (Tsobanoglous and Burton, 1991). Maas (1987) suggests that B concentration is considered toxic when it

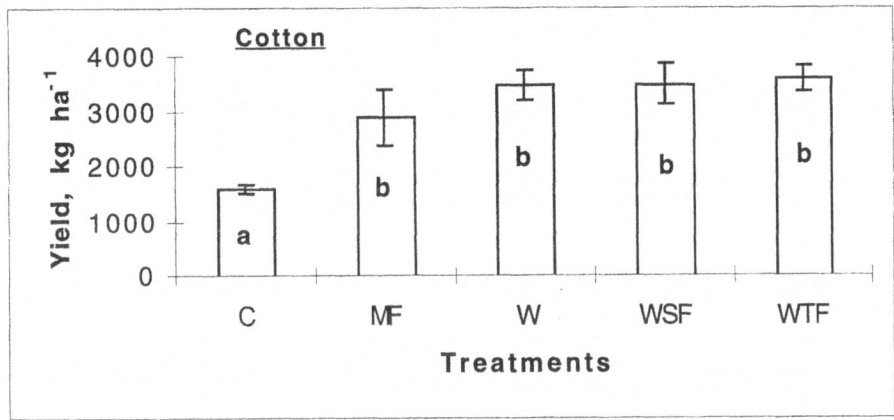

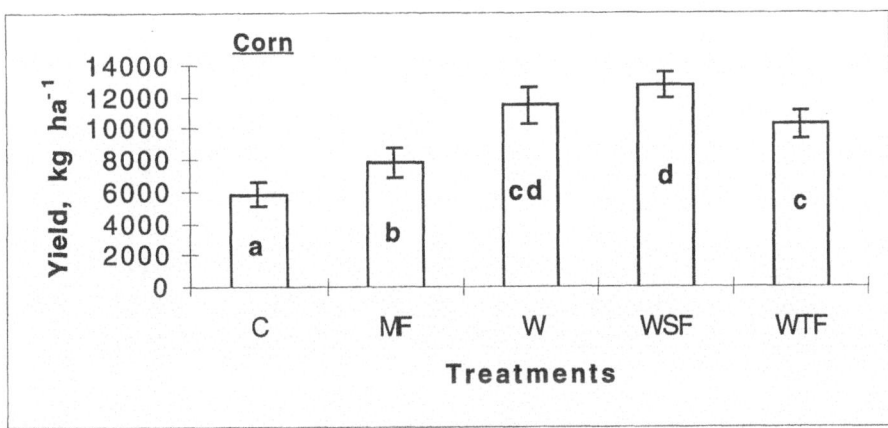

Figure 1. Influence of the treatments applied on cotton and corn yield. Bars with the same letter statistically were not different according to LSD test, vertical bars show the standard error of the mean. C: irrigation with natural water but no mineral fertilization; MF: irrigation with natural water plus complete mineral fertilizer; W: irrigation with wastewater only; WSF: irrigation with wastewater plus partial mineral fertilization; WTF: irrigation with wastewater plus complete mineral fertilizer.

is above 0.5, 1.0, 2.0, 4.0 and 6.0 mg l^{-1} for very sensitive, sensitive, moderately sensitive, moderately tolerant and tolerant crop species, respectively. According to the same classification, cotton and corn are considered moderately and very tolerant crops, respectively.

Irrigation with wastewater increased the yield of both crops (Figure 1). This is attributed to the appreciable quantities of N and P contained in the wastewater. In the case of cotton, the increase in yield caused by irrigation with wastewater without addition of mineral fertilisers (treatment W) was similar to the increase caused by the traditional mineral fertilisation and irrigation with natural water (treatment MF). Yield increase was similar in the remaining treatments (Figure 1). In the case of maize, the influence of irrigation on its yield was more complex. Irrigation with wastewater with-

out addition of mineral fertiliser (treatment W) caused higher increase in maize yield than the application of the usual mineral fertiliser rate and irrigation with natural water (treatment MF). The highest increase was observed in the treatment WSF (irrigation with wastewater, partial mineral fertilisation). The lower yield obtained in the treatment WTF (irrigation with wastewater, complete mineral fertilisation) in comparison to the treatment WSF (Figure 1) may be attributed to the lower amount of water applied in this treatment due to technical problems. Higher yield of the treatments WSF and WTF in comparison to the treatment MF may be explained by the fact that the usual mineral fertiliser rate was lower than that required for obtaining maximum maize yield.

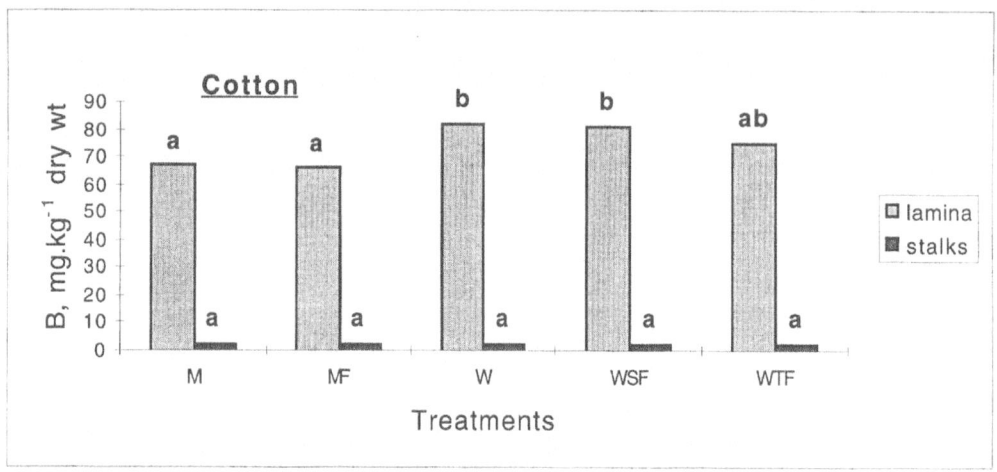

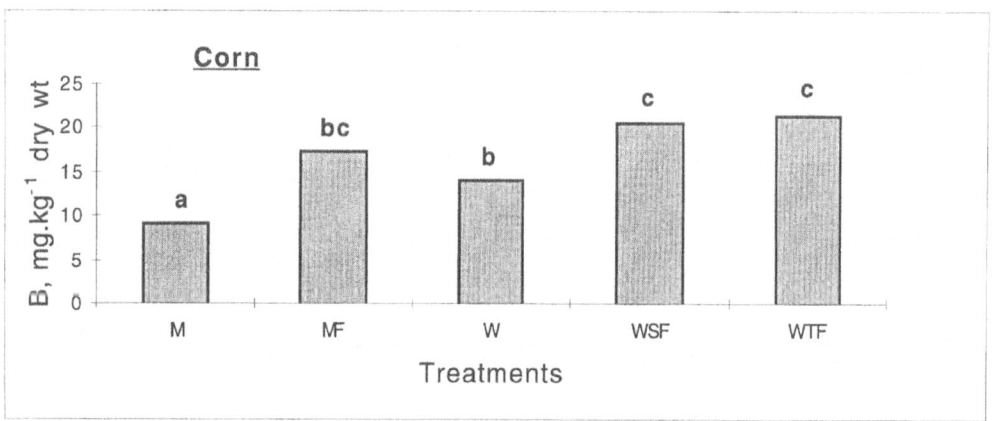

Figure 2. Influence of wastewater on boron (B) concentration of cotton and corn leaves. See Figure 1 for a description of treatments.

Irrigation with wastewater also significantly increased B concentration in leaves of both crops. For cotton, B concentration in the leaf laminas increased from 67 mg kg^{-1} in the control plot to 82 mg kg^{-1} in the W treatment (irrigation with wastewater, no mineral fertilization). In the stalks, B concentration was much lower than that in the laminas and was not significantly influenced by the irrigation with wastewater, with a concentration of 3 mg kg^{-1} in all the various treatments (Figure 2). Similarly B concentration in corn leaves increased from about 9 mg kg^{-1} in the control plot to 21 mg kg^{-1} in the treatment (Figure 2). For cotton, leaf B concentration is considered sufficient when it is between 20 to 60 mg kg^{-1} and high when it is above 60 mg kg^{-1}. The respective values for corn are between 5 to 25 mg kg^{-1} and above 25 mg kg^{-1} (Jones et al., 1991). In our experiments in both crops, leaf B

concentration was high without any negative influence on crop yield. In contrast, for corn a statistically significant positive relationship between B concentration in the leaves and the yield (r=0.63**) was observed. This response may be attributed to the tolerance of cotton and maize to B. Maize is characterised as moderately tolerant and cotton as very tolerant to B (Pescot, 1992).

Hot water soluble B was strongly affected by the treatments applied. In both experiments, Hws-B was raised from <1.0 mg kg^{-1} in the control treatments to above 2.0 mg kg^{-1} in some treatments containing irrigation with wastewater (Figure 3). In general, Hws-B concentration >1 mg kg^{-1} is considered high for crops sensitive to high B (Reisenauer et al., 1973). From the results of this study it seems that irrigation of soils with municipal wastewater may cause B contamination. Since B is adsorbed by soils, its removal

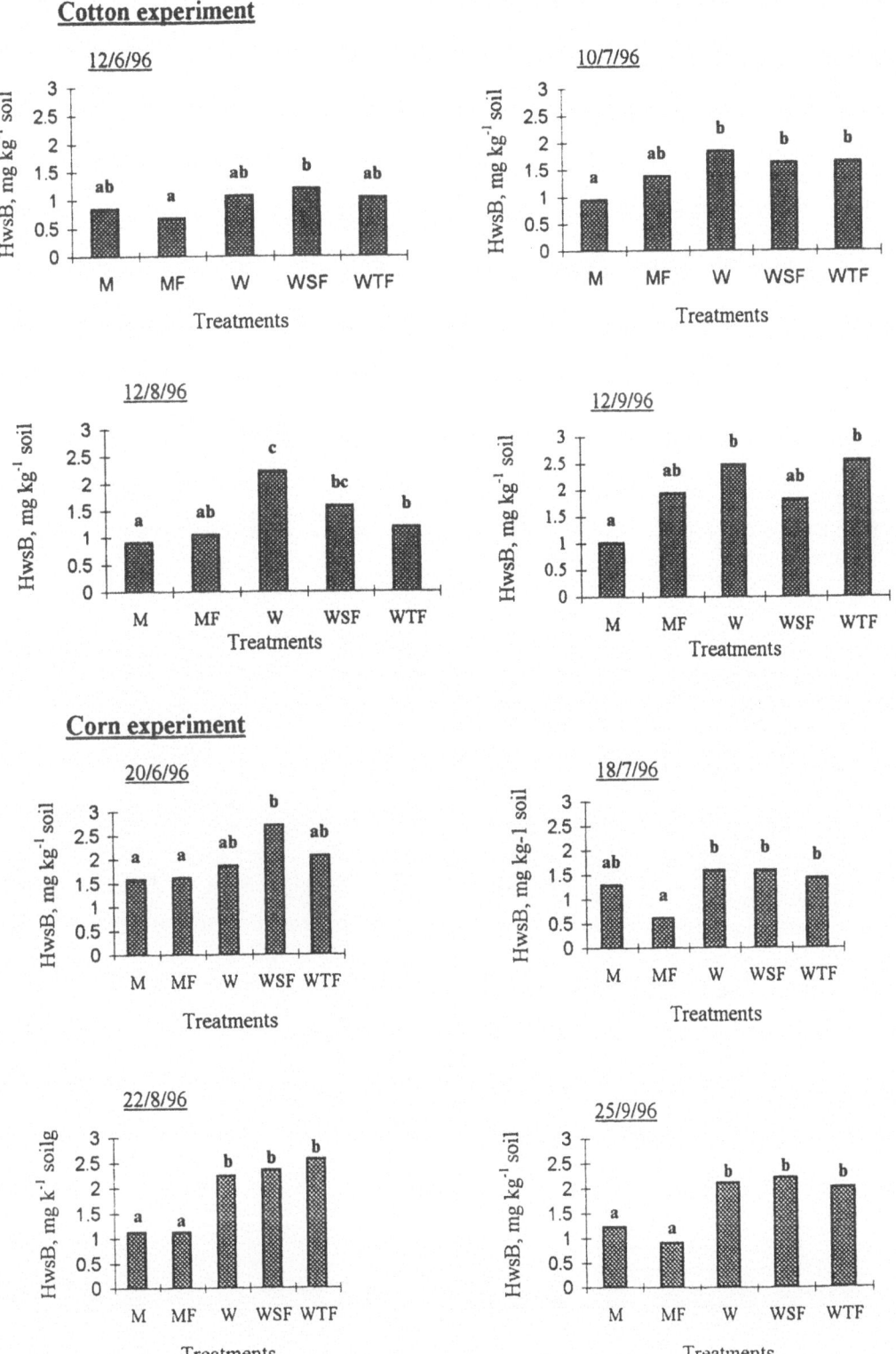

Figure 3. Influence of wastewater on hot water soluble boron (HwsB) in both experiments. See Figure 1 for a description of treatments.

270

from soils is much slower than of non-adsorbed salts such as chlorides. It is estimated that three times as much water is required to reduce B content of a soil to less than 20% of its initial value as was required to reduce the total soluble salt content by the same proportion (Leyshon and Jame, 1993). On the other hand, chemical methods trialled for detoxifying B toxic soils were often unsuccessful and of high cost (Hutchinson and Viets, 1969). Therefore the use of this water must be managed very carefully. Hot water soluble B was not significantly correlated with leaf B concentration in both crops.

Acknowledgments

This study is a part of the research program EPET II No 267 financed by the Ministry of Research and Technology of Greece and the European Union. The author would like to thank Mrs E. Mahera, V. Gianni and Mr Ch. Balais for their laboratory assistance.

References

Bingham F 1982 Boron. *In* Methods of Soil Analysis, Part 2 Chemical and Microbiological Properties. Ed A L Page, pp 431–447. Agronomy Society of America, Soil Science Society of America, Agronomy 9, Madison, Wisconsin, USA.

Hutchison G L and Viets F G Jr 1969 Detoxification of boron in plants with triopropanolamine. Soil Sci. 108, 217–221.

Gupta U C, Jame Y W, Cambell C A, Leyshon A J and Nicholaichuck W 1985 Boron toxicity and deficiency: a review. Can. J. Soil Sci. 65, 381–409.

Jones J B Jr, Wolf B and Mills A H 1991 Plant Analysis Handbook A Practical Sampling, Preparation, Analysis, and Interpretation Guide. Micro-Macro Publishing, USA.

Leyshon A J and Jame Y W 1993 Boron toxicity and irrigation management. *In* Boron and its Role in Crop Production. Ed U C Gupta, pp 207–226. CRC Press, Boca Raton, Florida.

Maas E V 1987 Salt tolerance of plants. *In* Handbook of Plant Science in Agriculture, Vol 2. Ed B R Christie, pp 57. CRC Press, Boca Raton Florida.

Page A L, Miller R H and Keeney D R 1982 Methods of Soil Analysis Part 2, Chemical and Microbiological Properties, 2nd Edn. Agronomy 9 Agronomy Society of America, Soil Science Society of America. Madison, Wisconsin.

Pescot M B 1992 Wastewater Treatment and Use in Agriculture. FAO Irrigation and Drainage Paper 47, FAO, Rome.

Reisenauer H M, Walsh L M and Hoeft R G 1973 Testing soils for sulfur, boron, molybdenum, and chlorine. *In* Soil Testing and Plant Analysis. Ed L M Walsh and J D Beaton, pp 173–200. Soil Science Society of America. Madison, Wisconsin.

Tsobanoglous G and Burton F L 1991 Wastewater Engineering. Treatment, Disposal and Reuse. Metfal & Eddy, Inc.

The manufacturer's authorised representative in the EU is Springer
Nature Customer Service Centre GmbH, Europaplatz 3, 69115 Heidelberg,
Germany. If you have any concerns regarding our products, please
contact ProductSafety@springernature.com

Printed and bound by CPI Group (UK) Ltd, Croydon, CR0 4YY

29/04/2026

02099554-0002